Treatise on Materials Science and Technology

VOLUME 21

Electronic Structure and Properties

TREATISE ON MATERIALS SCIENCE
AND TECHNOLOGY

VOLUME 21

ELECTRONIC STRUCTURE AND PROPERTIES

EDITED BY

FRANK Y. FRADIN

Materials Science Division
Argonne National Laboratory
Argonne, Illinois

 1981

ACADEMIC PRESS

A Subsidiary of Harcourt Brace Jovanovich, Publishers

New York   London   Toronto   Sydney   San Francisco

ACADEMIC PRESS, INC.
111 Fifth Avenue, New York, New York 10003

United Kingdom Edition published by
ACADEMIC PRESS, INC. (LONDON) LTD.
24/28 Oval Road, London NW1 7DX

Library of Congress Cataloging in Publication Data
Main entry under title:

Electronic structure and properties.

(Treatise on materials science and technology ;
v. 21)
Includes bibliographies and index.
Contents: Electronic structure of perfect and
defective solids / A. J. Freeman--Photoelectron
spectroscopy as an electronic structure probe /
B. W. Veal--Electronic structure and the electron-
phonon interaction / W. H. Butler--[etc.]
1. Electronic structure. 2. Free electron
theory of metals. I. Fradin, Frank Y. II. Series.
TA403.T74 vol. 21 [QC176.8.E4] 620.1'1s 81-2457
ISBN 0-12-341821-6 [530.4'1] AACR2

PRINTED IN THE UNITED STATES OF AMERICA

81 82 83 84   9 8 7 6 5 4 3 2 1

Contents

List of Contributors

Numbers in parentheses indicate the pages on which the authors' contributions begin.

W. H. BUTLER (165), Metals and Ceramics Division, Oak Ridge National Laboratory, Oak Ridge, Tennessee 37830

Z. FISK (297), Institute for Pure and Applied Physical Sciences, University of California, San Diego, La Jolla, California 92093

E. S. FISHER (223), Materials Science Division, Argonne National Laboratory, Argonne, Illinois 60439

A. J. FREEMAN (1), Department of Physics and Astronomy, Northwestern University, Evanston, Illinois 60201

P. JENA (351), Physics Department, Michigan Technological University, Houghton, Michigan 49931

S. G. STEINEMANN (223), University of Lausanne, Lausanne, Switzerland

B. W. VEAL (99), Materials Science Division, Argonne National Laboratory, Argonne, Illinois 60439

G. W. WEBB (297), Institute for Pure and Applied Physical Sciences, University of California, San Diego, La Jolla, California 92093

Foreword

Materials science has evolved to its present advanced state through contributions from numerous facets of the solid-state sciences. Physics and chemistry of materials, metallurgy, ceramics, and polymer science are all merging into more clearly defined common themes. This convergence is well exemplified by the topic to which this volume is addressed: electronic structure and properties.

Electronic-related properties have always been central to the growth of materials science. In recent years, new, fresh theoretical approaches and newly developed experimental techniques have allowed more realistic determinations to be made of band structure and energy states in pure systems and in a wide variety of solid solutions. Notable progress has been made in understanding and predicting phase stability, and this in turn has initiated new approaches to alloy design.

Simplified interpretations of complex and cross-linked phenomena, though required and intellectually satisfying, do not necessarily lead readily to reasonable predictions of behavior. To a large extent, the evolution of alloy theory had to wait for the introduction of very sensitive and high resolution experimental probes (e.g., photoelectron spectroscopy, positron annihilation), with an attendant appreciation for the theoretically based strengths and weaknesses of these techniques.

This volume, which has been edited by Frank Y. Fradin, has come at a propitious time, when experimental technology and electron theory are expanding significantly in a coordinated fashion. The near future holds great promise for studies of electronic structure of materials and for those applications that are bound to result from these activities.

H. HERMAN

Preface

There has been remarkable progress in the understanding of the electronic structure and properties of materials in the past two decades, due in large part to impressive advances in various spectroscopies for directly measuring the electronic structure, e.g., photoemission spectroscopy, as well as to advances in the theoretical methods of calculating the electronic structure of complex materials, e.g., the local density approach, supercell methods, and the coherent potential approximation. The improved understanding of the electronic structure has led to exciting new insights into a number of physical properties, e.g., superconductivity, phonon and lattice instabilities, magnetism, surface activity, energetics of point defects, and electron transport in complex materials. In this volume we present these new developments in electron theory and electron spectroscopies. The approach, while pedagogical in nature, avoids becoming bogged down in computational or experimental detail. The contributing chapters are exemplary rather than comprehensive, although sufficient references are given to help the interested reader find more detail. Emphasis in the chapters is on transition metals, alloys, and intermetallic compounds, with a number of illustrative examples from the oxides and silicates.

Although the electron theory of metals and alloys was developed by Mott and others as early as the 1930s, quantitative success of the theory was restricted to the simple alkali or noble metals. Of greater interest to the materials scientist are the alloys and compounds of technological importance for their interesting physical properties. As A. J. Freeman points out in Chapter 1, it has only been within the past 20 years that *ab initio* energy band structures have developed to the point that confidence could be placed on their predictions for highly complex structures. In his chapter on electronic structure, Dr. Freeman introduces the local density functional approximation and then describes the modern energy band methods. He also introduces the supercell method for treating the case of reduced (i.e., surfaces) or broken (i.e., defects, impurities) translational invariance and the coherent potential approximation for treating the disordered concentrated alloy problem. A number of illustrative examples are given in Chapter 1 that serve to relate the electronic structure calculations both to the various electron spectroscopies and to a number of physical phenomena discussed in the other chapters.

Photoelectron-emission spectroscopy is perhaps the most important spectroscopy to establish both qualitative and quantitative verification of the electron theory of solids. In Chapter 2, B. W. Veal introduces the ultraviolet and the x-ray photoemission techniques and describes the application of the techniques to study both core-electron and valence-electron structures. Dr. Veal includes illustrative examples of metals, alloys, oxides, hydrides, as well as amorphous materials.

The electron–phonon interaction is the central theme of many exciting phenomena in modern studies of metallic materials. In Chapter 3, W. H. Butler introduces the theory of the electron–phonon interaction. He then discusses the ramifications of the electron–phonon interaction for the superconducting transition temperature, lifetimes of phonons, lattice thermal resistivity, and interatomic force constants or phonon frequencies. Specific examples are given for the free electron metals, the transition metals, and the A-15 (e.g., V_3Si) type compounds.

The elastic properties of transition metals is the subject of Chapter 4 by S. G. Steinemann and E. S. Fisher. In this chapter emphasis is placed on the effect of deformation on the Fermi surface and the resultant systematic dependence of the elastic constants on the electronic structure. A heavy emphasis is given to the body-centered-cubic transition-metal alloys, which have received the most exhaustive theoretical and experimental attention. Other examples include the face-centered-cubic and the hexagonal-close-packed transition metals and the A-15 superconductors. A short section on the effects of interstitial hydrogen in niobium is also included.

The electrical resistivity of metals and alloys is the subject of Chapter 5 by Z. Fisk and G. W. Webb. In this chapter Drs. Fisk and Webb discuss the nature of defect-dominated electron scattering, including concentrated alloys and liquid metals, as well as contributions to the electrical resistivity due to the electron–electron interaction, the electron–phonon interaction, and magnetic effects. A discussion of the electrical resistivity changes occurring at various types of phase transformations is also included.

In the final chapter, P. Jena reviews the electronic structure of point defects in metals. In this chapter Dr. Jena surveys the various theoretical formulations for the point-defect problem. He then illustrates the theoretical results and makes contact with experimental probes such as nuclear quadrupole and nuclear magnetic resonance, impurity resistivity, muon spin rotation, and positron annihilation. Specific examples include hydrogenlike impurities in metals, including isotope effects, and vacancies in metals.

Electronic Structure of Perfect and Defective Solids

A. J. FREEMAN

Department of Physics and Astronomy
Northwestern University
Evanston, Illinois

I.   Introduction and Historical Perspective

The past decade has witnessed a coming of age of the energy band approach to the study of materials. During this time the energy band method has become an increasingly powerful and sophisticated tool of the solid-state physicist and materials scientist for theoretically studying the multitudinous and increasingly complex properties of matter in the condensed state. Thus today, information derived from energy band studies are increasingly attracting the interest of materials scientists for solutions and/or understanding of phenomena (including some in the more classical metallurgical domain) previously thought to be too complex for theoretical treatment.

The manifold successes of the band approach, which we now take for granted, are all the more remarkable when one recalls that only some 20 years ago very little confidence or physical meaning was attributed to ab initio energy band structures (i.e., eigenvalue phenomena). The status of band theory today may be viewed as a direct result of the fact that in the past decade new sophisticated experiments on both traditional materials and those having complex crystallographic structures have demanded not only theoretical descriptions of *eigenvalue* phenomena but also detailed and precise *wave functions* with which to determine the expectation values of different observable operators. Such a demanding test of the predictions of one-electron theory has the additional virtue in permitting, by their comparison with experiment, accurate determinations of the relative magnitude and importance of many-body effects in real solids. Thus there had developed considerable interest in the Hohenberg–Kohn–Sham (Hohenberg and Kohn, 1964; Kohn and Sham, 1965; Sham and Kohn, 1966) local density functional (LDF) formalism for investigation of various ground-state properties of solids, despite the usual difficulties of solving the associated one-particle equation characterized by a multicenter nonspherical potential. Aside from providing a rigorous basis and justification of the single-particle energy band description of the *ground-state* properties of materials, it has led to accurate, tractable, computational schemes for describing them from first principles.

In this chapter we describe and discuss some of these developments in the local-density theory energy band approach and indicate by means of selected examples some of the richness and variety of properties and pheonomena which are and can be studied. It will also give an overview of the detailed interaction that can now be made between ab initio theoretical calculations and experiment, it being well understood that only through this close interplay will a more fundamental understanding of electronic structure and properties emerge.

II. Theoretical Methods

The very success of the energy band approach and its recent great popularity (as attested to by the proliferation of such approaches and the large growth in the number of its practitioners) make it imperative that we discuss the basis of the method and its underlying assumptions and approximations in order to better understand the applicability (and reliability) of the results obtained and to assess their validity relative to experiment.

A. Complexities and Approximations Required for Treating the Many-Body Problem of a Solid in a Solvable, Tractable Form

The energy band method for calculating electronic eigenstates in crystals is based on a number of simplifying assumptions and approximations that reduce the many-body problem, involving the interactions between all the particles—electrons and nuclei—in the system, to a one-electron or independent electron model. (The reason for these approximations is clear: the many-body problem for the crystal entails the solution of Schrödinger's equation for 10^{23} nuclei and electrons and is a completely hopeless task.) It is important to keep these approximations in mind as they are not completely justifiable and may affect seriously some of the physical results obtained.

The first of these is the zero-order Born–Oppenheimer approximation, which essentially amounts to neglecting the electron–phonon interaction and reduces the problem to that of an interacting electron system in the field of fixed nuclear potentials. Actually, as discussed at length in Chapter 3, the electron–phonon interaction plays an important role in a number of physical phenomena, the most famous being its essential role as origin of superconductivity in the Bardeen–Cooper–Schrieffer (BCS) theory. In normal metals it can cause an enhancement of measured electron masses, electron velocities, g factors, the measured oscillator strength for optical transitions, density of states at the Fermi energy as obtained from specific heat measurements, etc. In general, it does not appear to affect measured Fermi surface dimensions, which therefore may be directly compared with predictions arising from the theoretical calculations.

The direct solution of the many-electron Hamiltonian with its 10^{23} variables is obviously not possible. Thus we must make an approximation that separates the variables to yield an effective single-particle equation that is the approximate Hamiltonian of the energy band model. There have been two "justifications" of this model: one is based on the Hartree–Fock formalism; the other arises from a more generalized many-body interacting homogeneous electron gas approach for treating the ground state of a system—the so-called local density functional theory.

The most recent justification of the band model is the local density functional formalism of Hohenberg, Kohn, and Sham, which we shall describe more fully in Section II,B (Hohenberg and Kohn, 1964; Kohn and

Sham, 1965; Sham and Kohn, 1966). Briefly, in this case one uses the theorem that for a nondegenerate ground state the particle density is the fundamental variable of the system, and the exchange-correlation functional is approximated using the results of many-body theory for the homogeneous electron gas. Each electron is assumed to experience, as it moves through the crystal, an *average* but spatially periodic potential $V(r)$ due to all the other electrons and nuclei in the crystal. The single-particle Hamiltonian, which replaces the complicated many-electron Hamiltonian, is given as the sum of a kinetic energy operator and a local potential energy $V(r_i)$ (i.e., the Bloch Hamiltonian). All the band methods thus reduce the problem into two parts: the choice of the exchange-correlation crystal potential and the accurate solution of the Schrödinger (or, in the relativistic case, the Dirac) equation with the assumed potential. High-speed computers, good mathematical analysis, and programming have made the second part of the problem relatively easy to solve and have led to considerable confidence in the numerical accuracy of the solutions obtained.

B. Local Density and Local Spin Density Functional Theories

The Hohenberg–Kohn–Sham local density formalism is based on the fundamental theorem that the ground-state properties of an inhomogeneous interacting electron system are functionals of the electron density $\rho(\mathbf{r})$ and that in the presence of an external potential $V_{ext}(\mathbf{r})$ the total ground-state energy $E[\rho(\mathbf{r})]$ in its lowest variational state can be written in terms of a universal functional of $\rho(\mathbf{r})$ and is *independent of the external potential* $V_{ext}(\mathbf{r})$. This theorem forms the basis of the approach to the electronic structure problem in that it provides an effective one-particle equation relating self-consistently the ground-state wave functions to the energy functionals (i.e., potential) of the electronic system. Identifying the external potential for a polyatomic system as the electron–nuclear and internuclear interactions and varying $E[\rho(\mathbf{r})]$ with respect to $\rho(\mathbf{r})$, one obtains an effective one-particle equation of the form

$$\left\{ -\frac{1}{2}\nabla^2 + \sum_m \frac{Z_m}{|\mathbf{r}_m - \mathbf{r}'|} + \int \frac{\rho(\mathbf{r}')}{|\mathbf{r} - \mathbf{r}'|}d\mathbf{r}' + \frac{\delta E_{xc}[\rho(\mathbf{r})]}{\delta\rho(\mathbf{r})} \right\} \psi_j(\mathbf{r}) = \varepsilon_j \psi_j(\mathbf{r}). \quad (1)$$

Here Z_m denotes the nuclear charge of the nucleus at site $\mathbf{R}_m$ and $E_{xc}[\rho(\mathbf{r})]$ denotes the total exchange and correlation energy of the interacting (inhomogeneous) electron system (square brackets are used to denote functional dependence). The eigenfunctions $\psi_j(\mathbf{r})$ are simply related to the total ground-state charge density of the occupied one-particle states. The total ground-state energy is then given by

$$E_{tot} = \sum_{j=1}^{\sigma_{oc}} \left\langle \psi_j(\mathbf{r}) \left| -\frac{1}{2}\nabla^2 \right| \psi_j(\mathbf{r}) \right\rangle + \int \rho(\mathbf{r}) \left\{ \sum_m \frac{Z_m}{|\mathbf{r} - \mathbf{R}_m|} + \frac{1}{2} \int \frac{\rho(\mathbf{r}')}{|\mathbf{r} - \mathbf{r}'|} d\mathbf{r}' \right\} d\mathbf{r}$$

$$+ \sum_{\substack{n,m \\ n \neq m}} \frac{Z_n Z_m}{\mathbf{R}_n - \mathbf{R}_m} + E_{xc}[\rho(\mathbf{r})], \tag{2}$$

where the first term represents the kinetic energy, the second and third terms are the total electrostatic potential energy, and the last term is the exchange and correlation energy.

Note that the LDF formalism in the form described above makes no claim on the physical significance of the eigenvalues ε_j in Eq. (1); hence we may concentrate only on ground-state crystal properties.

Retaining only the nongradient terms in the expansion of $E_{xc}[\rho(\mathbf{r})]$, the exchange and correlation potential becomes

$$\delta E_{xc}[\rho(\mathbf{r})]/\delta\rho(\mathbf{r}) \cong F_{ex}[\rho(\mathbf{r})] + F_{corr}[\rho(\mathbf{r})], \tag{3}$$

where the exchange potential has the well-known form

$$F_{ex}[\rho(\mathbf{r})] = \tfrac{4}{3}\varepsilon_x[\rho(\mathbf{r})] \equiv -((3/\pi)\rho(\mathbf{r}))^{1/3}, \tag{4}$$

which is $\tfrac{2}{3}$ of the value given by Slater (1951). The correlation energy of a uniform electron gas with local density $\rho(\mathbf{r})$ has been calculated from the many-body theory by authors using different techniques (Nozières and Pines, 1958; Pines, 1963; Singwi et al., 1970). The agreement between the most recent results lies within 5–8 mRy in the metallic density range. A particularly convenient form is to use the results of Singwi et al. (1970) fitted to an analytical expression by Hedin and Lundqvist (1971). As an example for the different contributions to the total potential, Fig. 1 shows the Coulomb $V(\mathbf{r})_{coul}$, the exchange $V_x(\mathbf{r})$, and the correlation potential $V_{corr}(\mathbf{r})$, in diamond.

In the local spin density functional (LSDF), which is of great interest and utility for treating properties of magnetic systems, the Kohn–Sham exchange potential [Eq. (4)] is replaced by a spin-polarized description such as that of von Barth and Hedin(1972). This description has been found to be more appropriate because comparison of calculations with the Kohn–Sham potential and experiment by Wang and Callaway (1977) indicates that the calculated magnetic moment and exchange splitting in ferromagnetic Ni are too large. The Kohn–Sham potential evidently overestimates the tendency toward ferromagnetism, and this overestimate needs to be reduced through

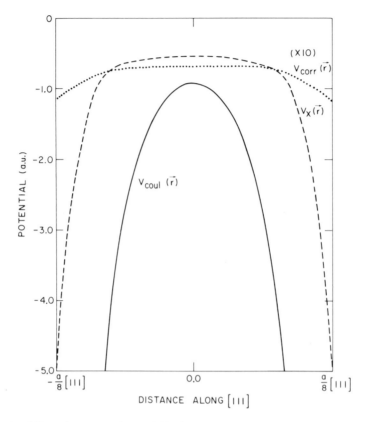

Fig. 1. Self-consistent crystal potential in diamond along the [111] direction. $V_{coul}(\mathbf{r})$, $V_x(\mathbf{r})$, and $V_{corr}(\mathbf{r})$ denote Coulomb, exchange, and correlation potentials, respectively. Vector quantities are indicated by overarrows in art and by bold letters in text.

the use of a potential which incorporates additional correlation effects (Wang and Callaway, 1977). The von Barth–Hedin potential for electrons of spin σ (↑ or ↓) apparently accomplishes this end.

C. Energy Band Methods

Traditionally, there have developed over the past 40 years a number of energy band methods for obtaining solutions to the Bloch equation for the motion of an electron in a periodic potential. The great simplification introduced by exploiting the high symmetry of the crystal lattice makes possible the accurate solutions of these otherwise complex equations for a given potential $V(\mathbf{r})$. Indeed, it was the discovery some 20 years ago that

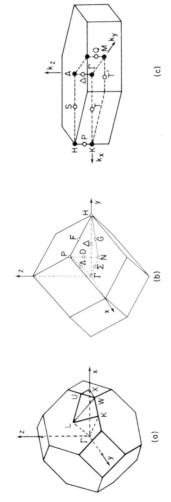

Fig. 2. The Brillouin zones for the (a) fcc, (b) bcc, and (c) hcp structures.

different band schemes applied to the same metal (Cu) using the same potential gave essentially identical energy bands that sparked interest in, and the development of, the modern computational band methods (Slater, 1974).

The well-known band methods [APW (augmented plane wave), KKR (Korringa–Kohn–Rostoker) or Green's function, OPW (orthogonalized plane wave), and LCAO (linear combination of atomic orbitals) or so-called tight binding] share the common objective of seeking solutions to the Bloch Hamiltonian. They each, of necessity, introduce a number of approximations to make the problem tractable. Since standard texts describe them in detail (Slater, 1972, 1974; Callaway, 1974), we only present here some basic elements in order to provide a framework for our discussion within the LDF formalism of recent developments.

The solution of the band Hamiltonian yields eigenvalues, $E_{n,\mathbf{k}}$ which, as labeled, depend not only on the particular band index n but also on the new "quantum" number $\mathbf{k}$, which measures the crystal momentum of the electron in the solid. Thus, in addition to the complication of having bands of allowed (and forbidden) energies, instead of the simple discrete electronic energy levels found in atoms and molecules, we need to know (in principle) the $\psi_{n,\mathbf{k}}$ and the $E_{n,\mathbf{k}}$ at all points in the k, or reciprocal lattice, space for the Brillouin zone (BZ) for the given crystal symmetry. The BZ for the face-centered-cubic lattice, for example, is shown in Fig. 2 with the principal symmetry directions labeled in the irreducible $\frac{1}{48}$ th of the zone. Also shown is the BZ for the body-centered-cubic and the hexagonal close-packed structures.

Of the various methods developed for the calculation of energy bands in solids, the most popular and most widely applied is the augmented plane wave (APW) method (Slater, 1937), which we now describe. Details and procedures involved in carrying out these calculations in either the nonrelativistic or the relativistic cases are available in the review articles by Loucks (1967), and Mattheiss et al. (1968). A review of the APW method and the results obtained using it for different solids may be found in the review article by Dimmock (1971).

1. THE APW METHOD

The APW method has considerable intuitive appeal; the one-electron states of the crystal are expanded in atomiclike functions for regions near the atomic sites and in plane waves for regions in the unit cell away from the atomic sites. This basis is physically reasonable and results from considering the crystal potential in the muffin-tin approximation.

Fig. 3. Schematic representation of a muffin-tin–like potential in a solid.

In this approximation the unit cell is separated into two regions by nonoverlapping spheres surrounding each atomic site. Inside each sphere the potential is assumed spherically symmetric and appears quite atomic-like; outside the spheres the potential is often assumed constant (cf. Fig. 3). It is easy to avoid this last assumption and include the general potential outside the spheres. This so-called warped muffin-tin (WMT) potential (Koelling *et al.*, 1970; Mattheiss, 1970–1972) is easy to incorporate into the APW formalism and has been used in a number of recent calculations. The construction of an approximate crystal potential generally follows a super-position procedure introduced by Mattheiss (1964); iteration to self-consistency is a tedious but straightforward procedure that is avoided whenever possible. The initial Coulombic potential is assumed to arise from overlapping atoms in the crystal and is calculated in a straightforward way, limiting the summation to a certain number of nearest neighbors.

Since we are after charge and spin densities, we need to understand the nature of wave functions in solids and the complications involved in their determination. An APW crystal wave function is expanded in augmented plane waves as

$$\chi_{\mathbf{k},E}(\mathbf{r}) = \sum_i A_i(\mathbf{k})\phi_{\mathbf{k}_i}(\mathbf{r}) \tag{5}$$

with the coefficients $A_i(\mathbf{k})$ to be determined variationally. The sum is over a set of reciprocal lattice vectors $\mathbf{K}_i$, where we have written $\mathbf{k}_i$ for $\mathbf{k} + \mathbf{K}_i$. An APW has the following representation:

Outside the spheres:

$$\phi_{\mathbf{k}_i}(\mathbf{r}) = \frac{1}{\sqrt{\Omega}} \exp(i\mathbf{k}_i \cdot \mathbf{r}), \tag{6}$$

where Ω is the volume of the unit cell;

Inside the νth sphere:

$$\phi_{\mathbf{k}_i}(\mathbf{r}) = \exp(i\mathbf{k}_i \cdot \mathbf{r}_\nu) \sum_{l,m} A_{lm}(\mathbf{k}) R_{l,E'}(\rho) Y_{lm}(\hat{\rho}), \qquad |\rho| = |\mathbf{r} - \mathbf{r}_\nu| < R_\nu,$$

$$(7)$$

where R_ν is the radius of the sphere and $\mathbf{r}_\nu$ is the vector to the center of the sphere. Figure 4 shows schematically an APW.

The A_{lm} are chosen so that each APW basis function is continuous at the sphere boundary. This guarantees that the crystal wave function will be continuous; there will remain, however, a slope discontinuity since we are limited to a finite expansion. The $R_{l,E'}(\rho)$ are the radial solutions to Schrödinger's equation inside the APW sphere. The wave-function coefficients $A_i(\mathbf{k})$ and eigenvalues can be determined by the Rayleigh–Ritz variational procedure as solutions of a standard eigenvalue problem, as discussed in Mattheiss et al. (1968), Loucks (1967), and Koelling (1970), by finding those E_i that satisfy

$$\det|H(E_i) - E\mathsf{S}| = 0. \qquad (8)$$

This procedure is useful for finding the eigenvalues, but is not as convenient for obtaining wave functions. One problem with obtaining eigenvectors is that the normalization is generally taken over the region outside the spheres and not over the unit cell. In order to obtain cell-normalized wave functions, it is best to employ the linearized form of the APW method as given by Koelling (1972) to solve

$$H(E_0)\mathbf{A} = E\mathsf{S}(E_0)\mathbf{A}, \qquad (9)$$

where S is the total overlap matrix. The procedure is to pick an E_0 and evaluate the resulting eigenvalues E_i. If one of the E_i is equal to E_0, then the wave function is evaluated. If E_0 does not equal any E_i, then E_0 is set equal to one of the E_i found, and an iteration performed. In practice this procedure converges very rapidly (Harmon et al., 1973).

Having thus obtained, at least formally, APW wave functions $\psi_{n,\mathbf{k}}(\mathbf{r})$ for the different bands (n) and at different k points in the BZ, the charge density is obtained by summing over the entire occupied volume of the BZ.

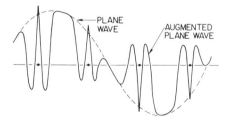

Fig. 4. Wave function in the APW method.

To simplify this task, one usually takes a finite (small) number of points into the summation with a weighting factor $S_{n,\mathbf{k}_i}$ for the $\mathbf{k}_i$th point, which is given by the local (in $\mathbf{k}$ space) density of states for band n. Thus

$$\rho(\mathbf{r}) = \sum_n \sum_{k_i} |\psi_{n,\mathbf{k}_i}(\mathbf{r})|^2 S_{n,\mathbf{k}_i}, \tag{10}$$

where

$$S_{n,\mathbf{k}_i} = \left[2\Omega_0 / (2\pi)^3\right]\left[\Delta\sigma_i^n / |\nabla_{\mathbf{k}_i} E_n|\right], \tag{11}$$

and Ω_0 is the volume of the unit cell and $\Delta\sigma_i^n$ denotes the element of surface area at the $\mathbf{k}_i$ point in the zone with energy E_n of the nth band.

2. ORTHOGONALIZED PLANE-WAVE AND PSEUDOPOTENTIAL METHODS

The orthogonalized plane-wave (OPW) and pseudopotential (PP) methods are very closely related. Both approaches make use of a plane-wave basis orthogonalized to a finite set of "core" states (assumed to be known). If the basis is made sufficiently large and all matrix elements are treated properly, of course, the two methods will produce identical results. In practice the OPW method as developed by Herman and co-workers (Herman, 1954; Herman and Skillman, 1961; Herman et al., 1963; Euwema et al., 1971) has been oriented toward first-principles band calculations within the framework of the HFS potential. On the other hand, Harrison (1966, 1970; Cohen and Heine 1970; Falicov and Cohen, 1963; Martinez et al., 1975a,b) and others have developed the PP scheme as a flexible method for generating effective potentials that provide an effective semiempirical procedure for fitting experimental data and interpolating band structures.

Generally, one can write the basis functions as

$$\phi_{j\mathbf{k}}^{(\mathbf{r})} = (1 - \mathcal{P})\exp\left[i(\mathbf{k} + \mathbf{K}_j)\cdot\mathbf{r}\right], \tag{12}$$

where $\mathbf{K}_j$ is a reciprocal lattice vector and

$$\mathcal{P} = \sum_{\nu,l} |\nu,l\rangle\langle\nu,l| \tag{13}$$

is a projection operator on the core states enumerated by l and located at site ν. In the PP approach a possibly energy-dependent parametrized form is often used (Cohen and Heine, 1970). When one writes out the matrix elements of the Hamiltonian $\langle\phi_i|h|\phi_j\rangle$ required for solution of the usual secular equation

$$\mathbf{HC} = \mathbf{SCE}, \tag{14}$$

terms involving the projection operator can be combined to define a nonlocal effective potential,

$$W_E(\mathbf{r}) = V(\mathbf{r}) + \sum_{\nu,l} (E - E_{\nu l})|\nu l\rangle\langle\nu,l|, \qquad (15)$$

where $E_{\nu j}$ are the core state eigenvalues. Thus one can consider the operator $W_E(\mathbf{r})$ as defining an effective Hamiltonian that, acting on pure plane-wave states, produces the correct band eigenvalues. Actually, it is well known that the pseudopotential of Eq. (15) is *not* unique and that a variety of forms can be generated for computational convenience.

One of the chief attractions of the PP method is that the pseudo–wave function obtained by solving the secular equation

$$\psi_n = \sum_j \exp\left[i(\mathbf{k} + \mathbf{K}_j)\cdot\mathbf{r}\right]C_{jn} \qquad (16)$$

can be smooth and slowly varying since the violent core-region oscillations have been largely removed. Thus perturbation calculations that describe valence and conduction electron properties are facilitated, and one hopes for rapid convergence in basis size.

In recent years, particularly through the efforts of Cohen (1979), there has been a resurgence of interest and effort in applying the PP method to a large variety of physical problems—including the electronic structure of solids and liquids, electron–lattice interactions, superconductivity, surfaces and interfaces, and chemisorption. A review of this work and a description of the various sophisticated approaches used to construct pseudopotentials and obtain solutions have been recently given by Cohen (1979).

3.   Other Approaches: KKR and LCAO

Two further approaches, namely, KKR (Korringa–Kohn–Rostoker) and LCAO (linear-combination-of-atomic orbitals), have been widely used in band theory. Both methods are capable of giving highly accurate solutions of the Schrödinger equation. Although the KKR, or Green's function, method is formulated as the solution of a multiple-scattering problem, it has many features similar to the APW technique. In the hands of Davis and others (Faulkner *et al.*, 1967; Davis *et al.*, 1968; Davis, 1971) it has been most useful in understanding a number of properties of metals and alloys. As will be discussed in Section II,G, this scheme has had also a profound impact on the coherent potential approximation (CPA) approach to disordered alloys and has led to a new simplified linearized method, the so-called LMTO (linear muffin-tin orbital) scheme of Andersen (1975), which is rapidly becoming the most popular scheme for treating complex materials.

After a long period of development the LCAO approach has become a quantitative method especially useful for studying semiconductors, insulators, and compounds. Since the muffin-tin potential approximation is avoided, covalent bonding effects can be adequately treated. Here we shall only mention a few applications. Lafon and co-workers (Chaney *et al.*, 1970; Lafon and Liu, 1966; Lafon *et al.*, 1971) have shown that x-ray structure factors for semiconductors like diamond and silicon are adequately represented. This work has been extended to study momentum densities and Compton profiles (Lafon, 1974). Similarly, Callaway and collaborators (Tawil and Callaway, 1973; Wang and Callaway, 1974, 1975, 1977; Ching and Callaway, 1975; Singh *et al.*, 1975) have performed self-consistent LCAO calculations on alkali and transition metals in order to interpret optical properties, *g* factors, magnetic, and Compton scattering data. Many other works have provided semiquantitative analysis of LCAO charge densities—for example, as interpolation schemes for more rigorous (but restricted in number of **k** points) band calculations (Maglic and Mueller, unpublished; Mattheiss, 1971).

Ellis and associates have developed a discrete variational approach using an LCAO basis to solve the energy band problem (Ellis and Painter, 1970). This approach has the virtue of avoiding the well-known problem of multicenter integrals that enters into the usual LCAO or type-binding scheme. The method, which provides a general variational scheme for efficiently calculating energy bands and charge densities in solids, can be viewed as a weighted local-energy procedure or, alternately, as a numerical integration scheme. This rapidly convergent procedure circumvents many of the difficulties associated with the evaluation of matrix elements of the Hamiltonian in an arbitrary basis and treats the general nonspherical potential with no more complication than the usual muffin-tin approximation. We discuss this DVM–LCAO approach in Section III in more detail when we describe more recent efforts to apply (and assess) the LDF formalism for describing electronic properties of solids.

4. LINEARIZED METHODS

Recently, several simplified energy band schemes have been proposed and shown to yield very good results. These methods, known as the linear-muffin-tin-orbital (LMTO) method [Andersen, 1975] and the linearized augmented plane wave (LAPW) method (Andersen, 1975; Koelling and Arbman, 1975) are "linearized" versions of the KKR and APW schemes, respectively. Their major virtues are that they avoid some of the computational complexities and high costs of treating complex (many atom per unit cell) systems inherent in the regular plane-wave based methods. These linearized methods retain relatively high accuracy and suffer little, if any, loss in computational speed compared to pseudopotential methods.

For example, the LAPW method is based on the augmented plane-wave method described above and thus combines the advantages of energy-independent muffin-tin (MT) Hamiltonian methods [fast root evaluation and rapid convergence for d-band metals as well as for nearly free-electron (NFE) crystals] with the simple matrix-element determination of the original APW method. The characteristic feature of the LAPW method is that a PW basis function in the interstitial region (where variations in the potential are relatively smooth) is augmented inside the MT spheres by functions constructed from the exact solutions in these spheres (where variations in the potential are large). The full one-electron potential is used and is constructed using the local-density approximation for exchange and correlation.

The main advantages of the LAPW method over the APW method are:

(i)   The secular determinant is linear in energy, which permits the simultaneous determination of both eigenvalues and eigenvectors by standard matrix diagonalization with very little loss of accuracy.

(ii)   Singularities in the matrix elements are eliminated, i.e., the asymptote problem of the APW method is avoided.

(iii)   The basis functions are everywhere continuous and differentiable.

In addition, the LAPW method retains such desirable features of the APW method as the ability to treat general potentials with no shape approximations, the ease with which relativistic effects can be included, and the fact that the basis size does not increase substantially for heavier elements.

Both the LAPW and the LMTO schemes are becoming increasingly popular for studies of complex systems. Examples will be given in Section III,D.

D.   LDF–Energy Band Approach

We have indicated briefly above some aspects of the traditional energy band computational methods for obtaining solutions of the Bloch equation. What may be said about the LDF formalism and its applicability to solids? As has been known for some time, applications of the LDF formalism to atoms (Lundqvist and Ufford, 1965) and molecules (Gunnarsson *et al.*, 1975) have yielded encouraging results. Similar applications for solids are complicated by (i) the need to consider both the short-range and the long-range multicenter crystal potential having nonspherical components, (ii) the difficulties in obtaining full self-consistency in a periodic system, and (iii) the need to provide a basis set with sufficient variational flexibility.

Hence, theoretical energy band studies of ground-state electronic properties of solids have used computational schemes that of necessity introduced a variety of well-known approximations *not inherent in the LDF formalism.* These additional approximations, invoked in order to simplify the solution of the one-particle equation, have hindered a meaningful comparison with experiment of the predictions of the LDF determined ground-state observables (i.e., total energy, lattice constants, Compton profiles, Fermi surface data, structure factors, etc.).

Several approaches have been developed recently that minimize the introduction of approximations not inherent in LDF. As an example of the sophistications inherent in these approaches, we describe briefly the approach of Zunger and Freeman (1977a,b) to the fully self-consistent (SC) solution of the one-particle equations in a periodic solid within the local density functional formalism. (Later on, we shall indicate some results they obtained compared with experiment.) It is designed and developed to incorporate special features with which to overcome difficulties encountered by other methods. Specifically, the method combines a discrete variational treatment of all potential terms (Coulomb, exchange, and correlation) arising from the superposition of spherical atomiclike overlapping charge-densities, with a rapidly convergent three-dimensional Fourier series representation of all the multicenter potential terms that are not expressible by a superposition model. The basis set consists of the exact numerical valence orbitals obtained from a direct solution of the local density *atomic* one-particle equations. To obtain increased variational freedom, this basis set is then augmented by virtual (numerical) atomic orbitals, charge-transfer (ion pair) orbitals, and "free" Slater one-site functions. The initial crystal potential consists of a non–muffin-tin superposition potential, including nongradient free-electron correlation terms calculated beyond the random-phase-approximation. The Hamiltonian matrix elements between Bloch states are calculated by the three-dimensional Diophantine integration scheme of Haselgrove (1961) and Ellis and Painter (1970), thereby avoiding the usual multicenter integrations encountered in the LCAO tight-binding formalism.

E. Problems of Itinerant versus Localized Electron Distributions and the Treatment of Non–Ground-State Properties

The LDF formalism permits solutions to be obtained for any nondegenerate ground-state system. By "solutions" one means in LDF any observable property that is a functional of the ground-state charge density. In many cases, however, it is clear that in order to study any system, we must perturb it. Thus we must know not only the ground state but excited states as well. Now strictly speaking, eigenvalues and eigenvectors determined from the local density Hamiltonian have no physical meaning in LDF theory and, in particular, excitation energies cannot be taken as the difference in eigenvalues, as is standard in band theory. This latter procedure for treating excitation energies relies on Koopmann's theorem assumption that, for an excitation of a single electron in the solid, the self-interaction, relaxation-polarization self-energy, and electron–hole interactions are small (of the order of $1/N$ times the equivalent atomic value, where N is the total number of particles in the system).

Let us consider a bit more closely this question from the local density-energy band versus many-electron points of view. We have seen how the LDF formalism yields a rather simple single-particle-effective Schrödinger (or its relativistic counterpart) equation

$$H = -\nabla^2 + V(\mathbf{r}),\tag{17}$$

where the potential $V(\mathbf{r})$ contains Coulomb, exchange, and correlation contributions. This simplified Hamiltonian is the familiar band description and differs from the many-electron Hamiltonian in not containing the well-known interelectronic Coulomb repulsion terms $(1/r_{ij})$ between electrons i and j. Thus in the band model, each electron may be interpreted to experience, as it moves through the crystal, an *average* but spatially periodic, now local, potential $V(\mathbf{r})$ due to all the other electrons and nuclei in the crystal. Obviously, this assumption involves an approximation in the treatment of the Coulomb interaction as it is the only coupling between electron coordinates. The energy band picture resulting from the eigenvalue spectrum will not be directly useful unless it is very nearly independent of the changes in occupation occurring for the experiment considered. The dominant term in the variation of the eigenvalues with occupation number is the intraatomic Coulomb interaction

$$U_{\mathrm{c}} = \left\langle \left\langle \psi(i)\psi(j) | (1/r_{ij}) | \psi(i)\psi(j) \right\rangle \right\rangle\tag{18}$$

discussed by Hubbard (1963, 1964). If U_{c} is large compared to the effective bandwidth W (which measures the effectiveness of band "hopping" or conduction), then the band picture is not valid and the energy of the system is strongly occupation-number-dependent (i.e., Koopmans's theorem does not hold).

Clearly U_{c} will be large if the orbitals $\psi(i)$ and $\psi(j)$ are localized in the same region of space. This is the situation found for the heavy rare earths where the intraatomic Coulomb interaction dominates (Freeman, 1972) and selects an f^n manifold within which crystal field effects operate. In this case, the only useful energy parameter is the total energy of the n f-electron system. A similar situation holds for any experiment involving transitions from the core states. At the same time, these localized (in space) electrons will not be able to move from atomic site to atomic site so their bandwidth W will be very small. Thus $U_c \gg W$, which is the case for the rare earths and the heavy actinides (Freeman and Koelling, 1974). At the opposite extreme, when the electron states are extended over the whole crystal, the Coulomb energy U_c will be very small and the bandwidth quite large ($U_c \ll W$). This is the case to which the energy band picture may be directly applied. Between these two extremes lies the much more complicated intermediate case ($U_c \approx W$). Examples of this case include the transition metal oxides as well as a large number of actinide systems.

F. Supercell Method for Systems Having Broken or Reduced Symmetry

When a foreign atom or any similar disturbance is introduced into an otherwise perfect crystal, the periodic potential of the crystal is broken. This resultant lack of symmetry removes the great simplifications inherent in solutions of the Bloch (periodic potential) equations of motion for the electron. Instead, a whole new range of approximations becomes necessary in order to obtain solutions for electrons in these perturbed lattices. A highly promising approach, which we describe here, is contained in the supercell method that is rapidly becoming popular for systems of broken symmetry (excited states, impurities, vacancy defects, and possibly disordered alloys, amorphous materials, and liquid metals) or reduced symmetry (surfaces, interfaces, chemisorption).

1. Disordered Alloys and Amorphous Materials

The determination of the electronic properties of disordered alloys, amorphous materials, and, we may add, liquid metals remains today one of the great challenges facing the solid-state theorist. Unlike the case of point defects, impurities or vacancies, which provide a highly localized degree of disorder to an otherwise perfect crystal, the high degree of disorder in these systems is the essential physical characteristic and a massive complication to the theory. (In a liquid, the disorder is structural and no long-range order is exhibited. Density correlations play an important role over distances on the order of several interatomic spacings.) This situation is in sharp contrast with the case of a crystalline solid whose most striking characteristic is its periodicity. The existence of this translational symmetry greatly simplifies the computational problem involved, as we have seen in earlier sections. For perfectly ordered alloys, the theoretical calculations can be done just as for pure materials. Disordered alloys require special techniques that go beyond the band methods discussed above. A major problem relates to the derivation of the appropriate alloy potential to be used to describe the disorder.

Beeby (1971) has given a particularly lucid account of the meaning of disorder and a discussion of the correct procedure for obtaining an alloy potential. The essential feature to be considered by the theorist is how best to describe the disorder by an averaging process. Beeby emphasized that a theoretical calculation for a disordered sample must of necessity proceed in ignorance of the actual positions of all its constituent atoms. The formal device that the theorist must use to offset this ignorance consists of an averaging procedure about which he makes the following comments:

(i) It is expected that all macroscopically *identically produced* samples will have (within experimental error) identical properties. Systems with large flucuations in their properties due to unavoidable variations in production need a different approach. Such fluctuations are usually due to variations in some macroscopic parameter not yet controlled in the production process.

(ii) The detailed microscopic order is therefore only important to the extent that certain macroscopic properties (e.g., order parameters) are satisfied.

(iii) Naturally the theoretician is, in these circumstances, at liberty to choose any one microscopic distribution that satisfies the macroscopic restraints. However, since all such distributions are equivalent, it is easier to average over them with a probability function specifying the chance that they occur.

(iv) Such an approach is well known in statistical mechanics and works for the same reason: the number of particles involved in the average is very large.

It is clear that any such averaging must be made only over observable properties of the sample using commonly accepted concepts and procedures of, for example, statistical mechanics. A virtue of this procedure is that the averaging makes for great simplification because one can in this way ignore all except the macroscopic features of the sample.

In the next subsection we describe the coherent potential approximation (CPA) for treating electronic properties of substitutionally disordered alloys. Here, as part of our discussion of the supercell method, we indicate its possible role in treating systems like disordered alloys and amorphous materials. The idea of a supercell is already found in the work of Butler and Kohn (1971), which provides a procedure for calculating the density of states of a disordered alloy. They consider a region in a material of a specific size, a so-called small neighborhood, possibly a cube of side L, and determine all possible relevant arrangements of atoms. Now each arrangement is characterized by a statistical probability of occurrence. For a given arrangement within this region, they define a hypothetical crystal made up as repeated periodic arrays of these regions. As Butler and Kohn point out, an energy band structure can be calculated for this "crystal" and, crudely speaking, one can describe the density of states (DOS) for the disordered system as an average of the DOS for the various periodic crystals that can be constructed in this fashion with each weighted by the statistical probability of occurrence. Further, they show that the errors of such a procedure can be made to vanish exponentially with the size of the neighborhood.

2. ELECTRONIC STRUCTURE OF SURFACES

Studies of surface electronic phenomena have made rapid advances in recent years mostly because of (i) the development of novel powerful experimental methods and theoretical computational schemes, and (ii) the strong interaction that has developed between them. On the experimental side, the development of high-vacuum techniques for the preparation of stable and well-controlled surfaces and the use of synchrotron radiation sources together with high-resolution spectroscopic and other methods (Feuerbacher et al., 1978; Hagstrum, 1966; Penn and Plummer, 1974; Soven et al., 1976; Dionne and Rhodin, 1976; Landolt and Campagna, 1977; Landolt and Yafet, 1978) for studying surface phenomena have provided (in many cases) a wealth of reproducible experimental data for a variety of materials, notably, the important free-electron metals, semiconductors, and transition metals. On the theoretical side, a variety of powerful and successful energy band methods for treating bulk structures have been adapted for the study of the electronic structure of surfaces (Appelbaum and Hamann, 1972, 1976; Alldredge and Kleinman, 1972; Kleinman and Caruthers, 1974; Caruthers et al., 1976; Cooper, 1973; Kohn, 1975; Kar and Soven, 1975; Kasowski, 1976a, b; Painter, 1978a, b; Schlüter et al., 1975; Louie et al., 1977; Louie, 1978; Krakauer and Cooper, 1977; Gay et al., 1977; Cooper, 1977; Wang and Freeman, 1978, 1979a).

Regardless of the geometrical model used to represent the surface (thin-film or slab geometry consisting of a finite number of atom planes, slab-superlattice geometry or the semiinfinite crystal), the many different theoretical approaches have one feature in common: they all face the necessity of treating large unit cells (supercells) containing many inequivalent atoms. For such calculations, reciprocal lattice or plane-wave (PW) based methods enjoy an important advantage, namely, simple matrix element determination and corresponding ease of programming. Of the PW methods, only pseudopotential (Appelbaum and Hamann, 1972; Alldredge and Kleinman, 1972; Schlüter et al., 1975; Louie et al., 1977) or supplemented-orthogonalized plane-wave (OPW) type (Kleinman and Caruthers, 1974; Caruthers et al., 1976; Louie, 1978) calculations have been performed to date. For treating d-band systems, these methods (Kleinman and Caruthers, 1974; Caruthers et al., 1976; Schlüter et al., 1975; Louie et al., 1977; Louie, 1978) have primarily relied on a slab-superlattice geometry in which the slab is periodically repeated and separated from adjacent slabs by several layer spacings of vacuum. In this manner periodicity is artificially retained normal to the slab, thus permitting the use of standard bulk electronic methods. Aside from the usual difficulties associated with pseudopotential methods, there are two disadvantages in this approach. Because of the large size of the perpendicular lattice parameter (the sum of the slab thickness plus the thickness of the vacuum region), convergence of the basis in reciprocal lattice space can be made significantly worse if many vacuum layers are required to prevent the slabs from interacting with each other. In addition, the PW-type basis is required to yield the correct behavior of the wave function inside the slab as well as to represent correctly the decay into vacuum.

We have described in Section II the linearized plane-wave based methods and their advantages for treating complex systems. Recently, Krakauer et al. (1979) have extended the LAPW bulk scheme in order to obtain a method for treating thin films that avoids the above-mentioned difficulties. As in the case of bulk systems, this method is very accurate, and suffers little, if any, loss in computational speed compared to pseudopotential methods. By extending the basic idea of the bulk LAPW method (a PW basis in the interstitial region augmented inside the MT spheres by functions constructed from the exact solutions inside these spheres) to the film geometry and using the film–muffin-tin (FMT) (Kohn, 1975; Kar and Soven, 1975; Krakauer and Cooper, 1977) potential, a suitably defined PW-basis function in the interstitial region of the slab is additionally augmented in the vacuum region by functions constructed from the exact solutions of Schrödinger's equation (SE) there. In the film-LAPW formalism, the two vacuum regions (above and below the film) are treated in a manner that is completely analogous to the MT spheres. There is no problem in optimizing the thickness of the vacuum region; to obtain the solutions of SE in this region we essentially integrate SE inward from $\pm\infty$. The FMT potential is used only for the purpose of constructing the film-LAPW basis functions. Once defined, however, this basis can be used to treat general potentials with no shape approximations.

To be more specific about the slab geometry and how it affects solutions of the LDF equations, consider the FMT potential schematically depicted in

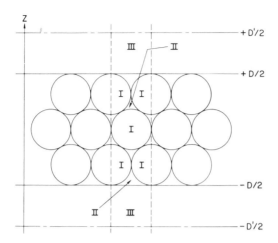

Fig. 5. Schematic representation of the film-muffin-tin (FMT) potential. The unit cell, indicated by dashed lines, extends to $z = \pm\infty$. Region I denotes the muffin-tin spheres, region II denotes the interstitial, and region III ($|z| > D/2$) denotes the vacuum region. The representation of the LAPW basis function in the interstitial region (II) involves a one-dimensional Fourier expansion over the interval $-D'/2 \leq z \leq D'/2$.

Fig. 5. (While we restrict ourselves to films with z reflection symmetry in this illustration, this is not a requirement of the method.) Since the film is periodic in the xy plane and nonperiodic normal to the film (the z direction), a unit cell can be defined extending to $\pm\infty$ in the z direction, as indicated by the dashed lines in Fig. 5. Inside the muffin-tin spheres (region I) the potential is spherically symmetric; in the interstitial region (region II) the potential is constant. Finally, in the exterior or vacuum region (region III is defined by planar boundaries at $\pm\frac{1}{2}D$), the potential is dependent only on z. As stated above, in analogy with the bulk LAPW method, a plane-wave–like basis function in the interstitial region is augmented by functions constructed from exact solutions in the MT regions (I) and the exterior regions (III) of Fig. 5. These functions must be matched onto the plane-wave–like basis function (and its derivative) at the MT sphere surfaces and at the boundary planes at $|z| = \frac{1}{2}D$. The resulting basis functions are then everywhere continuous and differentiable.

Another important example of LDF–energy band approaches to the study of the electronic structure of thin films is given in the work of Wang and Freeman (1979a, b). In this work the self-consistent numerical basis set LCAO method of Zunger and Freeman (1977 a, b) described above for bulk systems was generalized to the case of an unsupported thin film. As in the LAPW scheme just described, they consider a film of m layers with the origin of the system midway between the two surface layers and the z axis normal to the surfaces. The unit cell consists of a parallelepiped whose z dimension extends to $\pm\infty$. The Coulomb potential is formed as a non–muffin-tin superposition of spherical atomic potentials that can contain long-range ionic components to account for charge transfer in the film. A superposition of overlapping spherical atomic charge densities is used to construct the local density Kohn–Sham ($\alpha = \frac{2}{3})\rho^{1/3}$ exchange potential. Atoms up to 25 a.u. away from the atomic site were included in the two-dimensional direct-lattice sum to obtain the superposition potential and charge density. The long-range ionic component of the Coulomb potential accompanying the charge transfer near the surface is included through a generalized Ewald-type procedure.

3. INTERFACES AND CHEMISORPTION

We have seen that the superlattice representation just described also permits the treatment of the electronic structure of unsupported thin films (or slabs) by the same self-consistent LDF methods described earlier. This scheme provides a new and powerful approach to a variety of problems in that it makes a virtue out of the complexity introduced by the requirement that the supercell contain a large number of atoms. Since one has done away with periodicity in the z direction, there is no limitation on the type or nature of atom species to be placed in any layer (other than that there be periodicity within the layer). Thus it is a trivial extension of the concepts described above to apply the LDF slab (or superlattice methods) to the study of complex phenomena such as interfaces between dissimilar materials, modulated periodic structures of such materials, and chemisorption on surfaces. Increasingly, research in the near future will focus on these types of problems using the LDF schemes described in this section.

G. Coherent Potential Approximation (CPA) and KKR–CPA

Recent advances in the theoretical description of the electronic structure and properties of substitutional disordered alloys within the single particle approximation have attracted and received a good deal of attention. As is apparent from a number of reviews (Ehrenreich and Schwartz, 1976; Elliott *et al.*, 1974; Yonezawa and Morigaki, 1973; Economu *et al.*, 1974) dealing with their equilibrium transport and magnetic properties, the subject has a long history. Progress in the theoretical description of these systems has been made greatly within the framework of a multiple scattering description (Lax, 1951, 1952; Edwards, 1961; Beeby and Edwards, 1963; Beeby, 1964a, b). This approach usually involves the assumption that the effect of a given site on the propagation of an electron through the system can be decoupled from all other sites. In this way, the propagation of an electron depends on the average scattering from each site. This average scattering may then be determined within such a single-site decoupling scheme either beforehand, as in the average-t-matrix approximation (ATA) (Korringa, 1958; Soven, 1969; Beeby, 1967), or it may be determined self-consistently as in the coherent potential approximation (CPA) (Soven, 1967; Taylor, 1967). Until recently, theoretical investigations have been confined mainly to model Hamiltonians where the CPA is relatively easy to implement.

Briefly, the essential physical idea of the CPA approach is that the system can be described by an appropriate effective Hamiltonian defined by the condition that if it is employed there is, on the average, no further scattering from the individual sites. The Hamiltonian incorporates a potential that will permit a wave to propagate coherently through the alloy system. Unfortunately, this introduces the (necessary) complication that the effective Hamiltonian is both non-Hermitian and energy dependent.

Much of the theoretical effort has centered on the problem of the substitutional binary alloy $A_x B_{1-x}$ (Ehrenreich and Schwartz, 1976). The one-electron Hamiltonian for this alloy system is assumed to be of the form

$$H = H_0 + \sum_n V_n = H_0 + \mathcal{V}, \qquad (19)$$

where H_0 is a suitably chosen periodic unperturbed Hamiltonian and $\mathcal{V}$ is the total single-particle potential expressed as a sum of potentials V_n contributed by each site n. $\mathcal{V}$ and thus H are configuration-dependent since V_n takes the values V^A or V^B depending on whether atom A or B occupies site n. As discussed by Ehrenreich and Schwartz (1976), there are two approaches that have been followed in order to solve the quantum mechanical problem defined by H. In the first, the detailed construction of H for a piece of the alloy (considered as representative) is followed by an accurate numerical solution. As in molecular cluster approximations, the sample so constructed must be sufficiently large so that the calculated physical properties will not depend on boundary conditions. This requirement introduces the further complication of making the computations prohibitively costly.

The second approach considers the idea of configuration averaging as fundamental and so any macroscopic property of the system is described in terms of average over *all* configurations consistent with the structural definition of the alloy. In this approach, the need to use approximations for calculating these averaged quantities—typically consisting of decoupling procedures—introduces effects that are difficult to monitor. In principle, if solved with sufficient accuracy, the two approaches should yield the same results since a macroscopically large crystal may be visualized as made up of many smaller representative pieces that exhibit most, if not all, of the possible configurations.

The theory has now been well developed both physically and formally (Ehrenreich and Schwartz, 1976) beyond the early single-band models. Diagrammatic methods have been used and shown to be of importance in assessing the ranges of concentration, scattering strengths, and coordination numbers for which the theory is valid. Attention has been paid to the construction of alloy potentials (important in connection with both model Hamiltonian and alloy-band calculations) and to the incorporation of charge transfer between the alloy constituents that gives rise to important physical consequences.

The good results obtained with simple model calculations in recent years (Stocks *et al.*, 1971; Ehrenreich and Schwartz, 1976) has indicated that calculations based on first-principles crystal potentials similar to those used for perfect crystals were called for. Little progress was made along these lines because of computational complexities until recently when Stocks *et al.* (1978) presented the first complete solutions of the KKR–CPA equations for a nonoverlapping muffin-tin model of the crystal potential for three Cu_xNi_{1-x} alloys. In the muffin-tin model of the crystal potential they assign, within touching spheres, spherically symmetric potentials V^A and V^B. In the interstitial region the potential assumed is constant. They construct within CPA a system of effective scattering centers characterized by effective scattering amplitudes $f_{C,L}$ from which the configurational averaged properties of the real system may be calculated. The effective scattering amplitude is determined from a self-consistency condition that contains the difficult requirement of evaluating Brillouin zone integrals repeatedly for various guesses of the $f_{C,L}$s. Until the work of Stocks *et al.* (1978), such solutions appeared too difficult and it was proposed that the average *t*-matrix approximation (ATA)

$$f_2^{ATA} = xf_{A,L} + (1 - x)f_{B,L} \tag{20}$$

be used to calculate ensemble averaged observables. Stocks *et al.* [1978], showed that various proposed cluster CPA and so-called atomic sphere approximations (ASA) can be avoided with little extra computational effort and that the self-consistent equations can be directly solved. Their results, discussed later, generally agree with (and correct details of some of) the results obtained by Temmerman *et al.* (1978), who used a simplified version of the KKR–CPA scheme with ASA, but are qualitatively different from the results of the non-self-consistent form of the theory, i.e., ATA. The calculated densities of states are in good agreement with the results of photoemission experiments. Significantly, the aim of their KKR–CPA method goes far beyond simply obtaining densities of states. As they emphasize, once the effective scattering amplitude has been determined, most properties that are usually calculated for pure metals can be obtained without further complications and will be given in the future.

III. Illustrative Examples

We have emphasized earlier what may be taken as the dominant theme of this chapter, namely, that the current popularity of energy band theory stems from its successful application to the study of increasingly diverse problems in solid-state physics. Recent new sophisticated experiments on both traditional materials and those having complex crystallographic structures have demanded, however, not only theoretical descriptions of *eigenvalue* phenomena but also detailed and precise *wave functions* with which to determine not only charge and spin densities but also the expectation values of different observable operators. Such a demanding test of the predictions of one-electron theory has the additional virtue of permitting, by their comparison with experiment, accurate determinations of the relative magnitude and importance of many-body effects in real solids. Thus, as we have emphasized, there has developed considerable interest in applying the Hohenberg–Kohn–Sham local density functional (LDF) formalism, and its recent extension as a local "spin" density functional (LSDF) formalism to the investigation of various ground-state properties of solids. Indeed, it is the simplicity and universal applicability of the LDF that have made its application to a large variety of problems increasingly popular. In recent years, attention has been placed on testing the accuracy of ab initio LDF predictions of a variety of experimentally observed phenomena.

In what follows we discuss some of these tests and applications of LDF energy band theory by means of selected examples covering a variety of phenomena. No aim at completeness is possible, nor desirable. Our purpose is to provide a taste of the rich and diverse applications possible today. For practical and pedagogical reasons, and for greater clarity, the selection of examples will be mostly restricted to work in which we have been actively involved.

A. Charge and Spin Densities

1. CHARGE DENSITIES AND X-RAY SCATTERING FACTORS

Diamond has been long considered as a prototype for covalently bonded insulators and a great deal of experimental work has been done on its ground-state properties, including cohesive energy, lattice constant studies, x-ray scattering factors (Göttlicher and Wölfel, 1959; Renninger, 1955), charge density (Göttlicher and Wölfel, 1959), and directional Compton profile (Reed and Eisenberger, 1972; Weiss and Phillips, 1968). In addition, theoretical studies of its ground-state properties within the restricted Hartree–Fock (RHF) model are available (Euwema *et al.*, 1973; Surratt *et al.*, 1973; Wepfer *et al.*, 1974; Euwema and Greene, 1975), so comparison with the predictions of the LDF formalism is possible.

In an extensive study, Zunger and Freeman (1977b) have presented detailed results for the x-ray scattering factors, charge density, directional Compton profile, total energy, and equilibrium lattice constant using the LDF self-consistent LCAO–DVM scheme described in Section II, C, 3. In order to examine the effects of exchange and correlation on the ground-state charge density in diamond, they performed three fully self-consistent calculations; the first employed only the electrostatic electron–electron and electron–nuclear potential in the one-particle equation (electrostatic model), the second incorporated also the local exchange (exchange model), while in the third calculation the correlation potential was also considered (exchange and correlation model). All three calculations used an extended numerical set (1s, 2s, 2p, 3s, and 3p orbitals per carbon) and all lattice sums were performed to convergence. The lattice constant was fixed at 6.740 a.u.

Table I shows the calculated x-ray scattering factors in diamond at the three levels of local-density approximations, together with the experimental results (Renninger, 1955) and the canonical Hartree–Fock results of Euwema *et al.* obtained using an s/p Gaussian basis set (Surratt *et al.*, 1973; Euwema and Greene, 1975). It is apparent that exchange acts to increase the low-angle scattering factors quite dramatically, reflecting the increased localization of charge in the interatomic region, with the correlation effect being much smaller. In particular, the calculated (222) forbidden reflection (forbidden in the approximation in which the charge density is given as a superposition of spherically symmetric, possibly overlapping, atomic densities; last column in Table I) is increased by about a factor of 2 upon introducing exchange in the potential.

The general agreement between the calculated LDF results and experiment is reasonable. It is apparent from the comparison of the atomic and crystal values that, as expected, the scattering factors constitute a sensitive test for the details of the calculated charge density only for the first few reflections to which the density in the outer regions of the cell contribute. Similar conclusions can be drawn from comparing the exchange-model results with those obtained in the literature (Raccah *et al.*, 1970; Bennemann, 1964; Kleinman and Phillips, 1962; Goroff and Kleinman, 1967; Clark, 1964) by a number of approximations to the local exchange problem (Table I). Although these calculations employ various independent approximations, the results vary only in the first few reflections.

TABLE I

Calculated and Experimental X-Ray Scattering Factors for Diamond[a, b]

hkl	Electrostatic model	Exchange model	Exchange and correlation model	Experimental[c]	Hartree–Fock[e]	Superposition model
111	3.062	3.273	3.281	3.32	3.29	3.005
220	1.936	1.992	1.995	1.98	1.93	1.964
311	1.656	1.720	1.692	1.66	1.69	1.760
222	0.066	0.137	0.139	0.144[d]	0.08	0.0
400	1.470	1.494	1.493	1.48	1.57	1.585
331	1.625	1.600	1.605	1.58	1.55	1.519
422	1.411	1.423	1.408	1.42	1.42	1.432
511	1.347	1.385	1.392	1.42	—	1.387
333	1.346	1.381	1.392	1.42	—	1.387

[a] After Zunger and Freeman, 1977b.
[b] $f(\mathbf{K})_s = (1/\Omega) \int (1\mathbf{K}_x \mathbf{r})\rho_{cry}(\mathbf{r}) \, d\mathbf{r} \cdot f(0,0,0) = 6.0$. The "superposition model" results are calculated in the exchange model assuming the $1s^2 2s^2 2p^2$ configuration for the atoms.
[c] Göttlicher and Wölfel, 1959.
[d] Renninger, 1955.
[e] Euwema et al., 1973; Euwema and Greene, 1975.

Some further indication of the role of exchange versus correlation in LDF theory is given in the study of LiF, a prototype of an ionic solid, studied also by Zunger and Freeman (1977c). They considered the description of ground-state properties of the system, such as the band structure, charge density, x-ray scattering factors, cohesive energy, equilibrium lattice constant, and behavior under pressure, and compared the predictions of the LDF model with both experimental data and with available restricted Hartree–Fock (HF) results (Euwema et al., 1974a,b).

Figure 6 shows the total ground-state charge density calculated in (i) the exchange model, and (ii) in the exchange and correlation model, along the [100] direction in the unit cell. The position of their minima is given in Table II, and compared with the relevant experimental determination (Euwema et al., 1974a,b; Merisalo and Inkinen, 1966). The LDF results indicate that the effect of correlation is to *expand* the electropositive Li site at the expense of contracting the electronegative F. In this context it is interesting to note that Pauling's ionic radii (0.6 Å for Li and 1.36 Å for F^-) predict a much larger disproportion between the size of the lattice ions (15.3 and 34.7% for Li^+ and F^-, respectively) than do both the observed and the calculated values in the crystal.

The precise value of the minimum charge density in the unit cell is difficult to evaluate accurately from the experimental data, since small changes in the temperature parameters and structure factors introduce significant changes into this small quantity. The measurements of Krug

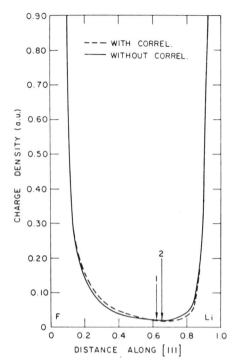

Fig. 6. Total ground-state charge density calculated in (1) the exchange and correlation model, and (2) the exchange model along the [100] direction in the unit cells. The arrows point to the positions of minimum density in the corresponding models. (After Zunger and Freeman, 1977c.)

TABLE II

THE DISTANCES FROM LI AND F SITES AT WHICH THE CHARGE DENSITY REACHES A MINIMUM EXPRESSED AS A PERCENTAGE OF THE LATTICE CONSTANT[a, b]

	Exchange and correlation model	Exchange model	Experimental[c]	Experimental[d]
R_{Li}	19.2	17.5	19.4	22.9
R_F	30.8	32.5	30.6	27.1

[a] $a = 4.01852$ Å.
[b] After Zunger and Freeman, 1977c.
[c] Merisalo and Inkinen, 1966.
[d] Krug et al., 1955.

et al. (1955) indicate a minimum density of 0.19 e/Å³ (0.028 e/a.u.³), while that of Merisalo and Inkinen (1966) show a minimum of approximately 0.15 e/Å³ (0.022 e/a.u.³). While the LDF calculation shows a minimal density of 0.155 e/Å³ (0.023 e/a.u.³) in good agreement with both measurements, the large uncertainties in the observed values may make this agreement fortuitous. It is interesting to note that Hartree–Fock *molecular* calculations (Bader and Bandrank, 1968) predict a much higher minimum along the Li–F bond, namely, 0.675 e/Å^3 (0.1 e/a.u.^{3}).

2. SPIN DENSITIES AND NEUTRON MAGNETIC FORM FACTORS

The electronic structure and properties of the transition (and, to a lesser extent, the rare-earth) metals have been measured using a variety of experimental methods that yield both macroscopic and microscopic information that have been related directly to the electronic band structure. By contrast, there are few experiments that yield direct information about the nature of wave functions in these metals. The determination of the neutron magnetic form factor in magnetically ordered systems, which gives the Fourier transform of the magnetization density, is perhaps the one outstanding exception. Here we describe several selected examples of spin densities and neutron magnetic form factors in metals.

a. Paramagnetic Sc Metal. In neutron measurements from a paramagnetic metal, one usually applies a uniform magnetic field and studies the paramagnetic scattering amplitude of the induced moment. Since the induced moment is small, only the electrons at the Fermi surface will be involved, as can be shown using linear response theory (Stassis *et al.* 1975; Freeman *et al.*, 1975, 1976). As shown by Stassis *et al.* (1975), for a uniform external field H, the Fourier transform of the induced moment is given as

$$M(\mathbf{G}) = H \int_{\text{F.S.}} \frac{ds_k}{|\nabla_k E_k|} \langle \psi_k | \exp(-i\mathbf{G} \cdot \mathbf{r}) | \psi_k \rangle, \qquad (21)$$

where the integration is restricted to the Fermi surface. Equation (21) may be rewritten as

$$M(\mathbf{G}) = H\rho(\mathbf{G}), \qquad (22)$$

where $\rho(\mathbf{G})$ is the Fourier transform of

$$\rho(\mathbf{r}) = \int_{\text{F.S.}} \frac{ds_k}{|\nabla_k E_k|} \psi_k^*(\mathbf{r}) \psi_k(\mathbf{r}). \qquad (23)$$

Here $\rho(\mathbf{r})$ [cf. Eq. (10)] is the spin density arising from the field-induced uncompensated states, and the factor multiplying the charge density, $\psi_k^*(\mathbf{r})\psi_k(\mathbf{r})$, is a local density of states weighting function that takes account of the detailed structure of the eigenstates at the point $\mathbf{k}$ (Freeman *et al.*, 1975, 1976). Thus the neutron form factor, i.e., the Fourier transform of the magnetization density, will be expected to provide highly useful information about the behavior of electrons at the Fermi surface. (The spin density at E_F was found using some 325 points in the $\frac{1}{24}$th Brillouin zone at which the wave functions were determined.)

In Fig. 7, we show the ρ_{LM} for some values of LM arising in the lattice harmonics decomposition of $\rho(\mathbf{r})$. We see that the main contributions to the spin density arise from the $L = 0$ and $M = 0$ term (i.e., the spherical part) and the $L = 3, M = 3$ term arising from interference between the $l = 1$ and $l = 2$ terms of the wave function. The $L = 2, M = 0$ and the $L = 4, M = 0$ contributions are considerably smaller. To better understand the spherical charge density ρ_{00}, Gupta and Freeman (1976a) further decomposed it into its respective $l = 0$, 1, or 2 character in order to determine its s, p, or d behavior. (Higher l components, i.e., $l > 2$ were not included since they were found to be rather small.) The induced spin density inside the spheres shown in Fig. 8 was found to be largely d-like except close to the sphere radius where it has considerable p character. This d-spin density itself is quite extended spatially in contrast to the atomic 3d density expected for the free atom/ion. Such a large expansion of the 3d density in Sc is in sharp contrast with the results obtained (Freeman *et al.*, 1975, 1976) for Pd metal — to be described shortly— and reflects the importance of solid-state boundary conditions: wave functions near the bottom of the d band are only required to be even under reflection across the Wigner–Seitz boundary, whereas the free atom wave function must approach zero at large distances.

What can be said about the relation of this theoretical spin density to experiment? The theoretical study described above was actually undertaken in close collaboration with the experimental work of Koehler and Moon

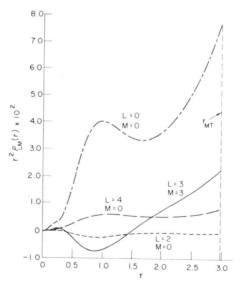

Fig. 7. The radial charge densities for some LM values arising in the lattice harmonics decomposition in Sc metal. (After Gupta and Freeman, 1976a.)

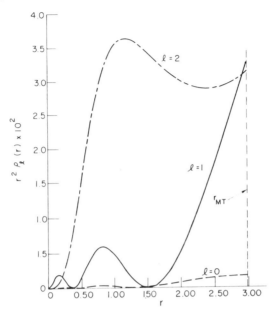

Fig. 8. Spherical harmonic contributions to the $L = 0$ charge density shown in Fig. 7 for Sc metal. (After Gupta and Freeman, 1976a.)

(1976). Thus it was hoped that a combined theoretical and experimental study would yield the distribution and nature of the conduction electrons of this important transition metal and (by analogy) the heavy rare-earth metals as well. As seen from Fig. 7, the theoretical APW magnetization density was found to be very *expanded* spatially and, hence, to be very different from that expected of a free atom/ion 3d density. The resultant form factor in Fig. 9 shows the interesting, but peculiar, feature of falling rapidly with $\sin\theta/\lambda$ then oscillating but yet remaining positive for the first ten reflections. This apparently large anisotropy is due to the asphericity in spin density and to interference, through the Fourier transform, between the inner and outer parts of the magnetization density. This predicted anomalous form factor is found to be in good agreement with the experiment of Koehler and Moon (1976). The orbital contribution to the form factor was not included because of the lack of knowledge about the gyromagnetic ratio for the metal. Exchange enhancement effects were also not included because, as found to be the case for chromium (Stassis *et al.*, 1975), the general shape of the curve is expected not to be greatly affected by such effects.

Koehler and Moon (1976) have also considered the nature of the experimental moment density shown in Fig. 10a as a contour map. The numbers shown are multiples of 10^{-3} μ_B/A^2 and the dark circles represent the projections of the atomic sites. In the projection the moment density is a

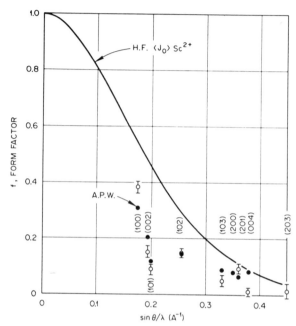

Fig. 9. Comparison of experimental (Φ) (Koehler and Moon, 1976) and theoretical (Gupta and Freeman, 1976a) APW results ($\circ$) for Sc metal along with the HF spherical free Sc^{2+} ion form factor.

maximum near the atomic sites but it is rather large, $\sim 0.3 \times 10^{-3} \mu_B/A^2$, and positive at the interatomic positions. They also prepared a similar map constructed from the Gupta and Freeman values. As seen from Fig. 10b, the theoretical results give quite good agreement with the experimental contour map shown in Fig. 10a. This good agreement gives added confidence to the theoretical results.

b. Paramagnetic Pd and Pt Metals. Even before this recent work on Sc metal, Cable *et al.* (1975) determined an accurate neutron magnetic form factor for paramagnetic Pd metal and showed that the magnetization was *contracted* by about 15% relative to the Hartree–Fock Pd^{2+} free-ion density function (which is already contracted by about the same amount relative to the neutral free atom). Freeman *et al.* (1975, 1976) obtained accurate APW solutions and, using Eq. (23) with wave functions calculated separately for the fourth, fifth, and sixth bands at 181 different points on the Fermi surface, obtained the results of the individual, spherically averaged, normalized band spin densities $\rho_j(r)$ shown in Fig. 11. Both the fourth and fifth bands have a localized distribution arising from the $l = 2$ (or d-like) electrons; by contrast, the sixth-band density is much more expanded

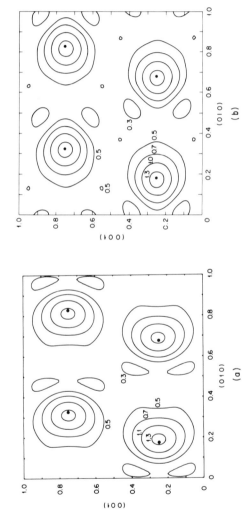

Fig. 10. Projected magnetic moment density in paramagnetic scandium and 100 K and 57.2 kOe. The contours are in units of 10^{-3} μ_B/A^2 with errors in the range 0.03–0.08. Atomic sites are indicated by solid circles. (a) Experimental results (Koehler and Moon, 1976); (b) theoretical results (Gupta and Freeman, 1976a).

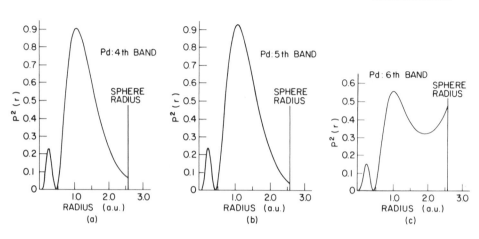

Fig. 11. The band 4, 5, and 6 contributions to the induced spin density in Pd. (After Freeman *et al.* 1975, 1976.)

indicating the large s- and p-like character of some of its wave functions. A surprising finding of these results is that the dominant fifth-band contribution yields a spin density that is *even more localized* than the Hartree–Fock spin densities obtained by Freeman and Watson (unpublished) for the free Pd^{2+} ion (also shown in Fig. 11). This is all the more remarkable when one recalls that the solid-state potential is derived from a superposition of neutral free-atom densities taken from the Hartree–Fock–Slater calculation with an assumed $4d^{10}$ configuration in which the 4d radial density is much more expanded spatially than the $4d^8$ ionic configuration.

The explanation for these highly contracted densities comes from considering the way the d orbitals are affected when placed in the solid (Freeman *et al.*, 1975, 1976). From an elementary viewpoint, the interaction among d orbitals on neighboring atomic sites can be thought of as producing bonding and antibonding states. The boundary condition for the lower energy-bonding states requires that the wave function be an even function between sites, whereas the antibonding states are required to have a node at the Wigner–Seitz cell boundary. In the solid, of course, a continuous range of radial densities is obtained. This is illustrated in Fig. 12, which shows the $l = 2$ or d-like radial probability density for energies near the bottom, middle, and top of the d bands. The antibonding wave functions at the Fermi energy are at the top of the d bands and the boundary condition that

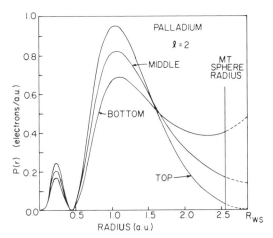

Fig. 12. Energy dependence of the $l = 2$ radial density in Pd metal. (After Freeman *et al.*, 1975, 1976.)

they go through zero at the Wigner–Seitz cell boundary (rather than at infinity as for atomic wave functions) causes them to be more contracted than the corresponding 4d wave functions in the free ion. It is thus clear why the free-ion model fails, and that solid-state effects are responsible for the highly contracted spin density observed in Pd.

Let us now consider the magnetic field induced form factors of Pd metal. The result of Freeman *et al.* (1976) for Pd is shown in Fig. 13. As expected from Fig. 11, the fourth- and fifth-band form factor contributions have the same angular dependence but differ markedly from the sixth band. As an estimate of the spherical part of the orbital contribution to the form factor, they have calculated the matrix elements of the atomic orbital operator from the spin-density functions for each individual band and for total contributions defined as in Eq. (21). The total magnetic form factor is taken as

$$f_{\text{tot}} = g^{-1}\big[2f_{\text{spin}} + (g - 2)f_{\text{orb}}\big] \qquad (24)$$

in order to compare with the experimental form factor of Cable *et al.* and the free-ion Hartree–Fock calculations. The final result is not sensitive to the approximate treatment of the orbital form factor, since with $g = 2.20$ from gyromagnetic measurements the orbital contribution to the form factor is only 10% of the total. We see from Fig. 13 that the agreement between the APW form factor and experiment is very good and is in much better agreement than that given by the Hartree–Fock Pd^{2+} calculations for which the experimental asphericity ($85\%\,t_{2g}$) has been assumed. The mean square deviation for f_{tot} compared with experiment is 0.023, whereas that for

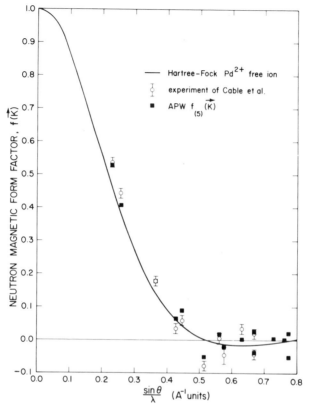

Fig. 13. Comparison between the experimental (Cable *et al.* 1975) form factor and the total band-5 form factor (Freeman *et al.* 1975, 1976); the solid curve is the free-ion Pd^{2+} Hartree–Fock form factor using the experimental asphericities.

f^{HF} is 0.035; the experimental mean square error is 0.017. In addition to the overall agreement between the theoretical and experimental form factors, there is also very good agreement between the measured and calculated asymmetry parameters (Freeman *et al.*, 1975, 1976). Exchange enhancement was found to have a small effect on the form factor and to mainly enhance the already dominant fifth-band contribution.

B. Cohesive Properties of Solids

As discussed in Section II, in LDF all ground-state properties of an interacting electron system are unique functionals of the ground-state charge density. This means that the very fundamental ground-state properties, such as equilibrium lattice constant, cohesive energies, and bulk mod-

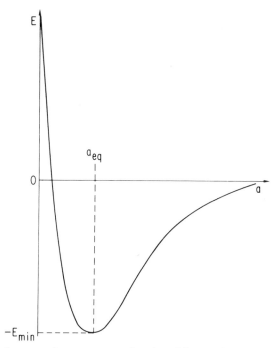

Fig. 14. Total energy of a system as a function of internuclear separation. a_{eq} denotes equilibrium separation.

ulus, are among those that can, in principle, be calculated with the least uncertainty. Here we indicate, by means of several examples, the degree of agreement obtained between LDF theory and experiment for these quantities.

The basic physical quantities are all obtainable from a determination of the total energy of the system as a function of internuclear separations. Figure 14 shows a typical curve with the zero of energy chosen to equal that of the ensemble of infinitely isolated atoms. The equilibrium internuclear separation is that distance a_{eq} at which the total energy has a minimum value $-E_{min}$, which in turn defines the cohesive or binding energy of the solid E_B. The bulk modulus corresponds to the curvature of the curve near the energy minimum. It is clear that these three features of the binding curve are rather sensitive quantities to inaccuracies in the theory and so provide stringent tests of any theoretical results.

1. COHESION IN COVALENTLY BONDED AND IN IONIC SYSTEMS

In Section II,C we described the fully self-consistent numerical basis set linear combination of atomic orbitals (LCAO) discrete variational method

(DVM) for treating ground- and excited-state properties of solids in the LDF approximation (Zunger and Freeman, 1977a). This scheme permits the treatment of general (i.e., analytic or numerical) basis functions and crystal potentials, and the determination of fully self-consistent solutions of the LDF one-particle equations without restricting the iterative path to a superposition of spherical charge densities or to muffin-tin models. It thus provides a highly accurate computational scheme for testing the predictions of LDF.

Zunger and Freeman (1978a) have studied the cohesive energy E_B and equilibrium lattice a_{eq} constants in covalently bonded solids (diamond and its heteropolar isoelectronic analog, cubic boron nitride) and a prototype ionic solid (LiF). It represents part of a detailed study into the ground-state properties of various solids, including the relative role of exchange and correlation. Table III represents their LDF results for E_B and a_{eq}, and, for comparison, those values calculated by the restricted Hartree–Fock (HF) model (Euwema *et al.*, 1974a,b; Surratt *et al.*, 1973) and the available experimental data (Hagstrum, 1947; Stull, 1965; Thewlis and Davey, 1968; Dworkin *et al.*, 1954; Berl and Wilson, 1961; Pauling, 1960; Gielisse *et al.*, 1967; Wentrof, 1957; Tosi, 1963; Fumi and Tosi, 1964). The values of the atomic total energies for C, B, and N were obtained from a spin-polarized local-spin density calculation (Gunnarsson, private communication) using the functional (Gunnarsson *et al.*, 1975) that approaches the Singwi *et al.* (1970) result in the nonspin-polarized limit. The corresponding values for the closed-shell Li^+ and F^- ions were obtained by a direct integration of the LDF total energy expression. The LDF model is seen to predict somewhat too low a binding energy (up to 7.5% for LiF) and too long a bond length (by 0.3–2%). The HF model, on the other hand, seems to yield too short a bond length. It is expected that those correlation corrections not present in both models will act to further stabilize the crystal over the noninteracting atoms, and to yield, thereby, better agreement of the LDF

TABLE III

Binding energies (E_B) and Equilibrium Lattice Constant (a_{eq}) of Diamond, Boron Nitride, and Lithium Fluoride Calculated by the Restricted Hartree–Fock (RHF) and the Present Local Density Formalism (LDF) Model[a]

	LDF		RHF		Experimental	
	E_B (eV/pair)	a_{eq} (Å)	E_B (eV/pair)	a_{eq} (Å)	E_B (eV/pair)	a_{eq} (Å)
Diamond	15.6	3.581	10.4	3.545	15.24	3.567
BN	12.8 ± 0.5	3.652	8.84	—	13	3.615
LiF	9.8 ± 0.5	4.09 ± 0.03	11.2	3.972	10.7	4.018 ± 0.03

[a] Zero-point energy corrections to the observed binding energies were done by means of a Debye formula. See Zunger and Freeman (1977e) for details and references.

results with experiment but too small a_{eq} values for the HF predictions. Both the HF and the LDF results show the predicted trend of decreasing binding and increasing lattice constant with increasing ionicity in the system (Pauling's electronegativity differences between the atoms in the cells are 0.0, 1.0, and 3.0 for diamond, BN, and LiF, respectively). Their charge analyses of the ground-state bands for these materials predict about 35 and 100% charge transfer in BN and LiF, respectively.

2. COHESION IN METALS

The transition metals have traditionally provided a rich and challenging testing ground for a variety of theoretical approaches. It is therefore not surprising that recent extensive investigations of the cohesive properties of metals by Janak, Moruzzi, Williams, and collaborators (Janak *et al.*, 1975; Janak and Williams, 1976; Moruzzi *et al.*, 1977; Williams *et al.*, 1979) have focused on the predictions of LDF in the 3d and 4d transition metals after an initial success with the simpler metals.

As is well known, the much discussed theory of cohesion in metals is based on the early semiquantitative estimates of Wigner and Seitz (1933, 1934) and Brooks (1963) for simple metals and the later work of Friedel and collaborators (1968) on the important role of d-band covalency in transition metals. More recently, self-consistent calculations performed either with Slater's $X\alpha$ method, i.e., variation of the exchange parameter α in Eq. (6) (Averill, 1972; Conklin *et al.*, 1972; Trickey *et al.*, 1973) or LDF (Janak *et al.*, 1975) demonstrated that a single theoretical formulation could describe cohesion in a broad range of metals from simple and transition metals to rare-gas solids. The analysis of Gelatt *et al.* (1977) of binding in the 3d and 4d series using the "renormalized atom" concept (Watson *et al.*, 1970) identified the role of configuration changes (i.e., occupation of s, p, and d states) and s–d hybridization in addition to that of d-band covalency.

In a set of state-of-the-art LDF calculations, Moruzzi *et al.* (1977) determined the lattice constant, cohesive energy, and bulk modulus for the 3d and 4d transition metals. As seen in Fig. 15, there is rather remarkable agreement of the LDF predictions with experiment. Not only is the trend with atomic number correctly described, but the absolute nuclear separations are predicted within a few percent. Since the calculations ignore magnetic effects, the actual closer agreement of the theory with experiment seen for the 4d series provides a clear indication of the importance of magnetic effects in the determination of these properties. It is seen that the sharp deviation between theory and experiment for the bulk modulus in the 3d series is particularly striking.

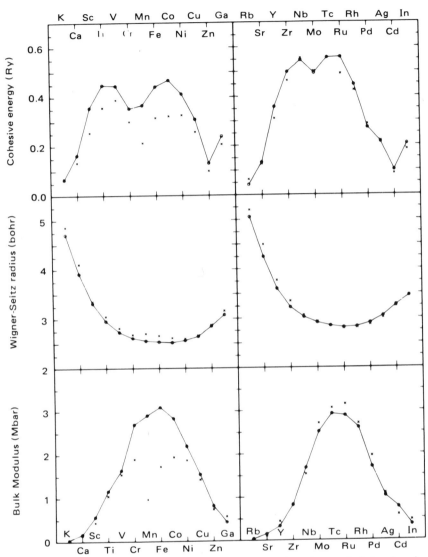

Fig. 15. Comparison with experiment of the results of self-consistent calculations based on the local-density approximation to exchange and correlation effects. The left- and right-hand sides of the figure present results for the 3d and 4d transition series, respectively. The middle row gives equilibrium lattice constants, or atomic densities, described by the corresponding Wigner–Seitz radius. The upper row gives cohesive energies and the bottom row bulk moduli. The calculations employ the muffin-tin approximation, ignore spin polarization, and are described in Moruzzi *et al.* (1977). Measured values are indicated by crosses, calculated values by the connected points. (After Williams *et al.*, 1979.)

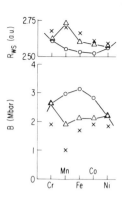

Fig. 16. Effect of magnetic order on the equilibrium lattice constant and bulk modulus. (After Williams *et al.*, 1979.) Crosses indicate measurements (Janak and Williams, 1976); circles indicate the results of paramagnetic calculations shown also in Fig. 15 and the triangles indicate the results of spin-polarized calculations.

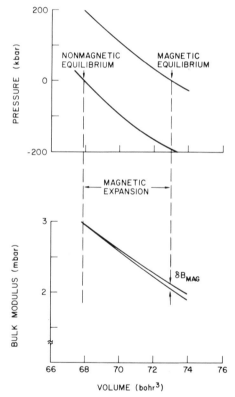

Fig. 17. Origin of anomalously small bulk moduli in magnetic transition metals. Variation with volume of hydrostatic pressure and bulk modulus in ferromagnetic and nonmagnetic Fe. Magnetic order causes the lattice to expand and the lattice expansion causes a large reduction of the bulk modulus. The intrinsically magnetic effect, i.d. the difference between the two curves in 2b, is seen to be relatively small. (After Williams *et al.*, 1979.)

The effect of magnetic order on the cohesive properties in the 3d series has been studied by Janak and Williams (1976) using the spin-polarized extension of the local density formalism. Their LSDF results are shown in Fig. 16. Janak and Williams explain the dramatic reduction of the bulk modulus due to magnetic order in a simple way. In Fig. 17 they show the effect of magnetic order on the pressure (i.e., the volume derivative of the total energy) and the corresponding effect on the bulk modulus. The tendency toward magnetic order is governed by a competition between the exchange and kinetic energy effects. Simply put, the exchange interaction causes a lowering of the energy by aligning spins. However, in a band scheme this is done by transferring electrons to states of relatively higher kinetic energy. Janak and Williams point out that this kinetic energy penalty is inversely proportional to the volume and can be reduced by a dilatation of the lattice. In other words, by expanding the lattice parameter the total pressure is increased at each volume and so magnetic order displaces the equilibrium volume to a larger value as is seen in Fig. 17.

The indirect, but larger, effect of the magnetic pressure is shown in the lower part of Fig. 17 where the magnetically induced lattice expansion is seen to result in a dramatic reduction of the bulk modulus. Figure 16 shows the magnitude of the resulting lattice dilatation for Mn, Fe, Co, and Ni. This figure shows that the lattice dilatation induced by magnetic order and the consequent effect on the bulk modulus account reasonably well for the discrepancies between theory and experiment seen in Fig. 15. Thus taken together, Fig. 15 and 17 indicate that the LDF self-consistent calculations do in fact contain all the essential ingredients of a theory of cohesion.

C. Electronic Structure of Metals

By far, the largest applications of energy band methods have been to the study of metals. Aside from their inherent interest to the materials science community, the calculation of their electronic structure has come to provide a good testing ground of the local density approach because the predictions of theory can be compared with a wide variety of experimental data. Several important examples of this have already been discussed. They represent but a very small fraction of the work done on electronic structure and properties of metals. Indeed the last two decades have witnessed remarkable advances in our understanding of metals and other materials sparked, in part, by the introduction and use of highly sophisticated experimental techniques that have brought about a mass of experimental knowledge. This latter has challenged previous (and often times simplified) theoretical models, and has brought about a more unified view of the electronic structure and behavior.

First the noble metals, then the transition metals and their compounds, and more recently the rare earths have each enjoyed a period of research on their electronic structure and properties that achieved a high degree of sophisticated understanding previously thought improbable. A similar period has most recently arrived for research on electronic structure and properties of actinide metals and compounds. Until recently, the problems of obtaining samples of sufficient purity and overcoming problems associated with their radioactive nature had resulted in slow progress and limited understanding.

Both the rare earths and the actinides have their own unique position among the elements. This arises from an unusual set of circumstances. In the transition elements, the s and d valence electrons form conduction bands that are responsible for their electric, magnetic, and optical properties. The rare earths are unique in that their 4f electrons are so highly localized that, although they determine the various exotic magnetic structures and properties of their metals, they have little effect on other chemical and physical behavior that arises from the transition metallike structure of the 5d and 6s valence electrons. By contrast the 5f electron actinide metals are not as well localized as the 4fs in rare earths but do have energies that are close to those of the 6d and 7s electrons. This produces in actinide systems the unusual condition of a strong "competition" between the 5f electron and the 6d and 7s electrons in determining their electronic structure and properties. The contrast between the results in the actinides and those in transition and rare-earths metals clarifies that the 5f electrons are a unique species.

Energy band calculations of both the rare-earth and the actinide metals have emphasized the importance of Coulomb correlations versus effective bandwidth in determining the itineracy (in light actinides) or localization of the 4f electrons in the rare-earth metals and of the 5f electrons in the heavy actinides. Thus one now has a description of the light actinides as a complex "transition metal" with overlapping s, d, and f bands all hybridized in the same region of energy space, and of the heavy actinides as a "second rare-earth series" (Freeman and Koelling, 1974). Thus in sharp contrast to the situation of the transition metals, the key question to be answered about the 4f and 5f electrons in the rare earths and actinides, respectively, is: Are the f electrons localized or itinerant?

It must be emphasized that the importance of these theoretical studies lies in the fact that they provide a means for undertaking quantitative correlations and interpretations of the electric, magnetic, and optical properties of these metals. Although experimental determinations of the electronic structures are still sparse (and here theory appears to have gotten ahead), what there is appears to agree with the more detailed predictions provided by band structure calculations. While still strictly within the framework of the one-electron model, the band results are also meaningful in revealing the nature and extent of many-body interactions in this interesting and important part of the periodic table. This development had, however, to await the availability of high-purity single crystals without which a wide variety of experiments cannot be made.

1. RARE-EARTH METALS

From the first determinations of the electronic band structure of the heavy rare-earth metals by Dimmock and Freeman (1964), there has emerged a picture of conduction bands that resemble closely those of the transition metals, but differ drastically from the free-electron model. Subsequent calculations, by these workers (Dimmock and Freeman, 1964; Keeton and Loucks, 1966; Duthie and Pettifor, 1977) and by Loucks and associates (Delley *et al.*, 1978; Young *et al.*, 1973; Williams and Mackintosh, 1968; Williams *et al.*, 1966), have confirmed this view and have yielded detailed results for the energy bands, density of states, and Fermi surfaces of both the heavy and light rare earths. The open f shell of electrons forms a narrow (localized in energy) band located well below the d and s conduction bands that are exactly like those of the transition metals (Freeman, 1972). These 4f electrons that are tightly bound to the atom do not overlap neighbors appreciably and, hence, form a very narrow energy band in the solid. Thus, 4f electron excitations may *not* be treated as band electrons because of the large errors (10–15 eV) introduced by neglect of the change in the intra-

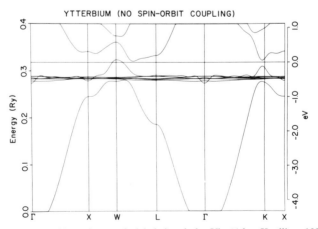

Fig. 18. The self-consistent relativistic bands for Yb. (After Koelling, 1980.)

atomic Coulomb correlation terms for the excited states (Freeman, 1972). As an example, we show in Fig. 18 the self-consistent relativistic bands for Yb (Koelling, 1980). The extreme narrowness in energy of the 4f states is evident.

Although these first calculations of Dimmock and Freeman were nonself-consistent, nonrelativistic, and non–spin polarized, nevertheless the general shape of the bands is surprisingly little affected by improved approximations. Typical of the early calculations, the 4f levels were ignored as much as possible. Later Loucks applied the relativistic APW method and published with Keeton (1966) the relativistic band structure of a number of rare earths. Two immediately noticeable effects caused by relativity are the net lowering of the s band relative to the d bands and the addition of spin–orbit coupling, which removes most of the band crossings.

These early calculations showed that as one went across the rare-earth series, the d bands rise in energy relative to the s–p band (Freeman, 1972). Recently Duthie and Pettifor (1977) were able to be more quantitative and showed that the number of d-like electrons decreases from about 2.5 in La to 1.9 in Lu. They managed to correlate the dhcp → Sm → hcp structural change along the series with this change in d-band occupancy. This trend is

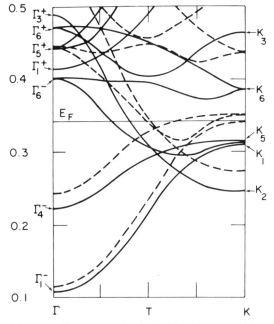

Fig. 19. The non–self-consistent bands of Gd. (Harmon and Freeman, 1974.)

discussed by Delley *et al.* (1978) in a recent publication dealing with the electrical resistivity of liquid rare-earth metals. It took 10 years after these initial calculations for precise experiments like the de Haas–van Alphen measurements of Young *et al.* (1973) to challenge the theory. Before this

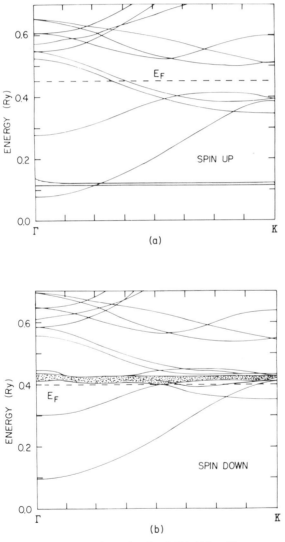

Fig. 20. (a) Self-consistent spin-up bands of Gd. (After Harmon *et al.*, 1978.) (b) Self-consistent spin-down bands of Gd. (After Harmon *et al.*, 1978.)

there were several experiments, notably positron annihilation (Williams and Mackintosh, 1968), which did not require the high-purity electrotransported samples and which did substantiate the general features of the theoretical calculations. Today, with the sample preparation problem overcome, accurate de Haas–van Alphen experiments (Young, 1979) are providing important information about both paramagnetic and ferromagnetic rare-earth metals and are providing the stimulus for carrying out more precise theoretical calculations. Recent efforts (Harmon, 1979) are being directed to the problem of how to treat properly the 4f bands and their interactions with the conduction electrons. This is a more difficult problem because the traditional band structure approach does not treat properly the atomiclike correlation of the localized 4f shells.

A good deal of attention has been paid to the problem of calculating the magnetic ground state of a ferromagnetic rare-earth metal like Gd. In such systems the inclusion of self-consistency can have important consequences. Figure 19 shows the result of the non–self-consistent calculation for ferromagnetic Gd (Harmon and Freeman, 1974). In the figure, the solid curves denote the spin-up bands. It is apparent that the s-band splitting is much smaller than the p-band splitting. In a self-consistent calculation, Harmon (1979) finds the s-band splitting is increased apparently because the d states, which have a net polarization from the 4f's, act in turn on the s states and cause a substantial s polarization. The self-consistent spin-polarized bands of Gd are given in Fig. 20a,b (Harmon *et al.,* 1978). It is seen that the conduction bands for the spin-up electrons look very similar to the original bands of Dimmock and Freeman (1964). The 4f bands (within the narrow shaded region shown in Fig. 20a) lie well below the Fermi energy. The situation is more complicated for the spin-down bands shown in Fig. 20b. Here the 4f bands lie in the shaded region just above E_F. As a result of hybridization, there are approximately 0.03 4f spin-down electrons per site and, hence, an extremely large density of states at the Fermi energy. Both results are unphysical and reflect the fact that the local-spin density formalism is not treating properly the atomiclike correlation that exists within the 4f shell (Harmon, 1979).

2. THE ACTINIDES

As expected from the electronic structure of their free atoms, the conduction bands of the actinides are derived from the 5f, 6d, and 7s electron states and so are more complicated than those of either the transition metals

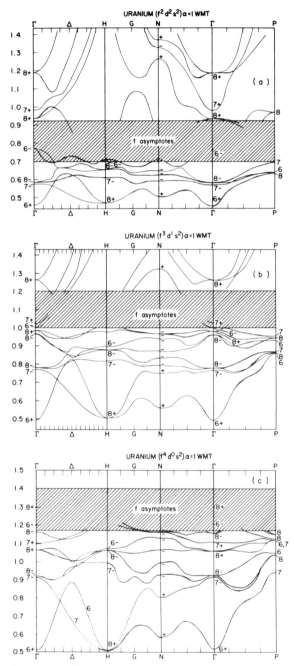

Fig. 21. Band structure of bcc uranium along high-symmetry directions. (After Koelling and Freeman, 1973.)

or the rare-earth metals. Perhaps the most important finding of the energy band work on the actinide metals is the apparent separation into two groups — the "lighter" (Th to Pu) and the "heavier" actinides (Am and beyond) (Freeman and Koelling, 1974). The 5f orbitals in the lower actinides must be considered from the itinerant or band point of view (since they do overlap strongly with their neighbors and hence are expected to form a fairly broad band). The band structure of bcc U is shown in Fig. 21 along high-symmetry directions (Koelling and Freeman, 1973). This is seen to be unlike the case of the 4f electrons described above, because the 5f electrons in the lighter actinides are very much part of the conduction band structure formed together with the 6d and 7s electrons. Their description by the band,* or itinerant, model is proper. By contrast, the 5f electrons in the heavier actinides, Am and beyond, form very narrow bands that do not hybridize greatly with the 6d and 7s itinerant bands. As an illustration, we show in Fig. 22 the band structure of Bk metal (Freeman and Koelling, 1974). This variation from the case of the light actinides is expected from the actinide contraction of the atomic 5f and 6d electrons, much as in the more familiar lanthanide contraction. This produces the sharply increased localization in energy of the 5f states in the heavier metals with increasing atomic number; i.e., greatly decreased overlap and rapid narrowing of their bandwidths.

Further, since the Coulomb correlation is large relative to the effective bandwidth, the itinerant (or band) description of excited state properties is *no longer valid*. Instead, the localized description of the ionic 5f electrons (i.e., large spin–orbit and Hund's rule coupling of L, S, and J—all maxi-

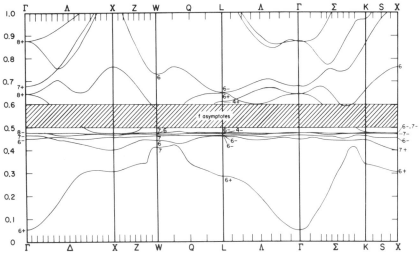

Fig. 22. The band structure of berkelium along high-symmetry directions. (After Freeman and Koelling, 1974.)

mized—interacting with their environment through the crystalline electric field and the exchange interactions via the conduction elecrons) appears more appropriate now, as in the case of the 4f electrons in the rare-earth metals. Indeed, an apt description of these metals is that they appear to form a second rare-earth series with all that it implies.

Why should the 5f orbitals be relatively less localized than the 4f orbitals? The primary factor is the requirement that they must be orthogonal to the 4f cores. As is well known in pseudopotential theory, this acts like a repulsive potential. Thus the 5f orbitals extend outside the closed shell 6p orbitals, whereas the 4f orbitals are found inside the closed shell 5p orbitals. As a result, the 4f's are much better shielded than are the 5f's.

Because the localization of the 5f orbitals in the solid is of such significance, it is interesting actually to see the effect of the actinide contraction. To do this, we show in Fig. 23 the $\kappa = 3$ ($f_{5/2}$) radial charge density out to the half-nearest-neighbor distance for γ - U, δ - Pu, β - Am, and β - Bk for the lowest f state (Γ_7^-). This should be the most extended density for the "f bands" of each solid. In γ - U, one sees that the radial density is quite large and increasing at the large distances; in δ - Pu, it is still quite significant and increasing; in Am, however, it is a good deal smaller and decreasing. For Bk, the radial density has actually decayed to zero with a small tail that arises from a small amount of plane-wave admixture. This will be seen to agree quite well with the conclusions already drawn from an examination of the band structures alone. Clearly the γ-U f orbitals are quite well delocalized and will satisfy the requirement that the Coulomb correlation (U_c) [cf. Eq. (18)] is less than the effective bandwidth (W). On

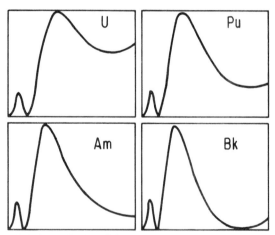

Fig. 23. The radial charge density of the $\kappa = 3$ ($f_{5/2}$) state for γ-U, δ-Pu, β-Am, and β-Bk for the lowest f state (Γ_7^-). (After Freeman and Koelling, 1974.)

the other hand, the Bk f orbitals are very compact and one will find $U_c \gg W$.

As an example of the current ability of the energy band scheme to describe the properties of complex systems, we turn to the recent work on α-U. The anomalous physical properties of α-U have made it a subject of great interest for experimental and theoretical study. Unfortunately, its complex structure (orthorhombic with 4 atoms/unit cell) and the resulting lack of symmetry, have made its theoretical study by conventional energy band methods exceptionally difficult and costly. Thus, while one was able to study the light and heavy actinide metals in the cubic (high temperature) structures (described above), the study of α-U awaited development of a computational scheme, which was not only rapid and efficient but which avoided the so-called asymptote problem that plagued earlier efforts.

Freeman *et al.* (1979a) and Koelling (1979) have reported results of relativistic energy band studies on α-U using a relativistic version of the linearized augmented plane-wave method described above. They determined the energy band structure, density of states (DOS), orbital angular momentum projected DOS [for use in analyzing the UPS experiments of Veal and Lam (1974)], detailed Fermi surface cross sections in close collaboration with de Haas–van Alphen work of Arko and Schirber [1979], wave functions and magnetic field–induced spin densities, neutron magnetic form factors [in close collaboration with and to understand the measurements of Maglic *et al.* (1978)], and generalized susceptibilities, $\chi(\mathbf{q})$ (for investigating possible electronically driven phonon anomalies and charge density waves). Because of space limitations, we are able only to give a brief indication of this extensive work here.

There is a good deal of structure in the total DOS in α-U arising from the lower symmetry and the hybridized set of s, p, d, and f bands. There is a rapidly increasing total DOS just above E_F which has relevance to specific heat and susceptibility determinations in alloys of uranium and to optical absorption measurement. Figure 24 shows the $l = 3$ angular momentum projected DOS that has been found to correlate well with the XPS data of Veal and Lam (1974), whereas the total DOS does not. This result is consistent with the fact that the higher angular momentum (more localized) f orbitals have much larger transition matrix elements. The large amount of $l = 3$ DOS above E_F indicates that this is the cause of the buildup in the total DOS in this same energy region. One might be tempted to attribute the occupied f character (i.e., the f DOS below E_F) to "tailing" caused by hybridization with the s, p, and d states. Although hybridization is present in this region of energy, many of the wave functions show nearly "pure" f character and so hybridization alone is not responsible for the observed occupied f states but involves the formation of a true f band.

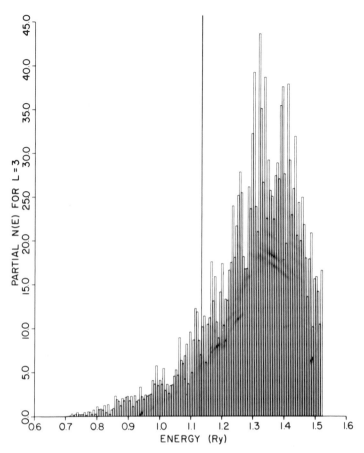

Fig. 24. The $l = 3$ angular momentum projected density of states in α-U. (After Freeman *et al.*, 1979a.)

D. Complex Systems

As an indication of the power and sophistication of the local density energy band approach dealing with the electronic structure and properties of complex systems, we briefly describe recent work dealing with magnetism and superconductivity in rare-earth ternary compounds. These materials are especially important because their discovery (Matthias *et al.*, 1977; Fischer *et al.*, 1975) has brought together the seemingly opposing collective phenomena inherent in magnetism and superconductivity. In particular, the observation of reentrant magnetism in $ErRh_4B_4$ (Ishikawa and Fisher, 1977) and $Ho_{1.2}Mo_6S_8$ (Fischer *et al.*, 1975), and, most recently, the coexistence of antiferromagnetism and superconductivity in several of the Chevrel ternaries have raised questions regarding the origin of these phenomena and our fundamental understanding of the interactions leading to magnetism and superconductivity.

1. TERNARY BORIDES

In the ternary boride MRh_4B_4 compounds (with M = a rare-earth metal) Matthias *et al.* (1977) found that either ferromagnetism (M = Gd, Tb, Dy, and Ho) or superconductivity (M = Y, Nd, Sm, Er, Tm, and Lu) existed at temperatures $T < 12$ K. (This behavior is in sharp contrast to that observed in the ternary Chevrel phase compounds $M_x Mo_6 S_8$ and $M_x Mo_6 Se_8$, where each rare-earth compound was found to be superconducting except for Ce and Eu). The sharp break in properties between the M = Ho and Er boride compounds is of especially great interest—with magnetism for Ho ($T_c^m = 6.56$ K) and superconductivity for Er ($T_c^s = 8.7$ K), even though the effective magnetic moments of both Re ions ($10.6\mu_B$ for Er) differ by only a small amount ($\sim 10\%$). Surprisingly, $ErRh_4B_4$, which becomes superconducting at 8.7, was found to become magnetic at $T = 0.9$ K with the return of the system to a normal conducting state ("reentrant magnetism" in a superconductor) (Fertig *et al.*, 1977).

Jarlborg *et al.* (1977, 1978) and Freeman and Jarlborg (1979) have obtained results of ab initio self-consistent (SC) energy band calculations carried out on three of these MRh_4B_4 systems, with M = Er, Y (an 11.3-K superconductor), and Ho in their paramagnetic states. The band calculations for the full 18 atom/unit cell structures were performed self-consistently using the linear muffin-tin orbital (LMTO) method and other related methods (Andersen, 1975). The calculational scheme is essentially the same as the one used earlier for a number of A-15 compounds (Jarlborg and Arbman, 1976) but extended to cover also ternary systems. [The LMTO method is a computationally rapid and efficient method for obtaining insight into the major physical aspects of energy band structure. By using the logarithmic derivative as a parameter, the usual energy dependence of the elements in the eigenvalue matrices can be neglected to a good approximation. Moreover, by replacing the Wigner–Seitz cell integrations by integration over overlapping Wigner–Seitz spheres, the matrix elements are very easily set up. All this makes the LMTO method much more rapid than the augmented–plane-wave and Kohn–Korringa–Rostoker methods without much sacrifice of accuracy (cf. Section II,C).]

Figure 25 shows the resultant total DOS for one spin for $ErRh_4B_4$, and the *l*-decomposed contributions to the DOS for the two Er and eight Rh atoms in a unit cell. (The DOS for the B atoms, not shown, shows a

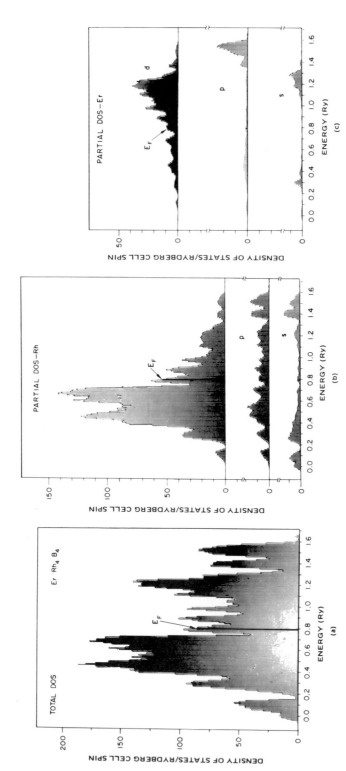

Fig. 25. The calculated (a) total DOS and *l*-decomposed partial DOS for (b) Rh and (c) Er sites in ErRh$_4$B$_4$. (After Jarlborg *et al.*, 1977, 1978.)

moderate p contribution of ~ 11.7 states/Rydberg cell spin at E_F, but a negligible s contribution.) This figure provides a qualitative understanding of the properties of the alloys. We focus on the fact that the Fermi energy E_F falls at a peak in the DOS and that this peak structure arises from the structure in the Rh 4d contribution. Since E_F did not occur at a peak in the DOS in the non–self-consistent calculations, this indicates the importance of the charge-transfer effects that are taken into account in the self-consistent calculations. The large transition-metal DOS from the Rh contribution indicates that a crucial requirement for the occurrence of superconductivity is satisfied. Further, a crude estimate of the electron–phonon coupling parameter λ obtained using the calculated bare total DOS and the measured electronic specific heat (Matthias *et al.*, 1977) yields $\lambda = 0.96$ and indicates that in strong-coupling theory a large T_c^s value would result.

The Er 5d DOS at E_F, ~ 8.9 states/Rydberg cell spin, is quite a bit smaller ($\sim 1/3$) than it is in the heavy pure rare-earth metals. Thus, through the exchange interaction between the localized 4f and rare-earth 5d electrons (hybridized with the Rh 4d's), the magnetic (RKKY type) coupling between the 4f local moments may lead to a still sizable magnetic ordering at a temperature T_c^m. For the alloys from Gd to Ho, the large effective spin moment leads to a magnetic interaction that dominates over the superconducting interaction and $T_c^m > T_c^s$. In the Er and Tm alloys— with their smaller spin moments, reduced 4f–5d exchange integrals (caused by the lanthanide contraction) and somewhat lower 5d DOS than in Ho—the "effective" magnetic-ordering temperature the system would have in the absence of the onset of the superconducting state is reduced, and so $T_c^s > T_c^m$. The observed value of T_c^m in Er is smaller than this effective ordering temperature because once the superconducting state has been achieved, the conduction-electron susceptibility is zero in principle, at least for the Rh electrons. Hence, for this compound, the RKKY interaction is largely ineffective compared to that in Ho (and earlier elements in the series); of course, dipole–dipole coupling may also contribute to the magnetic ordering. However, at lower temperatures, the RKKY interaction is still sufficiently strong to order magnetically the localized Er 4f moments that produce a sufficiently large exchange field to then destroy the superconducting state.

2. Chevrel Phase Compounds

One area of particular interest concerns the unusual behavior of rare-earth magnetic moments on the superconductivity of these compounds (Fisher, 1978). Contrary to observations on all other materials and the theory of Abrikosov and Gor'kov (1961), the superconducting T_c is only weakly depressed (if at all) by the magnetic moment through the pair-breaking effect of the exchange interaction. All the $(RE)_x Mo_6 S_8$ (or Se_8), with the exception of Ce and Eu, are superconducting, some with relatively high T_cs (Fischer et al., 1975; Shelton et al., 1976). The results for the system $Sn_{1.2(1-x)} Eu_x Mo_{6.25} S_8$ are particularly illustrative. As x increases to about 0.5, T_c is hardly changed with the depression occurring abruptly only at very high concentrations. The strength of the depairing interaction studied using ^{151}Eu Mössbauer experiments (Fradin et al., 1977; Bolz et al., 1977; Dunlap et al., 1979) show (1) a divalent Eu isomer shift (and hence a large magnetic moment of $7\mu_B$) typical for Eu^{2+} in an ionic compound without conduction-electron contributions, and (2) a spin-lattice relaxation rate that yields the product of the exchange coupling $\mathcal{J}$ and density of states $|\mathcal{J}N(E_F)|$ to be roughly one order of magnitude smaller (Fradin et al., 1977; Dunlap et al., 1979) than that measured in binary superconductors like Eu in $LaAl_2$.

Jarlborg and Freeman (1980) have presented results of the first ab initio self-consistent LMTO band studies of $SnMo_6S_8$, $SnMo_6Se_8$, $EuMo_6S_8$, and $GdMo_6S_8$ including all electrons in all 15 atoms in the unit cell. Figure 26 shows the structure of these materials. The charge densities from the core states as well as from partly occupied f states were recalculated in each iteration by using the actual MT potential.

The sharp structure in the total DOS shown in Fig. 27 for $EuMo_6S_8$ arises from the very flat nature of the energy band structure. (Very similar results were also found for the other compounds.) These flat bands yield low

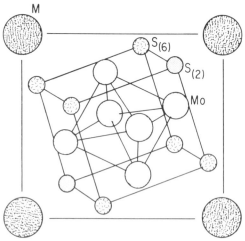

Fig. 26. The unit cell of the Chevrel phase structure for $SnMo_6S_8$, $SnMo_6Se_8$, $EuMo_6S_8$, and $GdMo_6S_8$.

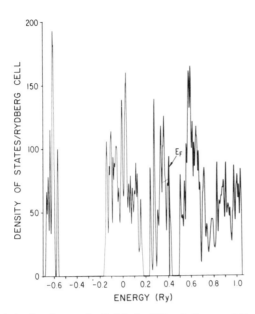

Fig. 27. Total density of states for $EuMo_6S_8$. (After Jarlborg and Freeman, 1980a.)

electron velocities at E_F and thus an unusually high contribution to the upper critical field from orbital effects $H^*_{c2}(0)$. There is a distinct gap in the DOS just above E_F, which falls in the middle of the Mo d bands and a smaller band gap between the Mo d and p states. In the "divalent" Sn and Eu compounds, E_F falls just two electrons below the middle of the bonding–antibonding "gap" of the metal d states. [This explains why the ternaries $Mo_2Re_4S_8$, which have two more electrons per cluster, are semiconductors (Fischer, 1978).] The additional electron contributed by Gd in $GdMo_6S_8$ raises E_F by one electron toward the gap and results in a much reduced $N(E_F)$.

The considerable structure in the total DOS, particularly around E_F, arises from the Mo 4d electrons. Usually for Mo compounds, the d bands are occupied up to the "gap" region where the DOS is low, but in the Chevrel compounds a large charge transfer from Mo to S (about 1 electron per Mo atom) was found to occur and E_F falls in a high DOS region below the "gap"; there is a high 4d DOS at E_F that is favorable for superconductivity. Indeed, the DOS per transition metal atom at E_F in the divalent systems is about 75% of that for the best superconducting A-15 compounds (Jarlborg, 1979).

Jarlborg and Freeman (1980) find that there is also a large charge transfer from the M site to the cluster; this gives Eu and Gd essentially no occupied conduction bands and Eu a typically divalent isomer shift in agreement with experiment. The Eu and Gd conduction-electron DOS and, hence, $|\oint N(E_F)|$ are reduced by an order of magnitude from their metallic state value. This results in a very weak coupling of the 4f electrons to the conduction electrons and hence only a very weak RKKY interaction leaving the dominant magnetic interaction to be the weak dipole–dipole interaction as surmised by Redi and Anderson (1979).

E. *Electronically Driven Phase Transitions and Phonon Anomalies*

One of the new exciting areas of investigation requiring accurate LDF–energy band solutions is the possible role of electronic structure on observed phase transformations and phonon anomalies. In particular, recent theoretical and experimental developments have called attention to the central role played by the generalized susceptibility function $\chi(\mathbf{q})$ in the understanding of many physical phenomena in solids. Since $\chi(\mathbf{q})$ measures the response of the system to an external (generally spatially inhomogeneous) perturbation, emphasis has been placed on possible anomalous behavior that may result from an instability in the conduction-electron gas as a result of a divergence in this response function. In linear response theory,

$$\tilde{\chi}(\mathbf{q}) = \chi(\mathbf{q})/[1 - I(\mathbf{q})\chi(\mathbf{q})], \tag{25}$$

where $I(\mathbf{q})$ is the electron–electron interaction and $\chi(\mathbf{q})$ is the bare susceptibility defined as

$$\chi(\mathbf{q}) = \sum_{n,n',\mathbf{k}} \frac{|M_{n,n'}^{\mathbf{k},\mathbf{k}+\mathbf{q}}|^2 f_{n,\mathbf{k}}(1 - f_{n',\mathbf{k}+\mathbf{q}})}{E_{n',\mathbf{k}+\mathbf{q}} - E_{n,\mathbf{k}}} \tag{26}$$

for bands n and n' and wave vector $\mathbf{k}$. The f's are the Fermi occupation numbers for occupied and empty states and the M's denote oscillator strength matrix elements.

Since the early work of Overhauser (1968, 1971), attention has been drawn to the role of Fermi surface "nesting" features, i.e., the existence of large parallel pieces of Fermi surface, in leading to a divergence in $\chi(\mathbf{q})$ and possibly to a divergence in $\tilde{\chi}(\mathbf{q})$ itself, depending on the value of $I(\mathbf{q})$. More recently, the possible role of so-called "volume" effects (parallel electron-hole bands crossing the Fermi energy) in producing divergences in $\chi(\mathbf{q})$ has been emphasized. The physical mechanism leading to either a spin density wave (SDW) or charge density wave (CDW) state is that a lower energy state results from the promotion of electrons from orbitals of one spin to orbitals of the same (or opposite) spin sustained by the Fermi surface, or band, geometry favorable to electron–hole excitations. Such a repopulation can also be achieved by the scattering of the electrons by phonons through the electron–phonon interaction. It can be shown that $\chi(\mathbf{q})$ plays an important role in the expression of renormalized phonon frequencies and that, under simplifying assumptions, a divergence in $\chi(\mathbf{q})$ can lead to a softening of the corresponding vibrational mode.

There are two steps required for the accurate determination of $\chi(\mathbf{q})$ in the constant matrix elements approximation: (1) the accurate determination of the underlying electronic band structure, and (2) a highly precise method for carrying out the phase-space (Brillouin zone) integral (summation) of a spectral Green's function [Eq. (26)]. The LDF–energy band method is now well accepted for successfully providing accurate eigenvalues and, hence, energy-related results in quite good agreement with experiment including complex metals and compounds. For the precise calculations of $\chi(\mathbf{q})$, it is now common to use the analytic tetrahedron linear-energy method (Rath and Freeman, 1975; Lindgård, 1975) derived as an extension of the work of Jepsen and Anderson (1971) and Lehmann and Taut (1972) on the density of states problem. In this method, using tetrahedrons as microzones with which to divide the BZ, a geometrical analysis is made of the occupied and unoccupied regions of any tetrahedron that reduces the problem of performing a volume integral over a tetrahedron with a linearized energy denominator. This procedure yields simple analytic expressions for the BZ integral that depend only on the volume of the tetrahedron and the differences of energies at its corners. The result is a computation scheme, not limited to constant matrix elements, which is highly accurate and, because of its simplicity, rapid to perform.

1. LATTICE INSTABILITY AND METAL–INSULATOR TRANSITION IN VO_2

The classic case of a metal–insulator transition occurs in VO_2 at 340 K. The crystallographic phase transformation from monoclinic rutile to the tetragonal rutile structure is accompanied by an abrupt jump in metallic conductivity ($\sim 10^5$) and a jump in magnetic susceptibility. VO_2 does not order magnetically at low temperatures. Gupta *et al.* (1977) have investigated whether the electronic properties, e.g., Fermi surface geometry and response function of the system, can account for the possible formation of a charge density wave (CDW), which could lead, through the electron–phonon coupling, to the renormalization of a phonon frequency, the corresponding phonon mode becoming overdamped at $T = T_c$, and driving the lattice to a new structural phase. This mechanism of a structural phase transformation proceeding via a soft phonon mode is well known in other materials.

First they have performed a first-principles energy band study of the metallic rutile phase of VO_2, using a general crystal potential and an

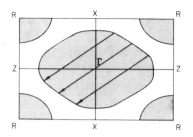

Fig. 28. The Fermi surface of metallic rutile phase of VO_2 showing nesting features corresponding to nesting vector $q = \Gamma R$. (After Gupta *et al.*, 1977.)

expansion of the Bloch functions in a linear combination of atomic orbitals. They obtained a large density of states at the Fermi energy. The Fermi surface was found to be determined by the two lowest d bands, at the bottom of the "t_{2g}" manifold which is split by the orthorhombic field; the lowest band Fermi surface possesses nesting features corresponding to a nesting vector $\mathbf{q} = \Gamma R$ (see Fig. 28).

Let us focus on the physical information contained in the generalized susceptibility function alone. The intraband contribution from bands 1 and 2 and the total interband contributions from the six bands belonging to the t_{2g} manifold are plotted separately in Fig. 29 along three high-symmetry directions $R(1, 0, 1)$; $P(\frac{1}{2}, \frac{1}{2}, 1)$, and $RP(\xi, \frac{1}{2} - \xi, \frac{1}{2})$. The total intraband value increases away from Γ and shows a sharp rise at the zone boundary R. The peak at R in the intraband function corresponds to a 29% increase from the value at Γ; this peak is due to the band 1 contribution and can be associated with the "nesting" features with wave vector $\mathbf{q} = (\frac{1}{2}, 0, \frac{1}{2})$ mentioned above. The magnitude of the band 1 contribution to $\chi(\mathbf{q})$ is generally 50% higher than the band 2 contribution throughout the Brillouin zone. This latter contribution shows in the ΓR direction a broad structure for $0.3 < q < 0.8$ associated with contributions from the FS piece around Γ, as the electron pocket around R has essentially the same characteristics for both bands.

The total interband contribution is remarkably flat and the resulting large background that it provides to the total $\chi(\mathbf{q})$ would be largely suppressed by the effect of the matrix elements. As no noticeable structure can be observed in the interband part, this contribution will be ignored. The results for $\chi(\mathbf{q})$ given in Fig. 29 determined along these several symmetry directions show an absolute maximum to occur at the point R in reciprocal space, which arises essentially from the FS nesting features of the lowest conduction band. Thus, the response function of the conduction electrons of the system shows an instability at the zone boundary R. The Fermi surface can sustain a charge density wave and, even though the phase transition for VO_2 is first order in nature, this instability could manifest itself by an overdamping of the corresponding phonon mode with wave vector $\mathbf{q} = (\frac{1}{2}, 0, \frac{1}{2})$, due to the renormalization of this mode through the electron–phonon coupling. This particular vector is compatible with the change from rutile to monoclinic

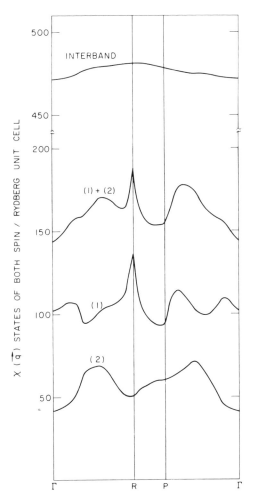

Fig. 29. The generalized susceptibility function $\chi(\mathbf{q})$ for VO_2. The intraband contribution from bands 1 and 2 and the total interband contributions from the six bands belonging to the t_{2g} manifold. (After Gupta *et al.*, 1977.)

structure and leads to a doubling of the unit cell in the low temperature phase. To test these predictions, Terauchi and Cohen (1978) carried out x-ray diffuse scattering studies as a function of temperature on VO_2 near the point R (0.5, 0, 0.5) using CuK_α radiation. They found that a lattice instability occurs at the R point with wave vector RR and polarization vector parallel to the "c" axis. This result is what is expected if a transverse phonon softens near T_c, whereas softening of a longitudinal phonon would cause diffuse scattering. Thus, the x-ray data are consistent with the softening of a transverse phonon near the R point.

2. Charge Density Waves in Layered Transition-Metal Dichalcogenides

Electron-diffraction studies (Wilson *et al.*, 1975; Williams *et al.*, 1974) have revealed that the previously observed (Thompson *et al.*, 1971) anomalous electromagnetic properties of the group-V B-layered transition-metal dichalcogenides may be related to the formation at some temperature of CDWs and their accompanying periodic lattice distortions. This anomalous behavior has been associated with a nearly two-dimensional Fermi surface supporting a CDW formation which, in turn, introduces a periodic lattice distortion generally incommensurate with the lattice (Wilson *et al.*, 1975; Williams *et al.*, 1974). The amplitude of these incommensurate waves grows from a high-temperature (below 550 K in $1T–TaSe_2$) down as far as the first-order transition temperature that represents the point of conversion (or "lock-in") to a commensurate geometry (Williams *et al.*, 1974).

Myron *et al.* (1977) reported results of detailed ab initio studies of $\chi(\mathbf{q})$ for the 1T polymorphs of both TaS_2 and $TaSe_2$. The band structures were determined by the KKR method. The Fermi surfaces (FS) (cf. Fig. 30) of

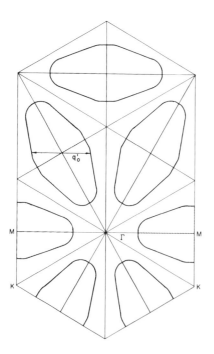

Fig. 30. Fermi surface of $1T–TaS_2$ in the basal plane showing the nesting vector q_0' in the ΓM direction. (After Myron *et al.*, 1977.)

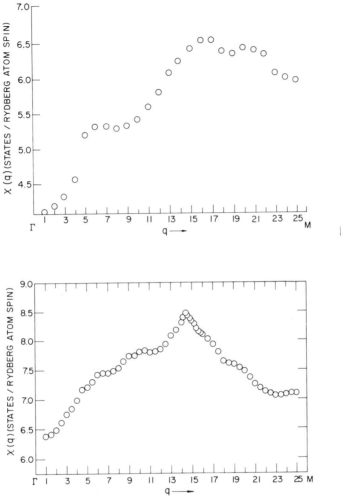

Fig. 31. (a) Generalized susceptibility of 1T–TaS$_2$ along ΓM direction. (b) Generalized susceptibility of 1T–TaSe$_2$ along the ΓM direction. (After Myron *et al.*, 1977.)

both metals were found to be very similar, having cross sections that are approximately constant in planes that are perpendicular to the *z* axis and which can be nested by approximately the same wave vector parallel to the ΓM direction. They also showed that the response function χ(**q**) contained not only information about the FS features (i.e., nesting) but also about band-structure effects for states just above and below the Fermi energy (volume effects). Both effects were found to contribute substantially to the important structure found in $\chi(\mathbf{q})$ as seen in Fig. 31. These accurate calculations showed major peaks to occur in the case of $TaSe_2$, at the $\mathbf{q}$ value corresponding to the observed CDW vectors. For $TaSe_2$, which shows *two* anomalies in the resistivity with temperature (whereas $TaSe_2$ shows only one), they find *two* peaks in $\chi(\mathbf{q})$ along the ΓM direction. Although a direct connection is perhaps somewhat tenuous, it is striking that this feature in $\chi(\mathbf{q})$ may well relate to the observed resistivity anomalies. The interested reader is referred to the original paper for details and a fuller discussion that appears to provide a detailed confirmation of the role of electronically driven instabilities as the origin of the observed CDW in these metals.

More recently, Zunger and Freeman (1977d) followed up these successful studies of the origin of the CDW in TaS_2 and $TaSe_2$ with a highly accurate self-consistent study of TiS_2 done with the LDF numerical basis set self-consistent linear-combination of atomic-orbitals-discrete-variational method (SC–LCAO–DVM) described above. Their results showed unambiguously that TiS_2 is an indirect gap (0.25 eV) semiconductor and thus resolved several controversies concerning this important material. The LDF results showed good agreement with optical properties for energies below 16 eV. Contrary to previous expectations, a *small* indirect gap (0.2–0.3 eV) was found to occur at the points L and M in the Brillouin zone, with a larger direct gap (0.8 eV) at Γ. A recent comprehensive analysis of transport, magnetic susceptibility, Hall effect, resistivity, Seebeck, etc., data (Wilson, 1978) yielded a gap of 0.2–0.5 eV, in agreement with these predictions.

These good results called attention to the need for a similarly accurate study of its sister compound, $TiSe_2$. Recent interest in the nature and unusual properties of $TiSe_2$ has centered on the unusual $2a_0$ by $2c_0$ superlattice which forms in $TiSe_2$ at low temperature. Wilson (1978) has suggested that this might be an example of the so-called excitonic state rather than a charge density wave state (CDW) such as that found in the Group V layered dichalcogenides. Woo *et al.* (1976) studied the intensity of x rays diffracted by the superlattice, magnetic susceptibility, resistivity, and far-infrared reflectivity, as a function of temperature in nonstoichiometric samples. Their results showed that the transition observed at $T_0 = 145 \pm 5$ K is second order, and they concluded that it was apparently a normal-to-commensurate transition driven by the Fermi surface (FS), which has a highly two-dimensional character. Recent neutron studies on stoichiometric samples found $T_0 = 202$ K involving transverse atomic displacement with wave vector $\mathbf{q} = (\frac{1}{2}, 0, \frac{1}{2})$; this transition was interpreted as being driven by electron–hole coupling.

Since extrapolation of the results for TiS_2 to the case of $TiSe_2$ is crude at best, and since muffin-tin or semiempirical models are not able to resolve these questions, Zunger and Freeman (1978a) undertook an ab initio study of the electronic band structure of $TiSe_2$ in order to understand its various measured properties. Of direct interest was to compare theory with the first angle-resolved photoemission studies of Bachrach et al. (1976), which showed conduction-band overlap of the valence band along the ΓM direction by at least 0.5 eV.

The band structure was calculated in the local density functional (LDF) approach, using their numerical basis set LCAO discrete variational method. They find that $TiSe_2$ is a semimetal with an indirect *negative* gap (0.20 eV) between Γ and L; contrary to some previous expectations, no electron pocket was found at M. The dimensions of the electron pocket at L (0.20 LH, 0.25 La, and 0.5 LM) indicate the presence of $7-8 \times 10^{20}$ conduction-band carriers/cm^3 in the perfect crystal. This is in very good agreement with the most recent independently reported transport and angle-resolved photo-emission experiments of DiSalvo et al. (1976) and Traum et al. (1978) at Bell Laboratories.

What may be said about the origin of the CDW? The several mechanisms proposed involve one or another form of "nesting" of electron–electron or electron–hole Fermi surfaces. It is possible that large peaks in the generalized susceptibility, $\chi(\mathbf{q})$, arising from characteristic "volume effects" provides the overscreening in these strong electron–phonon couples systems that drives the observed instability. Parallel bands, not necessarily crossing E_F, are not peculiar to $TiSe_2$ alone but prevail in many similar compounds such as TiS_2, $TaSe_2$, and VSe_2. Indeed, a major peak in $\chi(\mathbf{q})$ was found in $1T-TaS_2$ and $1T-TaSe_2$ to be caused almost entirely by "volume" effects (Myron et al. 1977). It therefore seems highly likely that this same mechanism drives the CDW instability in $TiSe_2$. Expressed in terms of the bonding in the system, such an instability can be described as a partially screened interaction between the electrons at L (with their wave-function amplitudes predominantly on the Ti site) and the holes at Γ (with wavefunction amplitudes predominantly on the Se site) with a zone boundary phonon supplying the coupling momentum and carrying out the structural change (e.g., shortening of the Ti–Se bond). Finally, whereas the phenomenological correlation between the observed instability temperature and the c/a ratio (or lattice ionicity) works remarkably well for many-layered dichalcogenides, including $TiSe_2$, it predicts an instability temperature in excess of 100 K for TiS_2. The Zunger and Freeman (1978a) analysis also suggests an explanation of the failure to observe such an instability in TiS_2; although both TiS_2 and $TiSe_2$ fulfill the volume effect condition, slight differences in the ionicities between them, which are not reflected in a dramatic change in the c/a ratio, are sufficient to place this "nesting" in a *gap* region in TiS_2 and in a *semimetallic* region in $TiSe_2$.

3. Electronic Structure and Phonon Anomalies in Transition-Metal Carbides

The transition-metal carbides and nitrides of the metal IV, V, and VI series are well known for their remarkable physical properties like high melting points and hardness, which indicate strong covalent bonds between the transition metal and the ligands. However, their electrical properties are comparable to those observed in good metals. An important fact is that the stoichiometric (or nearly stoichiometric) compounds with nine valence elecrons (like NbC and TaC) or ten valence electrons (like MoC and WC) are superconductors, with high T_c values ranging from 10 to ~ 17 K. In connection with the occurrence of superconductivity, an extensive experimental study of the phonon dispersion curves of the carbides has been carried out by Smith (1972) and by Smith and Gläser (1970). This study has shown that the carbides with high T_c temperatures possess anomalies in their phonon spectrum (NbC, TaC) that are not present in the spectra of the low T_c compounds (ZrC, HfC). The question of how phonon spectra develop anomalies is a complicated one involving details of the electron–phonon interaction, screening effects, and so on. This has been the subject of a great deal of theoretical work recently (Sinha and Harman, 1975; Varma and Weber, 1979; Ashkenazi et al., 1976; Hanke et al., 1976). The subject has recently been reviewed by Sinha (1980). Space does not permit us to discuss all these approaches, but to concentrate instead on the effect of the generalized susceptibility $\chi(q)$.

Gupta and Freeman (1976b) studied the possible origin of these anomalies by means of accurate calculations of $\chi(\mathbf{q})$. Without going through the exact derivation of the dynamical matrix, it can be shown that large maxima in the $\chi(\mathbf{q})$ function are related to a strong screening of the motion of the ions by the electrons; the ion–ion interaction is then expected to be strongly modified by an anomalous response of the electrons, and the phonon modes may be softened at those $\mathbf{q}$ values where the susceptibility function shows an anomalous behavior.

The remarkable result of these calculations for both NbC and TaC is the existence of strong structure in the contribution from band 4 to $\chi(\mathbf{q})$. More specifically, the maxima in $\chi(\mathbf{q})$, which occur at $\mathbf{q} = (0.6, 0.0, 0.0)$; $\mathbf{q} = (0.55, 0.55, 0.0)$, $\mathbf{q} = (0.5, 0.5, 0.5)$ in $2\pi/a$ units for the NbC match exactly with the positions of the dips in the longitudinal acoustic branches of the measured (Smith and Gläser, 1970; Smith, 1972) dispersion curves in the [100], [110], and [111] symmetry directions. For TaC, the positions of the maxima in the calculated $\chi(\mathbf{q})$ band-4 contribution found at $\mathbf{q} = (0.63, 0.0, 0.0)$, $\mathbf{q} = (0.55, 0.55, 0.0)$, and $\mathbf{q} = (0.5, 0.5, 0.5)$ are also in excellent agreement with experiment for the [110] and [111] directions. Even more remarkable, in the [100] direction, the small shift in the $\mathbf{q}$ value at which the anomaly occurs in going from NbC to TaC ($q_{th} = 0.60$ versus 0.63) reproduces the trend in the shift observed experimentally ($q_{exp} = 0.60$ versus 0.65).

One should also emphasize the large magnitude of the calculated maxima in $\chi(\mathbf{q})$. For band 4, the peaks in [100], [110], and [111] directions show an increase of 33.3, 55.2, and 91.7% for NbC and an increase of 32, 53, and 90 for TaC from the value of the function at $q = 0$ (which is the contribution from band 4 to the DOS at E_F). A striking feature of these results is that the ratios of these maxima also match closely to the magnitude of the depth of the mode softening, as can be seen by calculating $\omega^2(\text{ZrC}) - \omega^2(\text{NbC})$ or $\omega^2(\text{HfC}) - \omega^2(\text{TaC})$ at those q values where anomalies occur for NbC or TaC (note that ZrC and HfC do not possess any phonon anomalies). The calculated peaks are broad, as are the dips observed in the dispersion curves. Gupta and Freeman (1976b) also studied an arbitrary off-symmetry direction Γ to W defined by $(\xi, \xi/2, 0)$ with $0 \leq \xi \leq 2$. Their results show that the $\chi_4(\mathbf{q})$ function has a maximum at $\xi \sim 0.67$ for NbC and TaC in the Γ–W directions. Taken together, these calculations show that the maxima of $\chi_4(\mathbf{q})$ lie on the surface of a warped cube in $\mathbf{q}$ space, centered at Γ, and of approximate dimension $\sim 1.2(2\pi/a)$. This result is in striking agreement with the soft mode surface proposed by Weber (1973) starting from an entirely different (and phenomenological) formulation.

As a further test of the role played by structure in $\chi(\mathbf{q})$ in the occurrence of phonon anomalies, these authors also calculated $\chi(\mathbf{q})$ for the eight valence-electron compounds ZrC and HfC, by applying the rigid-band model to the NbC and TaC energy band results, respectively. Unlike the case of NbC and TaC, they find for ZrC and HfC that both the individual and the total interband contributions to $\chi(\mathbf{q})$ decrease in all directions from the value at $\chi(\mathbf{q}) = 0$. This results in a maximum at Γ and, consequently, in an overscreening at the zone center. These results correlate very well with the experimental results (Smith and Gläser, 1970; Smith, 1972) where the optic modes at Γ in ZrC and HfC have a significantly lower value than in NbC and TaC; the decrease at Γ is from ~ 17 to ~ 13 THz from NbC to ZrC. The general decrease of $\chi(\mathbf{q})$ away from Γ is also consistent with the fact that no anomalies are present in this low T_c compound. Thus, Gupta and Freeman conclude from this study of the nine and eight valence-electrons TMC that the phonon anomalies of these compounds can be explained essentially as being due to an anomalous increase in the response function of the conduction electrons, resulting in a strong screening of the corresponding phonon modes.

F. Surfaces, Surface Magnetization, and Chemisorption in Metals

As indicated in Section II,F,2 recent developments in experimental and theoretical methods have provided unique information about the electronic structure and properties of surfaces. Here we indicate the present state of the art in theoretical LDF capabilities as applied particularly to the transition metals.

1. TRANSITION-METAL FILMS: NI (001)

As in the case of bulk transition-metal studies, which have taken place over the past 40 years, the difficulty of treating localized d electrons along with the itinerant s–p electrons has provided the challenge and impetus for developing the sophisticated theoretical methods necessary for accurately determining the electronic structure of transition-metal surfaces. For bulk systems, considerable progress has been made in the last few years in this direction (Wang and Callaway, 1977; Callaway and Wang, 1977; Laurent *et al.*, 1979; Janak and Williams, 1976; Moruzzi *et al.*, 1977; Janak, 1977). Band-structure calculations of the electronic properties of transition-metal surfaces within the thin-film model include multiple scattering (Cooper, 1973, 1977; Kasowski, 1974a, 1975; Kar and Soven, 1975; Gurman, 1975), tight binding (Sohn *et al.*, 1976a–c; Dempsey and Kleinman, 1977a,b; Dempsey *et al.*, 1975, 1976a,b, 1977; Desjonqueres and Cyrot-Lackman, 1975), and supplemented orthogonalized plane-wave methods (Caruthers *et al.*, 1976; Caruthers and Kleinman, 1975). While providing valuable information about surface properties, early calculations are restricted by the arbitrariness in choosing a non–self-consistent (SC) potential, tight binding, or other parameters. In a recent study of the effects of different non-SC potentials on surface states in Fe, Caruthers and Kleinman (1975) concluded that SC is important for transition-metal film calculations in that both the existence and the symmetry of some surface states depend crucially on the details of the potential.

The first self-consistent calculation for a transition-metal surface was made for Nb (100) using a pseudopotential scheme (Louie *et al.*, 1976, 1977). Due to the localized nature of the d electrons, approximately 1000 plane waves ranging over an energy of 10.2 Ry were required to form the basis set. Convergence tests for bulk Nb (Ho *et al.*, 1977) indicated that the s–p levels and the d-band width were converged to 0.01 eV, and that the d level may shift relative to s–p levels by as much as 0.2 eV when additional plane waves up to an energy of 16 Ry were included by a perturbation technique. More recently, ab initio SC linear combination of atomic Gaussian orbitals (LCGO) calculations for three-layer Cu (100) and Ni (100) films were reported. While these required a much smaller basis set (Gay *et al.*, 1977; Smith *et al.*, 1978), the variational flexibility of their basis set may be limited by the neglect of the atomic 4p states. Both the pseudopotential and LCGO methods achieved self-consistency using the symmetrized plane-wave (SPW) expansion of the charge density. However, due to the lack of periodicity along the surface normal direction, this procedure is very cumbersome.

Extensive SC–LDF thin film energy band studies have been reported by Wang and Freeman (1979a,b) using the numerical SC–LCAO discrete variational method described in Section II,F. Their first application was to the investigation of a Ni (001) film. In this work, self-consistent LDF calculations were made for a single film slab of m layers with the origin of the system midway between the two surface layers; the z axis is normal to the surfaces. With the unit cell a parallelepiped (whose z dimension extends to $\pm \infty$), one has periodicity only in the two dimensions of the film. The LCAO Bloch basis set consists of 3d, 4s, 4p valence orbitals orthogonalized to the frozen core wave functions.

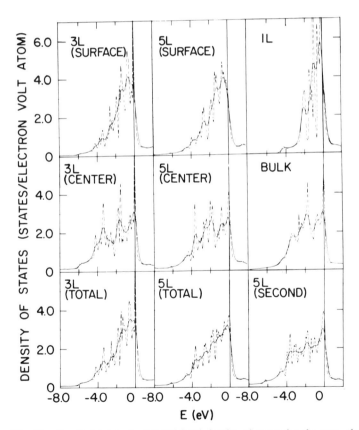

Fig. 32. Density of states and projected local density of states for the one-, three-, and five-layer Ni (001) films in units of electron states/atom electron volts. The dashed curves are the raw curves smoothed by a Gaussian broadening function of 0.1-eV FWHM; the solid curves are the result of a similar broadening of 0.5 eV. The bulk DOS for paramagnetic Ni of Wang and Callaway (1980) is shown in the right-hand center panel.

Their band structures are too complex to be easily understood. Instead, as an indication of the results obtained, as well as to demonstrate their convergence with thickness of the film, we show the layer projected density of states (DOS) for one-, three-, and five-layer paramagnetic Ni (001) films in Fig. 32. These DOS were calculated using a linear analytical triangle scheme based on a sampling of 15 k points in $\frac{1}{8}$ of the BZ before being smoothed by a Gaussian broadening function of 0.1 (dashed lines) and 0.5 eV (solid lines) full width at half maximum (FWHM). Smoothing is employed here because the continuous energy bands for wave vectors normal to the surface of a semi-infinite solid are replaced by discrete energy levels in a thin film model that leads to spikes in the DOS. Furthermore, the increasing number of accidental degeneracies in the film energy bands also makes the convergence in the k space integration rather slow. Note that (i) the center plane (CP) projected DOS of three and five layers converge rapidly toward that of paramagnetic bulk Ni, and (ii) the corresponding surface plane (SP) DOS show consistent narrowing of the d-band width and shifting of the peak positions from the CP DOS. Differences between the SP and CP DOS (and in the five-layer case also with the second plane DOS) may be important for interpreting results obtained with surface-sensitive spectroscopies.

In order to understand surface and finite thickness effects, we compare the Wang and Freeman (1979c) projected surface band structure with that of the bulk system as proposed by Caruthers *et al.* (1973). Note that the two-dimensional unit cell is rotated 45° in the xy plane away from the edge of the cube with length $a/\sqrt{x}$. To construct the projected band structure, we show in Fig. 33a the projection of a three-dimensional fcc unit cell defined by direct-lattice vectors $a(\frac{1}{2}, \frac{1}{2}, 0)$, $a(\frac{1}{2}, -\frac{1}{2}, 0)$, and $a(0, 0, 1)$ that is commensurate with that of the two-dimensional surface unit cell. The corresponding projection of the three-dimensional BZ defined by reciprocal-lattice vectors $(\mathbf{K}) = (2\pi/a)(1, 1, 0)$, $(2\pi/a)(1, -1, 0)$, and $(2\pi/a)(0, 0, 1)$

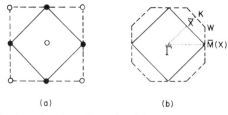

(a)   (b)

Fig. 33. (a) Projection of a three-dimensional fcc unit cell defined by the direct lattice vectors $a(\frac{1}{2}, \frac{1}{2}, 0)$, $a(\frac{1}{2}, -\frac{1}{2}, 0)$ and $a(0, 0, 1)$. The solid lines represent the base of the commensurate unit cell. (b) The corresponding projection of the three-dimensional BZ defined by the reciprocal-lattice vectors $[(\mathbf{K} = (2\pi/a)(1, 1, 0), (2\pi/a)(1, -1, 0), \text{ and } (2\pi/a)(0, 0, 1)]$.

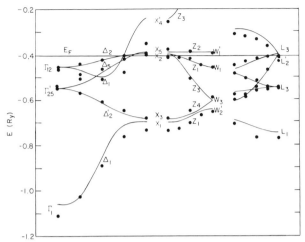

Fig. 34. Reconstruction of the three-dimensional band structure along the Δ, Z, and L to $(\pi/a)(\frac{1}{2}, \frac{1}{2}, 0)$ directions using the two-dimensional energy eigenvalues at the $\overline{\Gamma}$, $\overline{M}$, and $\overline{X}$ points, respectively. The bulk band structure for paramagnetic NI (solid lines) of Wang and Callaway (unpublished) are shown for comparison. (After Wang and Freeman, 1979a.)

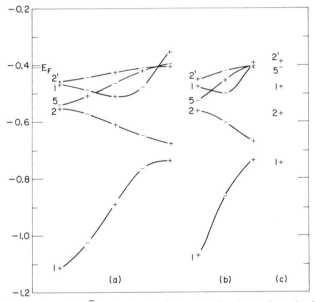

Fig. 35. Projection of the $\overline{\Gamma}$ level onto the Δ direction for the (a) five-, (b) three-, and (c) one-layer films. The $(+)$ and the $(-)$ symbols are used to denote even or odd parity with respect to reflection about the central plane. (After Wang and Freeman, 1979a.)

are shown in Fig. 33b. Symmetry points in the two-dimensional BZ are denoted by an overbar. Now in the limit of infinite film thickness the energies at a point $\bar{k}$ in the two-dimensional BZ span all the points at $(\bar{k}, k_z) + \mathbf{K}$ in the three-dimensional BZ, where $-\pi/a \leq k_z \leq \pi/a$. The two-dimensional energy levels at the $\bar{\Gamma}$, $\bar{M}$, and $\bar{X}$ points are used to reconstruct the three-dimensional energy bands in Fig. 34 along Δ, Z, and L to (π/a) $(\frac{1}{2}, \frac{1}{2}, 0)$ directions, respectively. The unpublished band-structure results of Wang and Callaway for paramagnetic bulk Ni are also included for comparison (solid lines, but with the origin of the bulk energy bands shifted in order to align the Fermi energies). Note that states of $\Delta_2(x^2 - y^2)$, $\Delta_2'(xy)$ symmetry agree very well with that of the bulk results, while states of $\Delta_5(xz, yz)$ symmetry, whose lobes at the surface atoms point toward the missing nearest-neighbor atom sites, deviate slightly more. A rather large downward shift of 0.05 Ry was found in the Δ_1 states of predominantly sp character relative to the d-band complex. The convergence with respect to the film thickness may be understood from the results of Fig. 35 where the projection of the $\bar{\Gamma}$ level onto the Δ direction for the one-, three-, and five-layer films are compared. Note that whereas the d-band complex can be described rather well by a three-layer film compared with the five-layer result, the plane-wave states (Δ_1) in the three-layer film are narrower. As the film thickness increases, the splitting of the levels, which for three-dimensional symmetry points are degenerate, gradually disappears. In the extreme case of a monolayer, the widths of both the d band and the sp states are much narrower and the resulting center of the d bands is found to lie relatively high in energy.

2. SURFACE STATES, SURFACE MAGNETIZATION, AND SPIN POLARIZATION IN FERROMAGNETIC 3D METALS

The study of surfaces has added a new and exciting dimension to the field of magnetism. This is clear from the wealth of experimental information that has recently been forthcoming to challenge the theoretical understanding of the role of surfaces in magnetism in general, and their importance relative to bulk contributions, in particular. Of particular interest have been the relative roles of bulk and surface contributions, since in several important cases agreement between experiment and bulk self-consistent (SC) calculations within the local spin density functional formalism is lacking (Feuchtwang et al., 1978). The theoretical transition metal surface studies share the common problem with bulk studies of treating localized d electrons along with the itinerant s–p electrons, but require, in addition, the treatment of larger numbers of atoms per unit cell.

For the case of ferromagnetic Ni, Wang and Freeman (1979c, 1980a,b) determined the spin-polarized SC band structure of a nine-layer Ni (001) film that is thick enough to accurately determine the energy dispersion and spatial character of SS and their effects on the surface spin polarization, charge distribution, and layer-projected DOS. The LSDF self-consistent potential (with von Barth–Hedin local density exchange-correlation) was obtained iteratively within the superposition of overlapping spherical atomic charge density model with the atomic configurations treated as adjustable parameters and a sampling of 15 equally spaced points in the irreducible Brillouin zone. The spin density is obtained self-consistently by minimizing the integrated root mean square difference between the crystal and superposition spin density.

Among their major results, they find a pair of majority spin $\overline{M}_3$ surface states (SS), which split away from the bulk bands and cross the Fermi energy E_F. This creates a majority spin d hole and decreases the surface layer spin magnetization (0.44 μ_B) and the exchange splitting (0.41 eV at $\overline{M}_3SS$) from their values (0.58 μ_B and 0.63 eV) for bulk ferromagnetic Ni (Wang and Callaway, 1977). This slight reduction in surface-layer magnetic moment is consistent with field emission (Bergmann, 1978; Sato and Hirakawa, 1975; Liebermann et al., 1970) experiments. No evidence was found for magnetically "dead" layers on Ni (001) surfaces. These results may be important (together with escape depth information) for interpreting recent angle-resolved photoemission (ARPE) experiments (Eastman et al., 1978).

Nearest-volume integration yields a practically neutral charge density around each atom (10.02, 9.97, 9.97, 10.02, and 10.05 electrons on surface and subsequent layers) and the spin magnetic moment values shown in Table IV. (Compared to the input superposition charge and spin densities, they are converged to within 0.03 electrons and 0.02 μ_β, respectively.) Note that the spin magnetic moments close to the center plane are in very good agreement with the experimental (bulk) value of 0.56 μ_β. The spin density shows the Friedel-type oscillations expected from a surface layer. Since the maximum magnetic moment occurs two layers below the surface, the SS responsible for the surface magnetism may be modified in a film less than or equal to five layers thick (Wang and Freeman, 1980a, b).

TABLE IV

LAYER SPIN MAGNETIC MOMENT IN μ_B IN 9-LAYER Ni (001) AND 7-LAYER Fe (001)[a]

	s	s − 1	s − 2	s−3	c	Bulk Theoretical	Bulk Experimental
Ni	0.44	0.59	0.62	0.56	0.54	0.58	0.56
Fe	3.01	1.69	2.13	—	1.84	2.16	2.22

[a]After Wang and Freeman, 1980a; s = surface layer and c = center layer.

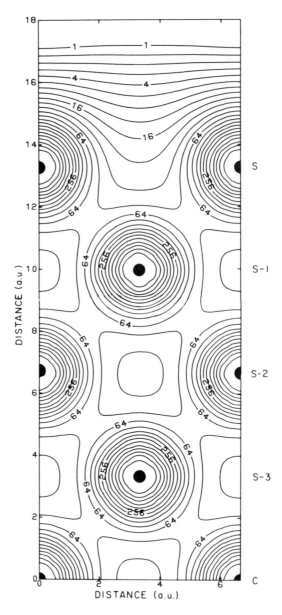

Fig. 36. The self-consistent charge density map for the nine-layer Ni (001) film on the (110) plane. (After Wang and Freeman, 1979c.)

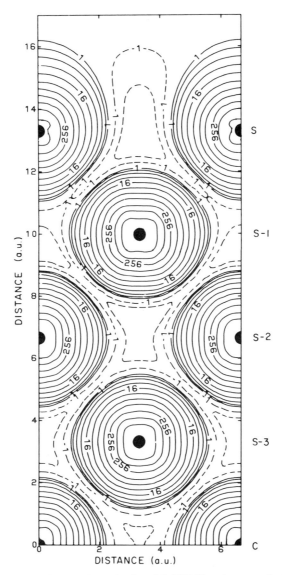

Fig. 37. The spin density map for a nine-layer Ni (001) film, shown on the face of the cube (vertical axis along 001). (After Wang and Freeman, 1980b.)

The SC valence electronic charge density map of the nine-layer Ni (001) film on the (110) plane with the surface normal along the vertical direction is shown in Fig. 36. The charge density is surprisingly bulklike starting at one layer below the SP. As one enters the surface region, the charge density gradually becomes smooth and parallel to the surface. Around each atom, there is a fairly large region where the charge density is spherically symmetric in confirmation of the often made muffin-tin approximation, except in the interstitial region on the SP where the charge density varies more rapidly due to the open-surface structure (Wang and Freeman, 1979c).

The spin density map on the face of the cube [vertical axis along (001)] is shown in Fig. 37 (Wang and Freeman, 1980b). In very good agreement with the bulk results (Wang and Callaway, 1977), the spin density is larger along the (110) than along the (001) direction. In the interstitial region (where the sp electrons dominate), the spin density, shown as dashed lines, is negative. This opposite polarization of the sp to the d electrons in the surface as well as in the bulk may be important in interpreting spin-polarized tunneling (Paraskevopoulos et al., 1977), field emission (Bergmann, 1978; Sato and Hirakawa, 1975; Liebermann et al., 1970), and electron capture (Eichner et al., 1971) experiments because the matrix elements for the extended sp electrons may be considerably larger than those for the localized d electrons.

Most recently, Wang and Freeman (1980a) carried out a similar spin-polarized self-consistent study of a ferromagnetic seven-layer Fe (001) film. Here they find a large number of SS in the middle of the d bands with the larger magnetic moment (2.15 μ_B) producing some difference between the two spin states. Comparing their layer-projected density of states (DOS) with bulk results, they find SS in the valley of the bulk DOS that seem to penetrate two layers below the surface for majority spin. In agreement with bulk results, the spin density for the film is largest along (001) directions and has similar magnitude along the (110) and (111) directions. Negative spin density was found along the (110) direction on the surface and center planes but not in the other planes.

The surface-layer spin density (cf. Table IV) shows an increase to ~ 3.0 μ_B compared to bulk. This result is only semiquantitatively correct because the Friedel oscillation penetrates deeper in Fe than was expected from the Ni results. Thus, a thicker film is required. The surface moment reduction in Ni (001) and its increase in Fe (001) may be consistent with the experiments of Bergmann (1978) showing that Ni films less than two layers thick possess no moment, whereas the first layer of an Fe film does. Here, Mössbauer or other hyperfine measurements may illucidate the surface magnetism in Fe.

3. CHEMISORPTION BONDING OF OXYGEN ON METAL SURFACES:
    OXYGEN C(2 × 2) ON NI (001)

Recent advances in spectroscopic measurements have provided important information about some fundamental features of chemisorption on transition-metal surfaces (Lapeyre et al., unpublished; Hagstrum, 1969; Becker and Hagstrum, 1972; Eastman and Cashion, 1971; Tibbetts et al., 1977; Yu et al., 1976; Andersson and Nyberg, 1975; Andersson, unpublished). Interest has centered on the nature of the chemisorption bonding of light atoms (especially O) and the chalcogens S and Se onto Ni (001) and other transition-metal surfaces and its proper theoretical description. Although a number of theoretical studies have been reported using molecular-cluster (or localized) models and energy band (or itinerant) methods, a full understanding of chemisorption bonding is still lacking. This is due to the inherent difficulties of performing self-consistent calculations on a realistic (but complex) system such as an ordered oxygen c(2 × 2) overlayer on the surface, say (001), of a transition metal, like Ni.

One central question of the theory has been the validity of LDF for describing chemisorption bonding. In the first of a number of theoretical studies of oxygen chemisorbed on Ni, Kasowski (1976a,b) studied an ordered p(1 × 1) O on Ni slab and found an O 2p nonbonding state at 5.4 eV below E_F and very little dispersion in the BZ. By contrast Liebsch (1977) found a rather large energy splitting and **k** dependence of the oxygen 2p bands for ordered p(1 × 1) oxygen on a semiinfinite Ni (001) film.

Since analysis of the low-energy electron diffraction (LEED) spectra (Demuth et al., 1973) deduced a c(2 × 2) structure on Ni (001), it became clear that a p(1 × 1) model overestimates the effect of interactions between adsorbates and between adsorbate and substrate. The qualitative feature of O 2p states split into bonding and antibonding states ($\sim$ 4 eV below E_F and above E_F, respectively) was found in a localized-orbital-pseudopotential calculation for a c(2 × 2) chalcogen overlayer on a three-layer Ni (001) slab (Bullett, 1977). The deviation of the position of the bonding state from experiment ($\sim$ 1.5 eV) is partly due to the neglect of self-consistency, of terms beyond second neighbors, and all three-center integrals in the tight-binding matrices. A more quantitative description of the bonds between the Ni 3d, 4s states, and O 2p states was reported recently by Liebsch in a layer Korringa–Kohn–Rostoker (KKR) muffin-tin calculation on the c(2 × 2) O on Ni (001) (Liebsch, 1978). Good agreement was found for the position of the O 2p bonding state with α (the value of the Slater exchange coefficient) chosen equal to 0.774.

A self-consistent chemisorption calculation by Smith et al. (1977) of N on Cu (001) found extremely large differences in the density of states (DOS) calculated using the self-consistent and the initially assumed non-SC potential. While the calculations found the antibondinglike N 2p resonance above the d bands observed in UPS, no obvious structure in the theoretical DOS can be associated with the UPS 2p peak found 5.6 eV below E_F (Tibbetts et al., 1977; Yu et al., 1976). This result may be due to the fact that the SC charge density for a p(1 × 1) overlayer of N on Cu (001) does not reflect directly the bonding of a c(2 × 2) structure found in the LEED analysis.

Wang and Freeman (1979b) have just reported the first ab initio SC–LDF study for oxygen c(2 × 2) on Ni (001). For this important prototype system they use the same LCAO scheme described above for an unsupported thin film and a numerical basis set (Ni: 3d, 4s, 4p; and O: 2s, 2p) orthogonalized to the (frozen) core wave functions.

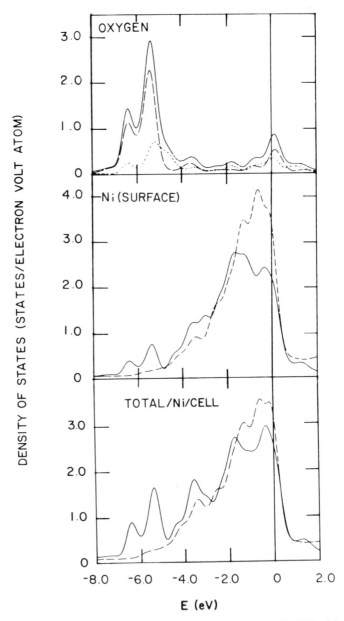

Fig. 38. Total (see text) and projected local DOS for three-layer Ni (001) slab (dashed lines) c(2 × 2) O on Ni (001) (solid lines) including the oxygen e (long dashes) and a_1 (dotted) orbital DOS. (After Wang and Freeman, 1979b.)

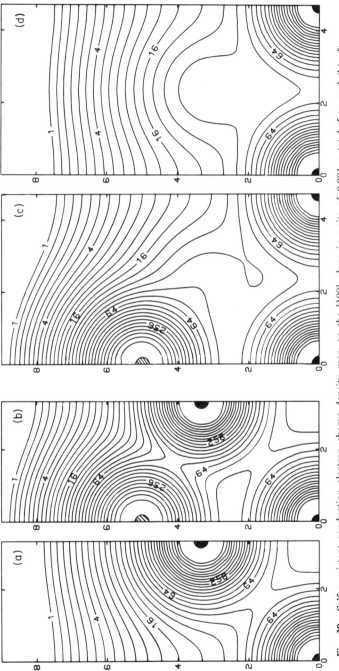

Fig. 39. Self-consistent conduction electron charge density map on the [100] plane in units of 0.001 e (a) before and (b) after oxygen chemisorption. Corresponding results on the [110] plane are shown in (d) and (c), respectively. (After Wang and Freeman, 1979b.)

Figure 38 compares their SC density of states results for c(2 × 2) O on a three-layer Ni (001) slab with that obtained for a pure three-layer Ni (001) slab. They were calculated by a linear analytical triangle method based on a sampling of 10 and 15 k points in $\frac{1}{8}$ of the irreducible BZ before being smoothed by a Gaussian broadening function of 0.5 eV full width at half maximum. In the top panel, in very good agreement with experiment, the projected local oxygen DOS shows bondinglike states 2 eV wide lying 5.5 eV below E_F and antibonding states right above E_F. The dotted and dashed lines are the $p_z(a_1)$ and $p_x, p_y(e)$ orbital density of states, respectively. The a_1 states lie slightly higher than the e states and are dominant in the region where they hybridize with the d bands of the substrate. This result agrees with Liebsch (1978) in that the ordering of the a_1 and e states is reversed at the Γ point; this is also consistent with recent angle-resolved photoemission measurements (Lapeyre, unpublished) for c(2 × 2) O on Ni (001). Compared with the clean Ni DOS (dashed lines) in the second panel, the surface plane Ni DOS has its main peak below E_F broadened by the addition of the adsorbate and shows induced states by (hybridization) having both sp and 3d contributions in the region around -5.5 eV, which corresponds to the position in energy of the O 2p bonding states. [Some of these effects persist into the center plane DOS between the corner Ni atom below the oxygen site (solid line) and the center Ni atom (dotted lines). Some of these changes may be exaggerated by the surface dipole moments on both sides of the center plane and should be reduced in a thicker film.] The bottom panel gives the total DOS divided by the number of Ni atoms per cell [i.e., 6 for c(2 × 2) O and 3 for clean Ni] to show the difference between the two cases.

Several features of the chemisorption bond are shown by the charge density contour maps in Fig. 39 for (110) and (100) planes with the vertical along the surface normal and contours plotted at $2^{1/3}$ times the previous value. Comparing the clean-Ni [(a) and (d)] and chemisorbed-Ni [(b) and (c)] maps shows the buildup of charge on the O site. The (100) plane contours (b) show more clearly the interaction of the O atoms with the surface plane Ni atoms in the formation of the adsorbate–substrate bond and the enhancement of the charge density in the interstitial region (Wang and Freeman, 1979b).

G. *Disordered Alloys*

In large part, the emphasis in this chapter has been on the current status of LDF-energy band theory for treating diverse phenomena in complex bulk systems and surfaces. In addition, we have indicated how these powerful methods can be adapted to systems of broken symmetry, including impurities and vacancy defects (to be treated extensively in Chapter 6) and in disordered systems and liquid metals. Here we give a brief indication of the status of theory and the type of results being obtained.

Theoretical studies of the electronic structure of disordered alloys have been largely confined to model Hamiltonians. These calculations do not yield detailed information such as Fermi surface topology of alloys that are now being measured. As indicated in Section II, G, the emphasis in recent years has been on using realistic Hamiltonians for studying the electronic properties of disordered alloys.

Here we discuss several recent calculations involving the coherent potential approximation (CPA) and its non–self-consistent form, namely, the average t-matrix approximation (ATA). The ATA has the advantage of being simpler than the CPA and hence easier to implement. Both the CPA and ATA are single-site approximations.

1. COMPLEX ENERGY BANDS IN α-BRASS

Bansil *et al.* (1974) have studied theoretically the electronic structure of the disordered alloy α-CuZn and its experimental consequences. They used the average t matrix approximation and evaluated complex energy bands over a range of Zn concentrations between 0 and 30 at.%. Such calculations of the complex bands within the ATA are very similar to those calculations of crystalline bands using a KKR approach. [This approach differs from the approach of Soven (1966) who calculates spectral density functions $\rho(\mathbf{k}, E)$ for a given $\mathbf{k}$ and E. The evaluation of $\rho(\mathbf{k}, E)$ not only involves the so-called KKR structure functions but also their derivatives.] Bansil *et al.* (1974) emphasize that by carrying over such concepts as the Fermi surface from the ordered crystals to the disordered alloys, complex bands provide a more natural means of comparing theory with different experiments than do the spectral density functions.

The calculations of Bansil *et al.* (1974) include the effects of charge transfer between the constituents of the alloy. Although these are known to affect the physical properties of many noble and transition metal alloys significantly, they had not been considered explicitly before. Previously neglected effects of the concentration dependence of the lattice constant of α-brass were also included in this work. The calculations are based on the KKR equations and on the use of atomic phase shifts obtained from renormalized atom muffin-tin potentials. Thus, this work considers realistic alloy potentials and predicts electronic level structure on the basis of the ATA. In discussing their calculated complex energy band, Bansil *et al.* (1974) find that the shifts of the real parts of various energy levels in going from the pure crystals to the alloy with and without charge transfer between the constituents are significant. This is particularly true of the copper d bands that are found to change significantly on alloying. Their results suggest that the virtual crystal approximation (Nordheim, 1931a, b) calculations (Amar *et al.*, 1966, 1967; Pant and Joshi, 1969) that hold them fixed artificially have a serious limitation. (Since the virtual crystal approximation yields only one set of d bands rather than the two characteristic, respectively, of Cu and Zn, a literal application would cause a rapid downward shift of a single d band in going from Cu to Zn.)

Interestingly, their results for the imaginary parts of the complex bands obtained from the ATA differ by about 30% at the Fermi surface from perturbation theory estimates. The copper d band damping within the ATA is either comparable in magnitude or larger depending on how charge transfer is introduced. As they explain, the Zn d-band damping vanishes in the ATA because these d states lie below the s muffin-tin zero of the alloy. Bansil *et al.* (1974) describe a useful decomposition of the damping of any state in the alloy into its angular momentum components. This is used for discussing the Ni anisotropy of the Fermi surface damping.

Bansil *et al.* (1974) also compared theory with experiment extensively in regard to both the real energy shifts and the damping of the states. Although they have adjusted the concentration dependence of the optical absorption edge to determine charge transfer, the changes in the 5-eV peak in the optical spectrum of Cu on alloying provide an independent test of the calculation. Figure 40 shows a comparison of their calculations with experiment. While details of this comparison are complex and are necessarily left to the original paper, an important conclusion of their work is that the study of optical structure as a function of concentration in various copper-based alloys may provide insight into the basic components contributing to the 5 eV and other peaks in this structure.

These authors have also examined in detail the Fermi surface properties of α-brass and have compared them with (i) positron annihilation measurements (Williams *et al.*, 1969; Murry and McGervey, 1970; Triftshäuser and Stewart, 1971; Becker *et al.*, 1971) for the radius of the neck orbit of α-brass, and (ii) low-concentration de Haas–van Alphen measurements (Chollet and Templeton, 1968; Templeton *et al.*, 1969; Coleridge and Templeton, 1971). The Fermi surface radii were obtained from a calculation of the Fermi energy that neglects the imaginary parts of the complex bands. Thus the effects of the broadening of the energy levels in the vicinity of the Fermi energy as well as those of the broadening of the copper d bands are neglected. As Bansil *et al.* (1974) emphasize, as in the case of a pure crystal, they assume each **k** state in the common conduction band in the alloy to have a weight of unity.

Figure 41 compares the neck and belly radii in the α-brass as a function of the Zn concentration both with and without including the lattice expansion effect. These results show that the various dimensions of the Fermi

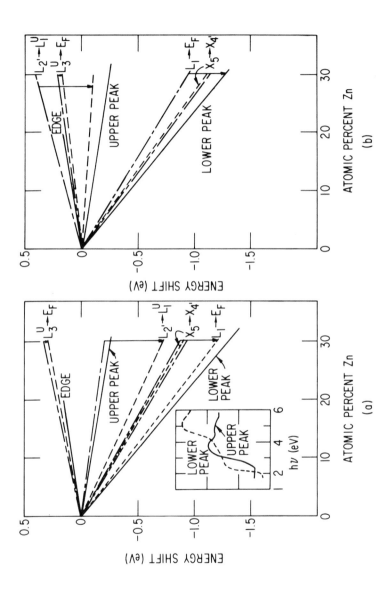

Fig. 40. Shifts with respect to Cu, in the dominant features of the experimental optical spectrum of α-CuZn, and the calculated changes in the energies of the principal optical transitions: (a) the CR model, and (b) the SMT model. The solid lines give the experimental results (Bionde and Rayne, 1959; Pells and Montgomery, 1970). The dot-dashed lines refer to the calculated values without lattice expansion; the dashed lines give the calculated values corrected for lattice expansion. The inset shows the experimental optical-absorption curves of Pells and Montgomery (1970) for Cu (dashed line) and α-Cu₀.₆₈Zn₀.₃₂ (solid line). (After Bansil *et al.*, 1974.)

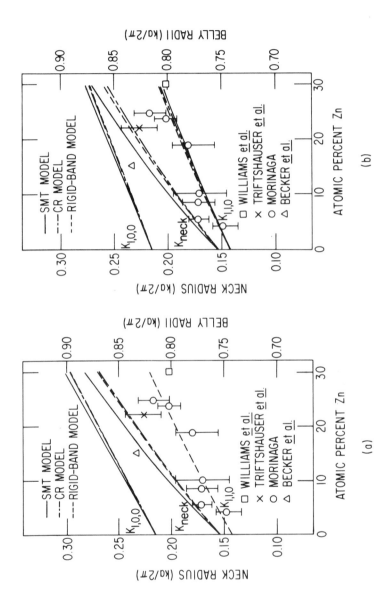

Fig. 41. Neck and belly radii in $\alpha\text{-}Cu_x Zn_{1-x}$ as a function of the Zn concentration: (a) without lattice expansion; and (b) including the lattice-expansion effect using the results of Davis *et al.* (1968). The solid lines and the dot-dashed lines refer to the SMT and the CS models, respectively, while the dashed lines give the results for the rigid-band model with the Cu density of states. The experimental data points for the neck radius are also shown. (After Bansil, *et al.*, 1974.)

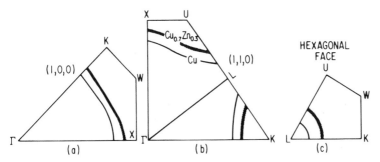

Fig. 42. Intersections of the Fermi surface of Cu (unshaded) and of α-$Cu_{0.7}Zn_{0.3}$ (shaded): (a) with a $(1,0,0)$ plane; (b) with a $(1,1,0)$ plane; and (c) with the hexagonal face of the Brillouin zone for the SMT model. The shading corresponds to six times the uncertainty in momentum on the Fermi surface of the alloy. (After Bansil *et al.*, 1974.)

surface appear to be insensitive to the choice of the model for charge transfer. The Fermi surface of copper is seen to expand rather uniformly in all the directions in **k** space as is shown in Fig. 42, which shows the intersections of the Fermi surface of copper with a few symmetry planes. The shading in this figure corresponds to six times the uncertainty and momentum of the Fermi surface of the alloy.

Finally, as further discussed by Bansil *et al.* (1974) the experimental information on the damping of the electronic states in α-brass is rather limited and the theoretical situation appears to be less satisfactory. Thus, for example, they find that the agreement between the experimental (Chollet and Templeton, 1968; Templeton *et al.*, 1969; Coleridge and Templeton, 1971) and calculated Dingle temperatures is within a factor of 2. In contrast with the real energy shifts, damping effects appear to be far more sensitive to the model for charge transfer and thus may be an indication that these results may be unreliable within their present scheme of calculation.

2.   KKR–CPA for CuNi Alloys

An important application of the KKR–CPA has been reported by Stocks *et al.* (1978). The authors have obtained results for the electronic states in disordered Cu–Ni alloys based on a complete solution of the CPA for a muffin-tin model of the alloy potential. They have shown that the approximations involved in the cluster CPA and ASA–CPA can be avoided with little extra computational effort. While their results generally confirm those obtained earlier by Temmerman *et al.* (1978), who used a simplified version of the theory in order to facilitate the calculation, they are quantitatively different from the results of the non–self-consistent form of the theory, namely, the ATA.

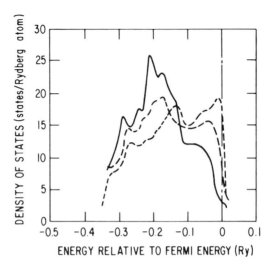

Fig. 43. Densities of states for three Cu–Ni alloys. $Cu_{0.77}Ni_{0.23}$, solid line; $Cu_{0.50}Ni_{0.50}$, dashed line; $Cu_{0.19}Ni_{0.81}$, dotted line. The zero of energy is the Fermi energy. (After Stocks *et al.*, 1978.)

The average density of states is shown in Fig. 43 for the three concentrations considered by Stocks *et al.* (1978). For all three concentrations the prominent structures in the density of states are in agreement with UV and x-ray photoemission experiments and also with the hybrid tight binding nearly free-electron calculations of Stocks *et al.* (1971). Note that in Fig. 43 the gradual switch from a split band regime at low nickel content to the modified virtual crystal behavior at high nickel content is clearly evident. As these authors point out "in the $Cu_{0.19}Ni_{0.81}$ alloy, calculation of the component densities of states shows that the two peaks at -0.01 and -0.12 Ry arise from Ni sites and are rounded versions of the two peaks in pure Ni, which occur at these energies. In $Cu_{0.50}Ni_{0.50}$ there is a broad structureless Ni peak that extends from E_F to 0.12 Ry below E_F and a more structured Cu peak that extends from 0.15 to 0.35 Ry below E_F. The Ni peak is centered around 0.04 Ry below E_F; this is in good agreement with the values of 0.04 and 0.20 Ry for the corresponding structures seen in photoemission experiments."

Stocks *et al.* (1978) have also emphasized the differences between their results and those of the ATA calculations of Bansil *et al.* (1974). This includes the fact that for the $Cu_{0.19}Ni_{0.81}$ the ATA gives a clearly resolved Cu-related structure at the bottom of the Ni band, whereas this structure is not present in the CPA. They attribute this to the fact that the CPA effective scattering amplitude has only a single "virtual crystal"-like resonance while the ATA scattering amplitude has a double resonance—one at the resonance energy of the Cu potential and one at the resonance energy of the Ni potential. The fact that both the other concentrations in the CPA results show the high-energy Ni band to be less structured than in the ATA is considered to result from intersite d–d interaction effects in the CPA, which broaden the Ni resonance in the effective scattering amplitude and push it to higher energies relative to the ATA. That the ATA omits this d–d repulsion effect also appears to account for the excessive erosion of the top of the copper d-band complex in $Cu_{0.77}Ni_{0.23}$.

Finally, as emphasized by Stocks *et al.* (1978) the aim of the KKR–CPA goes far beyond the density of states. They note that once the effective scattering amplitude has been determined, most properties that are usually calculated for pure metals can be obtained without further complications. Detail and comparison of the results with angle-resolved photoemission, soft x-ray emission, and Fermi surface experiments have been promised in future publications.

H. Microstructure Materials and Their Properties

In addition to the more traditional subjects facing the materials scientist discussed above, there are new exciting areas of research that are becoming increasingly active. The example, discussed below, may thus be a fitting way to close this chapter.

Recent scientific and technological developments have led to intense interest in (i) the study of phenomena possible with materials of submicron size, and (ii) the potential device applications inherent in microstructure phenomena. The range and variety of the phenomena that can be studied and understood pose an exciting challenge and opportunity for solid-state physicists and materials scientists. Much of the incentive for these techno-logical efforts lies in the universal recognition of the enormous economic potential inherent in large-scale applications of microelectronics based on silicon-integrated circuits. The extension of the technology to the submicron level promises to yield a new generation of ultrafast high-density logic and memory devices; here employment of compound (and possibly multilayer) semiconductors may play a significant role. Furthermore, with the move to fabricate cryogenic computer elements, utilizing the Josephson, and other phenomena, the further development of submicron superconducting tech-nology takes on increased significance.

In addition to the technological reasons for focusing attention on sub-micron problems, there are interesting phenomena to study that are of fundamental physical interest in themselves. In some sense we can see this field as a natural extension of a trend in the mainstream of condensed matter science; attention spreads from bulk properties to surfaces and the next natural extension is to obtain a better fundamental understanding of interfacial and reduced dimensionality phenomena.

A highly important development lies in the area of layered-synthetic microstructures. Work on such artificial superlattices was initiated by DuMond and Youtz (1940) and has been actively pursued in metallic systems by Hilliard and his associates over the last fifteen years (Hilliard, 1979). In general, the fabrication of layered synthetic microstructures permits a wide range of elements to be combined with layer thicknesses varying from a few angstroms to hundreds of angstroms. Thus it is clear that materials as microstructures, not found in nature, that are kinetically metastable are in fact realizable. It has been emphasized by a number of groups at different institutions that rather unique properties resulting from this submicron structural engineering capability are expected. Important examples include the enhanced superconductivity of metal layers when separated by, say, semiconducting layers (work of the Stanford group, following the pioneering work at Brookhaven by Strongin and co-workers), the enhanced magnetism and elastic modulus of Cu–Ni-layered microstructures (Hilliard and co-workers at Northwestern University), and enhanced conductivity in semiconductor heterojunctions (Dingle *et al.* at Bell Laboratories).

Energy band theory is in a good position to accept the challenge presented by experimental results on these systems. As should be apparent from the descriptions given in this chapter, the "supercell" local-density energy band scheme is ideally suited to determine accurately the electronic structure and properties of both unsupported thin films (slabs), sandwiches of dissimilar films, and the modulated structures of such thin films of metals/insulators. Theoretical calculations on modulated structures using both local density and local spin density approaches are being undertaken by Jarlborg and Freeman (unpublished) and are beginning to yield information of direct use in interpreting transport, magnetization, specific heat, and other measurements. For example, for Cu–Ni modulated structures consisting of three layers each of Cu and Ni planes of atoms, they find the layer projected DOS for the two types of Ni and two types of Cu atoms in the paramagnetic case shown in Fig. 44. Note that the DOS-labeled "bulk" refers to Cu (or Ni) planes that have two Cu (or Ni) planes as nearest neighbors and are close to the DOS for the bulk Cu (or Ni) metal with essentially three separate peaks. The "interface" DOS reflect the effect of the changed environment and have a smeared-out structure as a consequence. Finally, in the ferromagnetic case, they have carried out spin-polarized calculations for this modulated structure. Jarlborg and Freeman find that the Ni atoms on planes interfacing with Cu planes have their spin magnetic moments reduced to 0.36 μ_B and the Ni atoms on the center

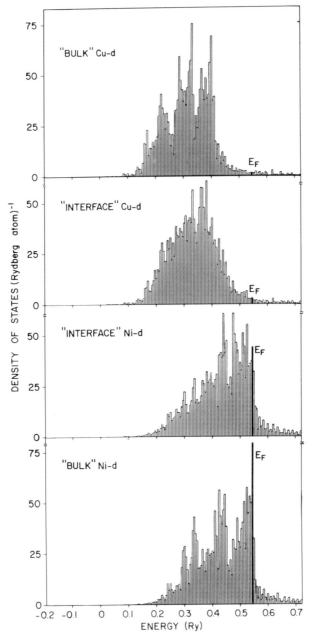

Fig. 44. Layer projected density of states for a Cu–Ni modulated structure, consisting of three layers each of Cu and Ni planes of atoms. (After Jarlborg and Freeman, unpublished.)

planes have their spin magnetic moments reduced to 0.5 μ_B (compared with the bulk value of 0.56 μ_B). Analysis of these results and the interpretation of the experiments are now being assessed.

Acknowledgments

I am grateful to my various colleagues for close collaboration on some of the research examples used here and to a number of authors for permission to reproduce their published figures. I am especially grateful to Susan Burge for her excellent help in producing the manuscript, and the National Science Foundation and the Air Force Office of Scientific Research for their support.

References

Abrikosov, A. A., and Gor'kov, L. P. (1961). *Sov. Phys. JETP* **12**, 1243.
Alldredge, G. P., and Kleinman, L. (1972). *Phys. Rev. Lett.* **28**, 1264.
Amar, H., Johnson, K. H., and Wang, K. P. (1966). *Phys. Rev.* **148**, 672.
Amar, H., Johnson, K. H., and Wang, K. P. (1967). *Phys. Rev.* **153**, 655.
Andersen, O. K. (1975). *Phys. Rev. B* **12**, 3060.
Andersson, S., and Nyberg, C. (1975). *Surf. Sci.* **52**, 489.
Andreoni, W., Altarelli, M., and Bassani, F. (1975). *Phys. Rev. B* **11**, 2352.
Appelbaum, J. A., and Hamann, D. R. (1972). *Phys. Rev. B* **6**, 2166.
Appelbaum, J. A., and Hamann, D. R. (1976). *Rev. Mod. Phys.* **48**, 479.
Arko, A. J., and Schirber, J. E. (1979). *J. Phys. C* **4**, A9.
Ashkenazi, J., Dacorogna, M., and Peter, M. (1979). *Solid State Commun.* **29**, 181.
Averill, F. W. (1972). *Phys. Rev. B* **6**, 3637.
Bachrach, R. Z., Skibowski, M., and Brown, F. C. (1976). *Phys. Rev.* **37**, 40.
Bader, R. F. W., and Bandrank, A. D. (1968). *J. Chem. Phys.* **49**, 1953.
Baldereschi, A., and Hopfield, J. J. (1972). *Phys. Rev. Lett.* **28**, 171.
Bansil, A., Ehrenreich, H., Schwartz, L., and Watson, R. E. (1974). *Phys. Rev. B* **9**, 445.
Bansil, A., Schwartz, L., and Ehrenreich, H. (1974). *Phys. Rev. B* **12**, 2893.
Bassani, F., and Yoshimine, M. (1963). *Phys. Rev.* **130**, 20.
Becker, E. H., Petijevich, P., and Williams, D. L. (1971). *J. Phys. F* **1**, 806.
Becker, G. E., and Hagstrum, H. D. (1972). *Surf. Sci.* **30**, 505.
Beeby, J. L. (1964a). *Phys. Rev.* **135**, A130.
Beeby, J. L. (1964b). *Proc. R. Soc. (London) A* **279**, 82.
Beeby, J. L. (1967). *Proc. R. Soc. (London) A* **302**, 113.
Beeby, J. L. (1971). "Electronic Density of States" (L. H. Bennett, ed.). NBS Special Publication, Washington, D.C.
Beeby, J. L., and Edwards, S. F. (1963). *Proc. R. Soc. (London) A* **274**, 395.
Bennemann, K. H. (1964). *Phys. Rev.* **133**, A1045.
Bergmann, G. (1978). *Phys. Rev. Lett.* **41**, 264.
Berl, G., and Wilson, W. E. (1961). *Nature* **191**, 380.
Bionde, M. A., and Rayne, J. A. (1959). *Phys. Rev.* **115**, 1522.
Bolz, J., Crecelius, G., Maletta, H., and Pobell, F. (1977). *J. Low Temp. Phys.* **28**, 61.
Born, M., and Huang, K. (1954). "Dynamical Theory of Crystal Lattices," pp. 219 and 321. Oxford Univ. Press, Oxford.

Brandow, B. H. (1975a). *Phys. Rev. B* **12**, 3464.
Brandow, B. H. (1975b). *J. Solid State Chem.* **12**, 397.
Brandow, B. H. (1976). *Int. J. Quant. Chem. S* **10**, 417.
Brener, N. E. (1975). *Phys. Rev. B* **11**, 1600.
Brooks, H. (1963). *Trans. Metal. Soc. AIME* **227**, 546.
Brown, F. C., Gahwiller, C., Kunz, A. B., and Lipari, N. O. (1970). *Phys. Rev. Lett.* **25**, 927.
Bullett, D. W. (1977). *Solid State Commun.* **23**, 893.
Butler, W. H., and Kohn W. (1971). "Electronic Density of States" (L. H. Bennett, ed.), p. 465. NBS Spec. Publ., Washington, D.C.
Cable, J. W., Wollan, E. O., Felcher, G. P., Brun, T. O., and Hornfeldt, S. P. (1975). *Phys. Rev. Lett.* **34**, 278.
Callaway, J. (1974). "Quantum Theory of the Solid State." Academic Press, New York.
Callaway, J., and Hughes, A. J. (1967). *Phys. Rev.* **156**, 860.
Callaway, J., and Wang, C. S. (1977). *Phys. Rev. B* **16**, 2095.
Caruthers, E., and Kleinman, L. (1975). *Phys. Rev. Lett.* **35**, 738.
Caruthers, E. B., Dempsey, D. G., and Kleinman, L. (1976). *Phys. Rev. B* **14**, 288.
Caruthers, E., Kleinman, L., and Alldredge, G. P. (1973). *Phys. Rev. B* **8**, 4570.
Chaney, R. C., Tung, T. K., Liu, C. C., and Lafon, E. E. (1970). *J. Chem. Phys.* **52**, 361.
Chaney, R. C., Liu, C. C., and Lafon, E. E. (1971). *Phys. Rev. B* **3**, 459.
Ching, W. Y., and Callaway, J. (1975). *Phys. Rev. B* **11**, 1324.
Chollet, L. F., and Templeton, I. M. (1968). *Phys. Rev.* **170**, 656.
Clark, H. (1964). *Phys. Lett.* **11**, 41.
Cohen, M. L., and Heine, V. (1970). *In* "Solid State Physics" (H. Ehrenreich, F. Seitz, and D. Turnbull, eds.), p. 38. Academic Press, New York.
Cohen, M. L. (1979). *Phys. Today* (July), 40.
Coleridge, P. T., and Templeton, I. M. (1971). *Can. J. Phys.* **49**, 2449.
Collins, T. C., Kunz, A. B., and Ivey, J. L. (1975). *Int. J. Quant. Chem. Symp.* **9**, 519.
Conklin, Jr., J. B., Averill, F. W., and Hattox, T. M. (1972). *J. Phys. (Paris) Suppl. C3* **33**, 213.
Cooper, B. R. (1973). *Phys. Rev. Lett.* **30**, 1316.
Cooper, B. R. (1977). *Phys. Rev. B* **16**, 5595.
Danese, J. B. (1974). *J. Chem. Phys.* **61**, 3071.
Danese, J. B., and Connolly, J. W. D. (1973). *Int. J. Quantum. Chem. Symp.* **7**, 279.
Davis, H. L. (1971). "Computational Methods in Band Theory," p. 183. Plenum, New York.
Davis, H. L., Faulkner, J. S., and Joy, H. W. (1968). *Phys. Rev.* **167**, 601.
Delley, B., Beck, H., Künzi, H., and Güntherrodt, H. J. (1978). *Phys. Rev. Lett.* **40**, 193.
Dempsey, D. G., and Kleinman, L. (1977a). *Phys. Rev. B* **16**, 5356.
Dempsey, D. G., and Kleinman, L. (1977b). *Phys. Rev. Lett.* **39**, 1297.
Dempsey, D. G., Kleinman, L., and Caruthers, E. (1975). *Phys. Rev. B* **12**, 2932.
Dempsey, D. G., Kleinman, L., and Caruthers, E. (1976a). *Phys. Rev. B* **13**, 1489.
Dempsey, D. G., Kleinman, L., and Caruthers, E. (1976b). *Phys. Rev. B* **14**, 279.
Dempsey, D. G., Kleinman, L., and Caruthers, E. (1977). *J. Phys. F* **7**, 113.
Demuth, J. E., Jepsen, D. W., and Marcus, P. M. (1973). *Phys. Rev. Lett.* **31**, 540.
Desjonqueres, M. C., and Cyrot-Lackman, F. (1975). *J. Phys. F* **5**, 1368.
Dimmock, J. O. (1971). *Solid State Phys.* **26**, 103.
Dimmock, J. O., and Freeman, A. J. (1964). *Phys. Rev. Lett.* **13**, 750.
Dionne, N. J., and Rhodin, T. N. (1976). *Phys. Rev. B* **14**, 322.
DiSalvo, F. J., Moncton, D. E., and Waszczak, J. V. (1976). *Phys. Rev. B* **14**, 4321.
DuMond, J., and Youtz, J. P. (1940). *J. Appl. Phys.* **11**, 357.
Dunlap, B. D., Shenoy, G. K., Fradin, F. Y., Barnet, C. D., and Kimball, C. W. (1979). *J. Mag. Magn. Matls.* **13**, 319.

Duthie, J. C., and Pettifor, D. G. (1977). *Phys. Rev. Lett.* **38**, 564.

Dutton, D. H., Brockhouse, B. N., and Miiller, A. P. (1972). *Can. J. Phys.* **50**, 2915.

Dworkin, A. S., Sasmar, D. J., and Van Artsdalen, E. R. (1954). *J. Chem. Phys.* **22**, 837.

Eastman, D. E., and Cashion, J. K. (1971). *Phys. Rev. Lett.* **27**, 1520.

Eastman, D. E., Himpsel, F. J., and Knapp, J. A. (1978). *Phys. Rev. Lett.* **40**, 1514.

Economu, E. N., Cohen, M. H., Freed, K. F., and Kirkpatrick, E. S. (1974). "Amorphous and Liquid Semiconductors" (J. Tauc, ed.), p. 1. Plenum, New York.

Edwards, S. F. (1961). *Phil. Mag.* **6**, 617.

Ehrenreich, H., and Schwartz, L. M. (1976). *Solid State Phys.* **31**, 149.

Eichner, S., Rau, C., and Sizmann, R. (1971). *J. Mag. Magn. Matls.* **6**, 204.

Elliott, R. J., Krumhansl, J. A., and Leath, P. L. (1974). *Rev. Mod. Phys.* **46**, 465.

Ellis, D. E., and Painter, G. S. (1970). *Phys. Rev. B* **2**, 7887.

Euwema, R. N., and Greene, R. L. (1975). *J. Chem. Phys.* **62**, 4455.

Euwema, R. N., and Stukel, D. J. (1970). *Phys. Rev. B* **1**, 4692.

Euwema, R. N., and Surratt, G. T. (1974). *J. Phys. C* **7**, 3655.

Euwema, R. N., Stukel, D. J., and Collins, T. C. (1971). *In* "Comp. Methods in Band Theory," p. 82, and many references therein. Plenum, New York.

Euwema, R. N., Wilhite, D. L., and Surratt, G. T. (1973). *Phys. Rev. B* **7**, 818.

Euwema, R. N., Surratt, G. T., Wilhite, D. L., and Wepfer, G. G. (1974a). *Phil. Mag.* **29**, 1033.

Euwema, R. N., Wepfer, G. G., Surratt, G. T., and Wilhite, D. L. (1974b). *Phys. Rev. B* **9**, 5729.

Falicov, L. M., and Cohen, M. H. (1963). *Phys. Rev.* **130**, 92.

Faulkner, J. S., Davis, H. L., and Joy, H. W. (1967). *Phys. Rev.* **161**, 656.

Fertig, W. A., Delong, L. E., Johnston, D. C., Maple, M. B., Matthias, B. T., and McCallum, R. W. (1977). *Phys. Rev. Lett.* **38**, 987.

Feuchtwang, T. E., Cutler, P. H., and Schmidt, J. (1978). *Surf. Sci.* **75**, 401.

Feuerbacher, B., Fitton, B., and Willis, R. F., eds. (1978). "Photoemission and the Electronic Structure of Surfaces." Wiley, London.

Fischer, Ø. (1978). *Appl. Phys.* **16**, 1.

Fischer, Ø., Treyvand, A., Chevrel, R., and Sergent, M. (1975). *Solid State Commun.* **17**, 721.

Fischer, Ø., Decroux, M., Chevrel, R., and Sergent, M. (1978). *In* "Superconductivity in d- and f-Band Metals" (D. H. Douglass, ed.), p. 175. Plenum, New York.

Fradin, F. Y., Shenoy, G. K., Dunlap, B. D., Alred, A. T., and Kimball, W. C. (1977). *Phys. Rev. Lett.* **38**, 719.

Freeman, A. J. (1972). "Magnetic Properties of Rare-Earth Metals" (R. J. Elliott, ed.), p. 245. Plenum, New York.

Freeman, A. J., and Jarlborg, T. (1979). *J. Appl. Phys.* **50**, 1876.

Freeman, A. J., and Koelling, D. D. (1974). "The Actinides: Electronic Structure and Related Properties" (A. J. Freeman and J. B. Darby, Jr., eds.), Vol. I, Chap. 2. Academic Press, New York.

Freeman, A. J., Harmon, B. N., and Watson-Yang, T. J. (1975). *Phys. Rev. Lett.* **34**, 281.

Freeman, A. J., Harmon, B. N., and Watson-Yang, T. J. (1976). *J. Mag. Magn. Matls.* **3**, 334.

Freeman, A. J., Koelling, D. D., and Watson-Yang, T. J. (1979a). *J. Physique* **C4**, 134.

Freeman, A. J., Watson-Yang, T. J., and Rath, J. (1979b). *J. Mag. Magn. Matls.* **12**, 140.

Friedel, J. (1968). *In* "The Physics and Metals" (J. M. Ziman, ed.), Vol. I. Cambridge Univ. Press, London.

Fumi, F. G., and Tosi, M. P. (1964). *J. Phys. Chem. Solids* **25**, 31, 45.

Gay, J. G., Smith, J. R., and Arlinghaus, F. J. (1977). *Phys. Rev. Lett.* **38**, 561.

Gelatt, Jr., C. D., Ehrenreich, H., and Watson, R. E. (1977). *Phys. Rev. B* **15**, 1613.

Gielisse, P. J., Mitra, S. S., Hendl, J. N., Griffis, R. D., Mansur, L. C., Marshall, R., and Pascoe, E. A. (1967). *Phys. Rev.* **155**, 1039.

Gurman, S. J. (1975). *J. Phys. F* **5**, L194.

Goroff, I., and Kleinman, L. (1967). *Phys. Rev.* **164**, 1100.

Göttlicher, G., and Wölfel, E. (1959). *Z. Electrochem* **63**, 891.

Gunnarsson, O., Johansson, P., Lundqvist, S., and Lundqvist, B. I. (1975). *Int. J. Quant. Chem.* **S9**, 83.

Gupta, M., and Freeman, A. J. (1976a). *Phys. Rev. B* **14**, 5205.

Gupta, R. P., and Freeman, A. J. (1976b). *Phys. Rev. Lett.* **36**, 613.

Gupta, M., Freeman, A. J., and Ellis, D. E. (1977). *Phys. Rev. B* **16**, 3338.

Gurman, S. J. (1975). *J. Phys. F* **5**, L194.

Hagstrum, H. D. (1947). *Phys. Rev.* **72**, 947.

Hagstrum, H. D. (1966). *Phys. Rev.* **150**, 495.

Hagstrum, H. D. (1969). *J. Appl. Phys.* **40**, 1398.

Hanke, W., Hafner, J., and Bilz, H. (1976). *Phys. Rev. Lett.* **37**, 1560.

Harmon, B. N. (1979). *J. Phys.* **40**, C5 65.

Harmon, B. N., and Freeman, A. J. (1974). *Phys. Rev. B* **10**, 1979.

Harmon, B. N., Koelling, D. D., and Freeman, A. J. (1973). *J. Phys. C* **6**, 2294.

Harmon, B. N., Schirber, J. E., and Koelling, D. D. (1978). *Int. Phys. Conf. Ser. No. 39, Transition Metals* (N. J. G. Lee, J. N. Perz, and E. Fawcett, eds.), p. 47. Institute of Physics, Bristol, England.

Harrison, W. A. (1966). "Pseudopotentials in the Theory of Metals." Benjamin, New York.

Harrison, W. A. (1970). "Solid State Theory." McGraw-Hill, New York.

Haselgrove, C. B. (1961). *Math. Compt.* **15**, 373.

Hedin, L., and Johansson, A. (1969). *J. Phys. B* **2**, 1136.

Hedin, L., and Lundqvist, B. I. (1971). *J. Phys. C* **4**, 2064.

Herman, F. (1954). *Phys. Rev.* **93**, 1214.

Herman, F., and Skillman, S. (1961). "Semiconductor Physics Conference, Prague," p. 20. Academic Press, New York.

Herman, F., Kuglin, C. D., Cuff, K. F., and Kortum, R. L. (1963). *Phys. Rev. Lett.* **11**, 541.

Hilliard, J. E. (1979). *AIP Conf. Proc.* **53**, 407.

Ho, K. M., Louie, S. G., Chelikowsky, J. R., and Cohen, M. L. (1977). *Phys. Rev. B* **15**, 1755.

Hohenberg, P., and Kohn, W. (1964). *Phys. Rev.* **136**, 864.

Howard, C. J., and Khadake, R. G. (1974). *Acta. Crystallogr. A* **30**, 296.

Hubbard, J. (1963). *Proc. R. Soc. (London) A* **276**, 238.

Hubbard, J. (1964). *Proc. R. Soc. (London) A* **281**, 401.

Ishikawa, M., and Fischer, Ø. (1977). *Solid State Commun.* **23**, 37.

Jarlborg, T. (1979). *J. Phys. F* **9**, 283.

Jarlborg, T., and Arbman, G. (1976). *J. Phys. C* **10**, 3107.

Jarlborg, T., and Freeman, A. J. (1980). *Phys. Rev. Lett.* **44**, 178.

Jarlborg, T., Freeman, A. J., and Watson-Yang, T. J. (1977). *Phys. Rev. Lett.* **39**, 1032.

Jarlborg, T., Freeman, A. J., and Watson-Yang, T. J. (1978). *J. Mag. Magn. Matls.* **7**, 296.

Janak, J. F. (1977). *Phys. Rev. B* **16**, 255.

Janak, J. F., and Williams, A. R. (1976). *Phys. Rev. B* **14**, 4199.

Janak, J. F., Moruzzi, V. L., and Williams, A. R. (1975). *Phys. Rev. B* **12**, 1257.

Jaros, M., and Ross, S. F. (1973). *J. Phys. C* **6**, 1753, 3459.

Jepsen, O., and Andersen, O.K. (1971). *Solid State Commun.* **9**, 1763.

Jepsen, O., Madsen, J., and Andersen, O. K. (1978). *Phys. Rev. B* **18**, 605.

Johnson, K. H. (1966). *J. Chem. Phys.* **45**, 3085.

Kar, N., and Soven, P. (1975). *Phys. Rev. B* **11**, 3761.

Kasowski, R. V. (1974a). *Phys. Rev. Lett.* **33**, 83.

Kasowski, R. V. (1974b). *Phys. Rev. Lett.* **33**, 1147.

Kasowski, R. V. (1975). *Solid State Commun.* **17**, 179.

Kasowski, R. V. (1976a). *Phys. Rev. Lett.* **37**, 219.
Kasowski, R. V. (1976b). *Phys. Rev. B* **14**, 3398.
Keeton, S. C., and Loucks, T. L. (1966). *Phys. Rev.* **146**, 429.
Keown, R. (1966). *Phys. Rev.* **150**, 568.
Kervin, P. W., and Lafon, E. E. (1973). *J. Chem. Phys.* **58**, 1535.
Killean, R. C. G., Lawrence, J. L., and Sharma, V. C. (1972). *Acta Crystallogr. A* **28**, 405.
Kittel, C. (1967). "Introduction to Solid State Physics," 3rd ed. Wiley, New York.
Kleinman, L., and Caruthers, E. B. (1974). *Phys. Rev. B* **10**, 3213.
Kleinman, L., and Phillips, J. C. (1962). *Phys. Rev.* **125**, 819.
Koehler, W. C., and Moon, R. M. (1976). *Phys. Rev. Lett.* **36**, 616.
Koelling, D. D. (1970). *Phys. Rev. B* **2**, 290.
Koelling, D. D. (1972). *J. Phys. Chem. Solids* **33**, 1335.
Koelling, D. D. (1979). *J. Phys. C* **4** 117.
Koelling, D. D. (1980). To be published.
Koelling, D. D., and Arbman, G. (1975). *J. Phys. F* **5**, 2041.
Koelling, D. D., and Freeman, A. J. (1973). *Phys. Rev. B* **1**, 4454.
Koelling, D. D., Freeman, A. J., and Mueller, F. M. (1970). *Phys. Rev. B* **1**, 1318.
Kohn, W. (1975). *Phys. Rev. B* **11**, 3756.
Kohn, W., and Luttinger, J. M. (1955). *Phys. Rev.* **97**, 1721.
Kohn, W., and Sham, L. J. (1965). *Phys. Rev. A* **140**, 1133.
Korringa, J. (1958). *J. Phys. Chem. Solids* **7**, 252.
Koster, G. F., and Slater, J. C. (1954). *Phys. Rev.* **95**, 1167.
Krakauer, H., and Cooper, B. R. (1977). *Phys. Rev. B* **16**, 605.
Krakauer, H., Posternak, M., and Freeman, A. J. (1979). *Phys. Rev. B* **19**, 793.
Krug, J., Witte, H., and Wölfel (1955). *Z. Phys. Chem.* **4**, 36.
Kunz, A. B. (1975). *Phys. Rev. B* **12**, 5890.
Kunz, A. B., Mickish, D. J., and Collins, T. C. (1973). *Phys. Rev.* **31**, 756.
Lafon, E. E. (1968). *MIT Solid State and Molecular Theory Group Quarterly Progress Report No. 69*, p. 66.
Lafon, E. E. (1974). *Bull. Am. Phys. Soc.* **19**, 201.
Lafon, E. E., and Liu, C. C. (1966). *Phys. Rev.* **152**, 579.
Lafon, E. E., Chaney, R. C., and Liu (1971). *In* "Comp. Methods in Band Theory" (P. M. Marcus, J. F. Janak, and A. R. Williams, eds.), p. 284. Plenum, New York.
Landolt, M., and Campagna, M. (1977). *Phys. Rev. Lett.* **38**, 663.
Landolt, M., and Yafet, Y. (1978). *Phys. Rev. Lett.* **40**, 1401.
Laurent, D. G., Wang, C. S., and Callaway, J. (1978). *Phys. Rev. B* **17**, 455.
Laurent, D. G., Callaway, J., and Wang, C. S. (1979). *Phys. Rev. B* **20**, 1134.
Lax, M. (1951). *Rev. Mod. Phys.* **23**, 287.
Lax, M. (1952). *Phys. Rev.* **85**, 621.
Lehmann, G., and Taut, M. (1972). *Phys. Status Solidi* **54**, 469.
Liebermann, L., Clinton, J., Edwards, D. M., and Mathon, J. (1970). *Phys. Rev. Lett.* **25**, 232.
Liebsch, A. (1977). *Phys. Rev. Lett.* **38**, 248.
Liebsch, A. (1978). *Phys. Rev. B* **17**, 1653.
Lindgård, P. A. (1975). *Solid State Commun.* **16**, 481.
Loucks, T. L. (1967). "The Augmented Plane Wave Method." Benjamin, New York.
Louie, S. G. (1978). *Phys. Rev. Lett.* **40**, 1525.
Louie, S. G., Ho, K. M., Chelikowsky, J. R., and Cohen, M. L. (1976). *Phys. Rev. Lett.* **37**, 1289.
Louie, S. G., Ho, K. M., Chelikowsky, J. R., and Cohen, M. L. (1977). *Phys. Rev. B* **15**, 5627.
Lubinsky, A. R., Ellis, D. E., and Painter, G. S. (1975). *Phys. Rev. B* **11**, 1537.

Lucovsky, G., Martin, R. M., and Burstein, E. (1971). *Phys. Rev. B* **4**, 1367.
Lundqvist, S., and Ufford, C. W. (1965). *Phys. Rev.* **139**, A1.
Maglic, R. C., Lander, G. H., Mueller, F. M., and Kleb, R. (1978). *Phys. Rev. B* **17**, 308.
Martinez, G., Schlüter, M., and Cohen, M. L. (1975a). *Phys. Rev. B* **11**, 651.
Martinez, G., Schlüter, M., and Cohen, M. L. (1975b). *Phys. Rev. B* **11**, 660.
Mattheiss, L. F. (1964). *Phys. Rev.* **133A**, 1399.
Mattheiss, L. F. (1970). *Phys. Rev. B* **2**, 3918.
Mattheiss, L. F. (1971). *In* "Comp. Methods in Band Theory," p. 355. Plenum, New York.
Mattheiss, L. F. (1972). *Phys. Rev. B* **5**, 290.
Mattheiss, L. F., Wood, J. H., and Switendick, A. C. (1968). *Meth. Comp. Phys.* **8**, 64.
Matthias, B. T., Corenzwit, E., Vandenberg, J. M., and Barz, H. E. (1977). *Proc. Nat. Acad. Sci. USA* **74**, 1334.
Menzel, W. P., Lin, C. C., Fouquet, D., Lafon, E. E., and Chaney, R. C. (1973). *Phys. Rev. Lett.* **30**, 1313.
Merisalo, M., and Inkinen, O. (1966). *Phys. Fen.* **207A**, 3.
Messmer, R. P., and Watkins, G. D. (1973). *Phys. Rev. B* **7**, 2568.
Mickish, D. J., Junz, A. B., and Collins, T. C. (1974). *Phys. Rev. B* **9**, 4461.
Miiller, A. P., and Brockhouse, B. N. (1971). *Can. J. Phys.* **49**, 704.
Morinaga, H. (1972). *J. Phys. Soc. Jpn.* **33**, 996.
Moruzzi, V. L., Williams, A. R., and Janak, J. F. (1977). *Phys. Rev. B* **15**, 2854.
Murry, B. W., and McGervey, J. D. (1970). *Phys. Rev. Lett.* **24**, 9.
Myron, H. W., Rath, J., and Freeman, A. J. (1977). *Phys. Rev. B* **15**, 885.
Neto, J. R. P., and Ferreira, L. G. (1973). *J. Phys. C* **6**, 3430.
Nordheim, L. (1931a). *Ann. Phys.* **9**, 607.
Nordheim, L. (1931b). *Ann. Phys.* **9**, 641.
Nozières, P., and Pines D. (1958). *Phys. Rev.* **14**, 442.
Overhauser, A. W. (1968). *Phys. Rev.* **167**, 691.
Overhauser, A. W. (1971). *Phys. Rev. B* **3**, 3173.
Painter, G. S. (1978a). *Phys. Rev. B* **17**, 3848.
Painter, G. S. (1978b). *Phys. Rev. B* **18**, 955.
Painter, G. S., Ellis, D. E., and Lubinsky, A. R. (1971). *Phys. Rev. B* **4**, 3610.
Pant, M. M., and Joshi, S. K. (1969). *Phys. Rev.* **184**, 635.
Paraskevopoulos, D., Meservey, R., and Tedrow, P. M. (1977). *Phys. Rev. B* **16**, 4970.
Pauling, L. (1960). "The Nature of the Chemical Bond." Cornell Univ. Press, Ithaca, New York.
Pells, G. P., and Montgomery, H. (1970). *J. Phys. C* **3**, 5330.
Penn, D. R., and Plummer, E. W. (1974). *Phys. Rev. B* **9**, 1216.
Pines, D. (1963). "Elementary Excitations in Solids." Benjamin, New York.
Raccah, P. M., Euwema, R. N., Stukel, D. J., and Collins, T. C. (1970). *Phys. Rev. B* **1**, 756.
Rasolt, M., and Geldart, D. J. W. (1975). *Phys. Rev. Lett.* **35**, 1234.
Rath, J., and Freeman, A. J. (1975). *Phys. Rev. B* **11**, 2109.
Redi, M., and Anderson, P. W. (1979). *Bull. Am. Phys. Soc.* **24**, 389.
Reed, W. A., and Eisenberger, P. (1972). *Phys. Rev. B* **6**, 4596.
Renninger, M. (1955). *Acta Crystallogr.* **8**, 606.
Sabin, J. R., and Trickey, S. B. (1975). *J. Phys. B* **8**, 7593.
Sambe, H., and Felton, R. H. (1975). *J. Chem. Phys.* **62**, 1122.
Sato, M., and Hirakawa, K. (1975). *J. Phys. Soc. Jpn.* **39**, 1467.
Schlüter, M., Chelikowsky, J. R., Louie, S. G., and Cohen, M. L. (1975). *Phys. Rev. Lett.* **34**, 1385.
Seth, A., and Ellis, D. E. (1977). *J. Phys. C* **10**, 181.

Sham, L. J., and Kohn, W. (1966). *Phys. Rev.* **145**, 561.

Shelton, R. N., McCallum, R. W., and Adrian, H. (1976). *Phys. Lett.* **56A**, 213.

Simmons, J. E., Lin, C. C., Fouquet, D. F., Lafon, E. E., and Chaney, R. C. (1975). *J. Phys. C* **8**, 1549.

Singh, M., Wang, C. S., and Callaway, J. (1975). *Phys. Rev. B* **11**, 287.

Singwi, K. Sjölander, A., Tosi, M. P., and Land, R. H. (1970). *Phys. Rev. B* **1**, 1044.

Sinha, S. K., and Harmon, B. N. (1975). *Phys. Rev. Lett.* **35**, 1515.

Sinha, S. K. (1980). "Dynamical Properties of Solids" (C. K. Hartin and A. A. Maradudim, eds.), Chap. 1.

Slater, J. C. (1937). *Phys. Rev.* **51**, 846.

Slater, J. C. (1951). *Phys. Rev.* **81**, 385.

Slater, J. C. (1960). "Quantum Theory of Atomic Structure." McGraw-Hill, New York.

Slater, J. C. (1965). *J. Chem. Phys.* **43**, 5228.

Slater, J. C. (1972). "Symmetry and Energy Bands in Crystals." Dover, New York.

Slater, J. C. (1974). "Self-consistent Field for Molecules and Solids: Quantum Theory of Molecules and Solids," Vol. 4. McGraw-Hill, New York.

Slater, J. C., and Johnson, K. H. (1972). *Phys. Rev. B* **5**, 844.

Slater, J. C., and Koster, G. F. (1954). *Phys. Rev.* **94**, 1498.

Smith, H. G. (1972). *In* "Superconductivity in d and f Band Metals" (D. H. Douglass, ed.), AIP, New York.

Smith, H. G., and Gläser, W. (1970). *Phys. Rev. Lett.* **25**, 1611.

Smith, J. R., Arlinghaus, F. J., and Gay, J. G. (1977). *Solid State Commun.* **24**, 279.

Smith, J. R., Arlinghaus, F. J., and Gay, J. G. (1978). *Inst. Phys. Conf. Ser.* **39**, 303.

Sohn, K. S., Dempsey, D. G., Kleinman, L., and Carruthers, E. (1976a). *Phys. Rev. B* **13**, 1515.

Sohn, K. S., Dempsey, D. G., Kleinman, L., and Carruthers, E. (1976b). *Phys. Rev. B* **14**, 3185.

Sohn, K. S., Dempsey, D. G., Kleinman, L., and Carruthers, E. (1976c). *Phys. Rev. B* **14**, 3193.

Soven, P. (1966). *Phys. Rev.* **151**, 539.

Soven, P. (1967). *Phys. Rev.* **156**, A09.

Soven, P. (1969). *Phys. Rev.* **178**, 1136.

Soven, P., Plummer, E. W., and Kar, N. (1976). *Solid State Sci.* **6**, 111.

Stassis, C., Kline, G. R., and Sinha, S. K. (1975). *Phys. Rev. B* **11**, 2171.

Stocks, G. M., Williams, R. W., and Faulkner, J. S. (1971). *Phys. Rev. B* **4**, 4390.

Stocks, G. M., Temmerman, W. M., and Gyorffy, B. L. (1978). *Phys. Rev. Lett.* **41**, 339.

Stull, D. R. (1965) (Ed.). "JANAF Tables of Thermochemical Data." Dow Chemical Co., Midland, Michigan.

Surratt, G. T., Euwema, R. M., and Wilhite, D. L. (1973). *Phys. Rev. B* **8**, 4019.

Tawil, R. A., and Callaway, J. (1973). *Phys. Rev. B* **7**, 4242.

Taylor, D. W. (1967). *Phys. Rev.* **156**, 1017.

Temmerman, W. M., Gyorffy, B. L., and Stocks, G. M. (1978). *J. Phys. F* **8**, 2461.

Templeton, I. M., Coleridge, P. T., and Chollet, L. F. (1969). *Phys. Kondens. Materie* **9**, 21.

Terauchi, H., and Cohen, J. B. (1978). *Phys. Rev. B* **17**, 2494.

Thewlis, J., and Davey, A. R. (1968). *Phil. Mag.* **1**, 409.

Thompson, A. H., Gamble, F. R., and Revelli, J. R. (1971). *Solid State Commun.* **9**, 981.

Tibbetts, G. G., Burkstrnd, J. M., and Tracy, J. C. (1977). *Phys. Rev. B* **15**, 3652.

Tosi, M. P. (1963). *J. Phys. Chem. Solids* **24**, 965.

Traum, M. M., Margaritondo, G., Smith, N. V., Rowe, J. E., and DiSalvo, F. J. (1978). *Phys. Rev. B* **17**, 1835.

Trickey, S. B., Green, F. R., Jr., and Averill, F. W. (1973). *Phys. Rev. B* **8**, 4822.

Triftshäuser, W., and Stewart, A. T. (1971). *J. Phys. Chem. Solids* **32**, 2717.

Varma, C. M., and Weber, W. (1979). *Phys. Rev. B* **19**, 6142.

Veal, B. W., and Lam, D. J. (1974). *Phys. Rev. B* **10**, 4902.
von Barth, U., and Hedin, L. (1972). *J. Phys. C* **5**, 1629.
Wang, C. S., and Callaway, J. (1974). *Phys. Rev. B* **9**, 4897.
Wang, C. S., and Callaway, J. (1975). *Phys. Rev. B* **11**, 2417.
Wang, C. S., and Callaway, J. (1977). *Phys. Rev. B* **15**, 298.
Wang, C. S., and Freeman, A. J. (1978). *Phys. Rev. B* **18**, 1714.
Wang, C. S., and Freeman, A. J. (1979a). *Phys. Rev. B* **19**, 793.
Wang, C. S., and Freeman, A. J. (1979b). *Phys. Rev. B* **19**, 4930.
Wang, C. S., and Freeman, A. J. (1979c). *J. Appl. Phys.* **50**, 1940.
Wang, C. S., and Freeman, A. J. (1980a). *J. Mag. Magn. Matls.* **15–18** 869.
Wang, C. S., and Freeman, A. J. (1980b). *Phys. Rev. B* **21**, May 15.
Watson, R. E., Ehrenreich, H. and Hodges, L. (1970). *Phys. Rev. Lett.* **24**, 829.
Weber, W. (1973). *Phys. Rev. B* **8**, 5082.
Weiss, R. J., and Phillips, W. C. (1968). *Phys. Rev.* **176**, 900.
Wentrof, R. H. (1957). *J. Chem. Phys.* **26**, 916.
Wepfer, G. G., Euwema, R. N., Surratt, G. T., and Wilhite, D. L. (1974). *Phys. Rev. B* **9**, 2670.
Wigner, B. P., and Seitz, F. (1933). *Phys. Rev.* **43**, 804.
Wigner, B. P., and Seitz, F. (1934). *Phys. Rev.* **46**, 509.
Williams, A. R., Gelatt, C. D., Jr., and Janak, J. F. (1979). "Proceedings of the Symposium on Theory of Alloy Phase Formation" (L. Bennett, ed.). AIME, New York.
Williams, D. L., Becker, E. H., and Petijevich, P. (1969). *Bull. Am. Phys. Soc.* **14**, 402.
Williams, P. M., Parry, G. S., and Scruby, C. G. (1974). *Phil. Mag.* **29**, 695.
Williams, R. W., Loucks, T. L., and Mackintosh, A. R. (1966). *Phys. Rev. Lett.* **16**, 168.
Williams, R. W., and Mackintosh, A. R. (1968). *Phys. Rev.* **168**, 679.
Wilson, J. A., DeSalvo, F. J., and Mahajan, S. (1974). *Phys. Rev. Lett.* **32**, 882.
Wilson, J. A., DeSalvo, F. J., and Mahajan, S. (1975). *Advan. Phys.* **24**, 117.
Wilson, J. A. (1978). *Phys. Status Solidi(b)* **86**, 11.
Woo, K. C., Brown, F. C., McMillan, W. L., Miller, R. J., Schaffman, M. J., and Sears, M. P. (1976). *Phys. Rev. B* **14**, 3242.
Yonezawa, F., and Morigaki, K. (1973). *Prog. Theor. Phys. Suppl.* **53**.
Young, R. C. (1979). *J. Phys.*, **40**, C5 71.
Young, R. C., Jordan, R. G., and James, D. W. (1973). *Phys. Rev. Lett.* **31**, 1473.
Yu, K. Y., Spicer, W. E., Lindau, I., Pianetta, P., and Lin, S. F. (1976). *Surf. Sci.* **57**, 157.
Zachariasen, W. H. (1968). *Acta Crystallogr. A* **24**, 324.
Zunger, A. (1974). *J. Phys. C* **6**, 72.
Zunger, A., and Freeman, A. J. (1974). *J. Phys. C* **6**, 72, 96.
Zunger, A., and Freeman, A. J. (1977a). *Phys. Rev. B* **15**, 4716.
Zunger, A., and Freeman, A. J. (1977b). *Phys. Rev. B* **15**, 5049.
Zunger, A., and Freeman, A. J. (1977c). *Phys. Rev. B* **16**, 2901.
Zunger, A., and Freeman, A. J. (1977d). *Phys. Rev. B* **16**, 906.
Zunger, A., and Freeman, A. J. (1977e). *Int. J. Quant. Chem. Symp.* **11**, 539.
Zunger, A., and Freeman, A. J. (1978a). *Phys. Rev. B* **17**, 1839.
Zunger, A., and Freeman, A. J. (1978b). *Phys. Rev. B* **17**, 2030.

2

Photoelectron Spectroscopy as an Electronic Structure Probe†

B. W. VEAL

Argonne National Laboratory
Argonne, Illinois

I. Introduction

Photoelectron spectroscopy (PES) refers to a number of specialized experimental techniques that depend upon the measurement and analysis of photoejected electrons. Ideally, the exciting photon flux is monochromatic. In this case, energy analysis of the emitted electrons can provide a great deal

---

†Work supported by the U.S. Department of Energy.

of information about the electronic properties of solids. Photoelectron spectroscopies can serve as atom-specific, spatially local probes (with deep core-level studies) or as direct probes of spatially delocalized valence-electron states. The spectroscopies are extremely versatile; the electronic properties of gases, liquids, and virtually any solid material that does not decompose in vacuum can, in principle, be studied with these techniques.

The spectroscopies are manifestations of the photoelectric effect, a phenomenon discovered in 1887 by H. Hertz (1887). The photoelectric effect played a central role in the early development of quantum theory. To explain the effect, Einstein in 1905 published the revolutionary postulate that photons were quanta of energy existing in integral units of $h\nu$ (Einstein, 1905). Yet is was not until the 1950s that advances in the understanding of the effect and the development of sophisticated instrumentation permitted exploitation of the effect as a spectroscopic tool by the materials scientist.

We have pointed out that the techniques, in their simplest form, involve the energy analysis of photoejected electrons stimulated by incident monochromatic photons. In fact, many experimental parameters can be controlled to obtain different kinds of electronic structure information. The photoelectron current is dependent upon the energy, emission angle, and spin angle of the emitted electron, and the energy, polarization, and incidence angle of the incident photon. The experimental techniques and theoretical framework needed to investigate the dependence of electronic properties upon these parameters are still in a rapid state of development. Even with its short life and the relatively limited experimental capability that is generally available, the PES field has generated scientific publications numbering in the thousands.

The early development (1950s) of PES was almost entirely devoted to angle-integrated measurements acquired in two extreme regions of the incident photon spectrum. One group of workers concentrated on the vacuum ultraviolet region (usually below the LiF cutoff at 11.6 eV) while another group explored the x-ray region. These spectral regions were chosen because they offered photon fluxes of adequate intensity to permit meaningful measurements to be made. Not surprisingly, the experiments provided significantly different, but certainly complementary, information. In recent years new advances have been made with the exploitation of synchrotron radiation from charged-particle storage rings. These machines produce an extremely high-intensity photon flux that spans much of the spectral region between the ultraviolet and soft x-ray energies. Thus, the gap between the ultraviolet and x-ray regions has been, for most purposes, completely bridged. Furthermore, as sophistication develops and techniques improve, more experimental parameters are coming under scrutiny. Of strong current emphasis are angular-resolved photoemission studies performed on single-crystal materials of known orientation. Studies of the spin polarization of photoemitted electrons are also providing an exciting new class of measurements, largely for ferromagnetic materials. Because of the extensive range of experimental parameter variation that is available with the photoemission techniques, a tremendous quantity of electronic structure information can be obtained for a given material. Since the techniques are also applicable to a great variety of materials, it is clear that they will continue to play an increasingly important role in the study of electronic structure of solids.

Since elastically scattered electrons (see Section II) have very short escape depths (generally less than 20 Å) the photoemission spectroscopies might well be used to probe the electronic properties of surfaces. At suitable photon excitation energies, surface effects are dominant and a rapidly growing field of surface science relies heavily on PES as a sensitive surface probe. However, the material in this chapter will be directed to studies of bulk electronic properties.

We will present some discussion of the various experimental approaches in order to provide a limited overview of the field of PES. In no instance will there be an in-depth discussion; instead, the reader will be referred to appropriate literature. Furthermore, we will rely heavily on the use of selected studies as examples and will make no attempt to balance the presentation among the various capabilities. In keeping with the interests of the author, the XPS technique will receive strongest emphasis.

Many excellent books, surveys, and conference proceedings are available to the serious student of PES. Recent books devoted entirely, or in part, to PES provide in-depth discussions of the photoemission technique, relevant theoretical formalism, selected case studies, and references to an extensive original (and review) literature. The literature cited here emphasizes the study of bulk materials properties reather than gas- or liquid-phase properties or surface applications of PES. The list of books includes Siegbahn *et al.* (1967) Cardona and Ley (1978), Ley and Cardona (1980), Brundle and Baker (1977–1979), Carlson (1975, 1978), Dekeyser *et al.* (1973), Buck *et al.* (1979), Spicer (1972), Craseman (1975), and Kane and Larrabee (1974). Numerous review articles, often directed to specific aspects of PES, are also available. A partial listing includes Watson and Perlman (1975), Campagna and Wertheim (1976), Jørgensen (1975), Veal and Lam (1980), Feuerbacher and Willis (1976), Campagna and Wertheim (1975), Shirley *et al.* (1977), Eastman (1972), and Spicer *et al.* (1976; Spicer, 1976). Conference proceedings (which also contain reviews) include Shirley (1972), Caudano and Verbist (1974), Fabian and Watson (1973), and Koch *et al.* (1974). The PES literature has been reviewed by Hercules (1972, 1976), Hercules and Carver (1974), Schwartz (1973), and Baker *et al.* (1978). The Journal of Electron Spectroscopy and Related Phenomena contains numerous PES publications.

II.   The Photoemission Techniques: Ultraviolet Photoemission Spectroscopy (UPS) through X-Ray Photoemission Spectroscopy (XPS)

A.   Description

In its simplest form (and the form best approximated by XPS), photo-emission spectroscopy provides a direct measure of the occupied density of electron states $n(E)$ in a solid. Figure 1 schematically illustrates the measurement. Here we show the valence-band density of states that one might find in a transition metal and, beneath it, two fully occupied core levels. A photon of energy $h\nu$ impinging on a sample surface can photoeject an electron into the vacuum. This electron departs the sample with a measurable kinetic energy

$$E_{ke} = h\nu - E_{\phi} - |E|, \tag{1}$$

where E_{ϕ} is the work function of the material and E is the initial energy or "binding energy" of the ejected electron. Since the photon energy is known, measurement of E_{ke} gives a unique measure of E, the distance (in energy) below E_F from which the electron came. Clearly, a photoelectron current at E_{ke} is observable only if occupied electron states exist at energy E. If the photoelectron intensity from the state E is proportional to the density of states at E, then measurement of the energy dependence of the photoemission intensity $I(E)$ provides a direct measure of $n(E)$. Since XPS is typically performed with photon energies in excess of 1000 eV, the core levels within about 1 keV of E_F can also be studied. Spectra are normally plotted against electron binding energy and, for metals, the Fermi energy is usually defined to be the energy zero. Figure 2 shows a segment of a 1000-eV scan for gold. Atomiclike core levels from 4s to 4f are labeled. The valence-band region is shown enlarged in the inset. Figure 2 is a very graphic picture of the electron states in a solid. The atomiclike core levels are isolated peaks whereas the outer electron states show the hybridized band states that characterize valence electrons in solids. These band states terminate abruptly at the Fermi energy; the rounding of the spectrum at E_F is a consequence of the limited instrumental resolution. As we shall discuss subsequently, the XPS $I(E)$ appears to be a very good measure of the occupied density of valence states.

The incident photon flux penetrates far into the sample (typically a few hundred angstroms) and may generate photoelectrons anywhere along its path. A small fraction of those electrons will leave the sample apparently without energy loss. At most, they experience *elastic* scattering processes

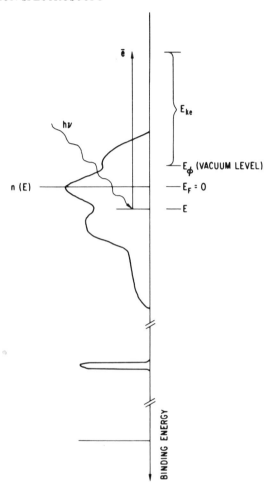

Fig. 1. Schematic of photoemission process. A monochromatic photon of energy $h\nu$ ejects an electron, with binding energy E, into the vacuum. The electron escapes the solid with kinetic energy E_{ke}. For the metallic solid illustrated here, E_F is the Fermi level, $E\phi$ is the work function, and $n(E)$ is the density of states. Using XPS, an approximate measure of $n(E)$ versus E_B can be obtained by measuring the photoemission intensity and the electron kinetic energy.

during photoejection. These elastically scattered electrons preserve "memory" of the initial energy E and thus carry the information of greatest interest to the photoelectron spectroscopist. However, escape depths for elastically scattered electrons are small, only a few atomic layers, so that a large inelastic loss current (at degraded kinetic energies) is always present. Thus, the spectrum of Fig. 2 is characterized by the resonant elastic peaks followed by the broad and featureless, but high- (integrated) intensity inelastic background.

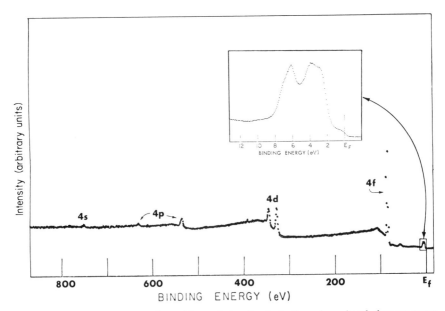

Fig. 2. An XPS spectrum for gold metal showing both the valence-band electron states (expanded in the inset) and *atomiclike* core levels.

A great deal of effort has been devoted to measuring elastic electron escape depths λ for a variety of materials at various photon energies (see Spicer, 1976; Shirley, 1978; Brundle, 1974). These studies have revealed that escape depths for most materials are well approximated by a single, universal curve. This curve, displayed in Fig. 3, shows that λ varies between about 3 and 30 Å, depending on photon energy. The very small λ means that only

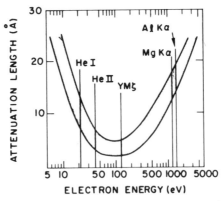

Fig. 3. The universal curve of electron attenuation length in various heavy metals, drawn as a band that encompasses most of the existing experimental data. The energies of several laboratory photon sources are shown for reference. (After Shirley, 1978.)

a few atomic layers of the sample are probed. This fact has stimulated concern that photoemission may not, in general, probe bulk electronic structure but rather might be sensitive to surface phenomena. The variable λ (with $h\nu$) enables the depth penetration, and hence the surface sensitivity, to be varied. The wealth of information that has recently been obtained appears to substantiate the view that at high $h\nu$, the bulk properties are reliably probed. At appropriate $h\nu$, surface states associated with the termination of Bloch functions may also be observed (Feibelman and Eastman, 1974; Feuerbacher and Willis, 1976).

B. *"Band Structure" and "Density of States" Regions*

The simplified description of the photoemission process that was given above (Fig. 1) is appropriate to angle-integrated XPS but totally inappropriate to the UPS spectral regime. Since photoemission involves the ejection of an electron from an occupied state to an empty excited state, it follows that both the initial and final states of that electron might affect the observed photoemission intensity $I(E,\nu)$. If the photon energy is small (UPS), then only valence electrons can be excited and they can only be elevated into conduction bands lying rather close to E_F. In this case, the character of the possible final states are as important to $I(E,\nu)$ as are the initial states. For UPS, $I(E)$ is considered to be a "joint density of states" function since $I(E)$ is a convolution of the valence and conduction states. If one considers the photoexcitation to be an isolated one-electron process and imposes energy and momentum conservation, then the electron transition must be **k** conserving, i.e., a "direct interband transition," since the photon supplies very little momentum. In this case, $I(E)$ is sensitive to the interband transition probabilities for electrons (excited from initial energy E) integrated over all points of the Brillouin zone. Changing $h\nu$ produces a new convolution of valence and conduction states in the $I(E)$ spectrum (see Cardona and Ley, 1978, for a summary of formalism). $I(E)$ in the UPS region is commonly called the energy distribution of the joint density of states. For UPS, curves of $I(E)$ are typically presented for a series of photon energies $h\nu$. Figure 4 shows $I(E,\nu)$ spectra for gold recorded for photon energies between 10.2 and 23 eV (Eastman and Grobman, 1972). These spectra show a remarkable richness of structure with features moving about as determined by the electron band structure of gold.

At x-ray energies, electrons are removed from valence-band states to final states far above E_F. Since MgKα (1283.6 eV) or AlKα (1486.6 eV) characteristic radiation is typically used in XPS, the valence electrons are excited to final states 1 kV or more above E_F. At these energies, band structure

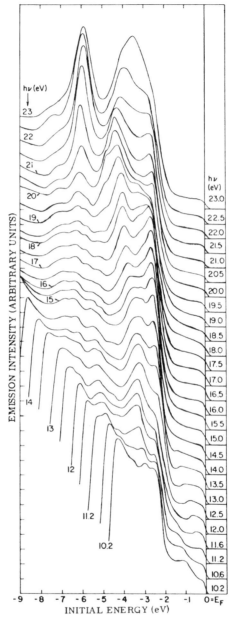

Fig. 4. UPS spectra for gold with $10.2 \leq h\nu \leq 23$ eV. Zero energy corresponds to the Fermi level. (After Eastman and Grobman, 1972.)

effects are not important, and the final density of states is high and free-electron–like. Furthermore, at x-ray energies, the photon momentum is no longer negligible relative to the Brillouin zone dimensions so that a larger region of the zone is available to the excited electron. Phonon-assisted processes also become more likely (Shirley *et al.*, 1977). In this case, all initial states are sampled with approximately equal probabilities. Thus, the x-ray measurement is sensitive to $n(E)$ and XPS falls in the density of states spectral region. [Matrix-element effects can modulate transition probabilities of electrons with different quantum character, however, so that XPS need not provide an exact measure of $n(E)$.] The UV end of the spectrum, where spectral features are dependent on the **k** dependence of valence- and conduction-electron interband transition probabilities, is called the band structure region (Feibelman and Eastman, 1974). The spectra shown in Figs. 4, 5, and 6 illustrate the transition between these two regions. For photon energies less than 30 eV (Fig. 4), the spectra vary dramatically with choice of $h\nu$. [In this display, for $h\nu < 15$ eV, the spectra are terminated by the

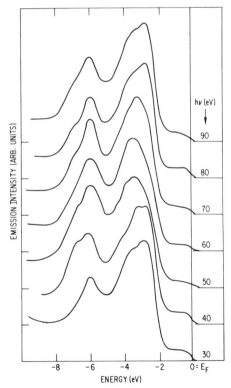

Fig. 5. Photoelectron energy distributions for an evaporated gold film for $30 \leq h\nu \leq 90$ eV. (After Freeouf *et al.*, 1973.)

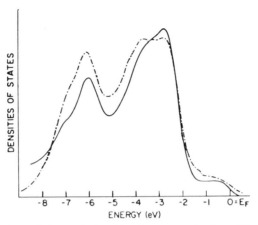

Fig. 6. Comparison of photoemission d-band spectrum for $h\nu = 80$ eV (——) with an x-ray photoemission spectrum (— · —). (After Freeouf *et al.*, 1973.)

work function barrier, Eq. (1).] However, for $h\nu \gtrsim 40$ eV, the spectra $I(E)$ become quite independent of the excitation radiation. These results are shown in Fig. 5 where $h\nu$ spans the range 30–90 eV (Freeouef *et al.*, 1973; Eastman, 1974), and in Fig. 6 where 80-eV spectra are compared with XPS results at 1487 eV. Thus, for gold, at 40 eV the transition to the density of states regime is largely complete. Other work (Shirley *et al.*, 1974) on the noble metals shows that the density of states region may not be fully

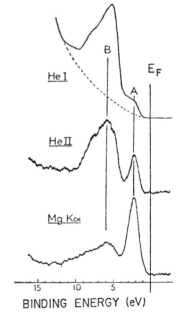

Fig. 7. Photoemission spectra for UO_2 taken at photon excitation energies of 21.2 eV (He I), 40.8 eV (He II), and 1.253 eV (MgKα). The U 5f electrons (peak A) have a very different dependence on photon energy from the O 2p's (peak B). (After Evans, 1977.)

realized until the photon energies exceed ~ 100 eV. The transition between the band structure and density of states regimes for angle-integrated photo-emission is discussed at length by Feibelman and Eastman (1974).

The cross section for photoemission of electrons of a given angular momentum character depends upon the energy of the exciting photons. This $h\nu$ dependence is a matrix-element effect (Eastman and Kuznietz, 1971) that can be exploited to obtain information about the angular momentum character of the occupied states. For example, Fig. 7 shows photoemission spectra for UO_2 taken at 21.2 eV (He I), 40.8 eV (He II), and 1253 eV (MgKα) (Evans, 1977). Peak B of Fig. 7 consists primarily of O sp-derived electrons while peak A corresponds to localized U 5f electrons that are in the $5f^2$ ground-state configuration (see Section V,C). We see that the 5f's are very dominant at x-ray energies, while the p's dominate at UV energies.

C. Direct and Indirect Transitions

The k-conserving, direct interband transition process discussed above appears to provide a good description of $I(E,\nu)$ below $h\nu \sim 25$ eV in certain well-studied cases. However, only a few detailed tests of the direct transition model (utilizing angle-integrated UPS) are available. The needed calculations are complicated and costly and only a few materials have been adequately studied experimentally. In these cases, impressive correspondence between the measured and calculated $I(E,\nu)$ has been obtained using the direct transition model.

Germanium is, of course, one of the most well-studied materials, and its electronic structure has been determined with high precision. The most demanding experimental tests of the numerous band structure calculations (see Grobman *et al.*, 1975, for a list of reviews) have traditionally been provided by optical spectroscopy measurements which now are impressively well reproduced by theory. A recent pseudopotential band calculation of Ge was also used to calculate $I(E,\nu)$, making use of the direct transition model (Grobman *et al.*, 1974, 1975). Photoemission was treated as a three-step process involving (1) optical excitation, (2) electron transport to the sample surface, and (3) electron escape to the vacuum. The calculated results are shown in Fig. 8, along with the corresponding experimental measurements. There is a remarkably strong similarity between these very complex experimental and theoretical results. Furthermore, the band calculation provides an accurate determination of the optical absorption $\varepsilon_2(\omega)$. Like $I(E,\nu)$ this function is also obtained from a direct interband transition analysis, but $\varepsilon_2(\omega)$ represents a very different convolution of conduction- and valence-electron states.

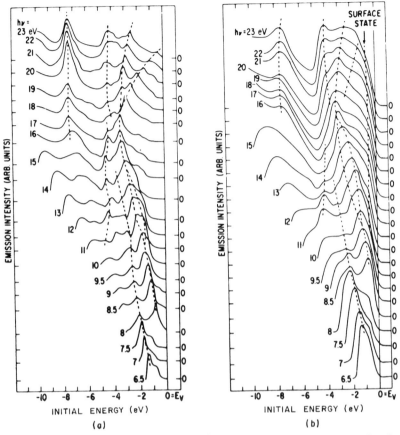

Fig. 8. (a) Calculated and (b) measured energy distribution curves for Ge for $6.5 \leq h\nu \leq$ 23 eV. (After Grobman *et al.*, 1974.)

Figure 9 shows a similar comparison of experimental and theoretical results for copper for the spectral range 9–26 eV (Eastman, 1974). Again, the direct transition model was used and good correspondence between theory and experiment was obtained. The Cu band calculation (Janak *et al.*, 1975) utilized the KKR method iterated to self-consistency. It included computation of interband transition probabilities (momentum matrix elements) to obtain $I(E,\nu)$ and $\varepsilon_2(\omega)$ as well as the Fermi surface. Agreement with all these experimental results is good.

As PES techniques developed, the view that UPS should be regarded as a direct transition process did not go unchallenged (see e.g., Spicer, 1967). Furthermore, recent measurements using angular-resolved photoelectron spectroscopy (see Section II,D) indicate that the matter is still not finally resolved (Grandke *et al.*, 1978).

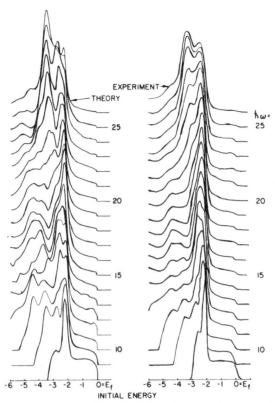

Fig. 9. Comparison of experimental and theoretical photoemission energy distributions for Cu for $8 \leq \hbar\omega \leq 26$ eV. (After Eastman, 1974.)

In the intermediate photon energy range (25–100 eV), a simple uncertainty-principle argument can be made to demonstrate that indirect transitions must become important. In the optical excitation process, an (elastically scattered) electron is created with some final-state momentum $\mathbf{k}_F$. The electron position can, at most, be specified to an accuracy of λ, the elastic escape depth. Then, if $\Delta p \Delta \chi \simeq \hbar/2$, we have $\Delta(\hbar k_F)\lambda \simeq \hbar/2$ or $\Delta k_F \simeq 1/(2\lambda)$. The uncertainty in the normal component of the final-state momentum thus becomes increasingly important as λ becomes smaller. With increasing $h\nu$, λ decreases to a minimum at $h\nu \simeq 100$ eV (Fig. 3) where Δk_F is approximately 10% of $2\pi/a$ (for copper). The uncertainty in the final-state momentum permits a larger sampling of the first Brillouin zone. Stöhr *et al.* (1976a) have demonstrated that this uncertainty effect is a dominant process for producing indirect transitions in copper (see also Shirley *et al,* 1977). Figure 10b shows angle-integrated measurements obtained using synchrotron source radiation of $50 \leq h\nu \leq 200$ eV along with the 1486.6-eV XPS spectrum. Figure 10a shows corresponding calculated

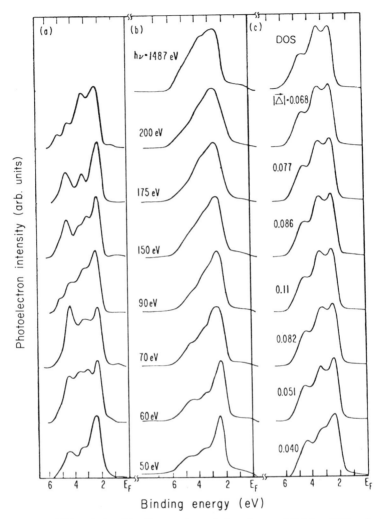

Fig. 10. (a) PEDs calculated for Cu assuming strict **k** and energy $[E_f(\mathbf{k}) - E_i(\mathbf{k}) - h\nu \leq 0.1$ eV] conservation. The calculated curves are convoluted with a 0.5-eV FWHM Gaussian function. (b) Experimental results for polycrystalline Cu. The original data have been corrected for inelastic background. (c) PEDs, convoluted with 0.5-eV FWHM Gaussian function, calculated for Cu assuming an uncertainty in the final-state momentum. The amount of uncertainty $|\boldsymbol{\Delta}|$ is in units of $2\pi/a$ where a (3.615 Å) is the lattice constant of Cu. The theoretical density of states (DOS) is shown at the top of column c. (After Shirley *et al.*, 1977.)

spectra obtained with a direct (strictly **k**-conserving) transition model. XA free-electron final-state and constant-matrix elements were used in the calculation. Initial-state energy bands $E_i(\mathbf{k})$ were taken from Smith (1971). Structure in the calculated $I(E,\nu)$ is strongly modulated by the choice of incident photon energy for the entire series. Experimentally, however, very little difference is found in $I(E,\nu)$ for $h\nu \gtrsim 70$ eV. The result is also found in the calculated spectra of Fig. 10c, where the uncertainty in the final-state **k** is included in the calculation. For $\Delta k_F = 1/\lambda \gtrsim 0.08$, appropriate to the 70-eV spectrum, the calculated $I(E,\nu)$ is essentially indistinguishable from the valence-band density of states (shown as the top curve in Fig. 10c). The DOS calculation accurately reproduces the XPS spectrum.

For $h\nu \gtrsim 100$ eV, the escape depth λ increases and the indirect processes associated with the uncertainty principle must correspondingly decrease. However, indirect transitions continue to dominate because of the increased importance of phonon-assisted processes (Shirley *et al.*, 1977).

D. Angle-Resolved Photoemission

As understanding of the photon energy dependence of photoemission has progressed, angular effects have begun to receive increasing attention. Angular-resolved photoelectron spectroscopy (ARPES) now constitutes a major area of the photoemission field (see Smith, 1978, for a recent review). These experiments, generally conducted at UV photon energies, promise to be extremely powerful tools for determining electron band structures of solids. These experiments probably represent the first opportunity to map out, on a point-by-point basis, complete $E(\mathbf{k})$ energy-dispersion curves both for the filled valence and the empty conduction bands. Problems with interpretation of results remain. Nonetheless, several variations of the ARPES technique are available, some of which provide quite unambiguous results.

With reference to Fig. 11, we briefly describe the technique. Let us consider a portion of $E(\mathbf{k})$ that lies along the ΓX symmetry line in the Brillouin zone of a face-centered-cubic crystal. For simplicity we show only

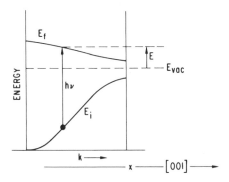

Fig. 11. Energetics of a direct transition occuring in the [001] propagation direction. E_{vac} is the vacuum level and E represents the kinetic energy of the photoemitted electron.

one filled and one empty band separated by the vacuum level designated E_{vac}. (For a metal, E_{vac} is separated from the Fermi level by the work function; see Fig. 1.) For incident photons of energy $h\nu$, electronic excitations occur between the initial $E_i(\mathbf{k})$ and final $E_f(\mathbf{k})$ energy bands along a surface in k space defined by

$$E_f(\mathbf{k}) - E_i(\mathbf{k}) = h\nu. \qquad (2)$$

The ARPES experiment provides a measure of the kinetic energy E (Fig. 11) and propagation direction of the emitted electron. For electrons in vacuum, $E = \hbar^2 K^2/2m$. Thus, both the magnitude and direction of $\mathbf{K}$, the free-electron momentum or wave vector are measured. Usually, the free electron $\mathbf{K}$ is then related to the bulk band structure by means of the direct interband transition model and the "three-step model" of photoemission. In the three-step model, recall that (1) optical transitions (considered to be the same as those associated with bulk optical absorption) produce photoelectrons, (2) the electrons propagate to the surface without energy loss, and (3) the electrons escape across the surface. During escape the parallel component of $\mathbf{K}$ is conserved:

$$\mathbf{K}_\| = \mathbf{k}_\| + \mathbf{G}_\|. \qquad (3)$$

$\mathbf{K}_\|$ is the component of the external (vacuum) electron momentum that is parallel to the surface, $\mathbf{k}_\|$ is the parallel component of the crystal momentum, and $\mathbf{G}_\|$ is the parallel component of a reciprocal lattice vector $\mathbf{G}$. The vector $\mathbf{G}$ must be considered since the wave function associated with the electron state in the solid, $E(\mathbf{k})$, is a Bloch wave of the form

$$\psi_{\mathbf{k}}(\mathbf{r}) = u(\mathbf{r})e^{i(\mathbf{k}+\mathbf{G})\cdot\mathbf{r}} \qquad (4)$$

whose wave vector is specified by the quantity k plus G, an integral multiple of reciprocal lattice vectors having magnitude $2\pi/a$. From Fig. 11 we see that the measured kinetic energy is

$$E = E_f(\mathbf{k}) - E_{vac} \quad \text{and} \quad E_i(\mathbf{k}) = E_f(\mathbf{k}) - h\nu, \qquad (5)$$

so that knowledge of E_{vac} and $h\nu$ specifies both E_f and E_i at the measured θ (the collection angle relative to the surface normal). Knowledge of θ then provides

$$\mathbf{k}_\| + \mathbf{G}_\| = \mathbf{K}_\| = |K|\sin\theta = \sqrt{2mE/\hbar^2}\,\sin\theta. \qquad (6)$$

Thus, the measurable parameters are the energies $E_f(\mathbf{k})$ and $E_i(\mathbf{k})$ that have $\mathbf{k}_\|$, the observed parallel component of k. Unfortunately, this does not completely specify the bands $E(\mathbf{k})$ since the perpendicular component ($k_\perp$) is not strictly conserved and hence cannot be readily measured. However, if

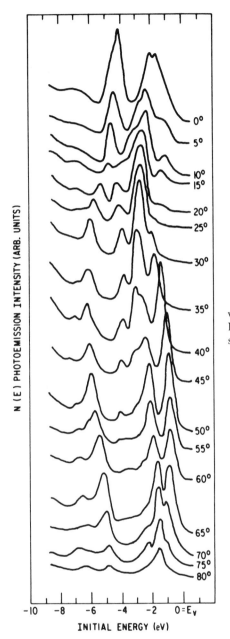

Fig. 12. Polar angle dependence (5° intervals) of photoelectron energy spectra along
$\Gamma M \Gamma$ azimuth with $h\nu = 18$ eV. (After Larsen *et al.*, 1977.)

conservation of $k_\perp$ is assumed, an energy-conservation equation relating $K_\perp$ and $k_\perp$,

$$\left(\hbar^2/2m\right)K_\perp^2 = \left(\hbar^2/2m^*\right)k_\perp^2 - V_0, \qquad (7)$$

can be used to advantage. Here m^* and m are the effective electron masses inside and outside the crystal and V_0 is the height of the potential step at the sample surface (Mahan, 1970).

Because of the indeterminacy of $\mathbf{k}_\perp$, early ARPES studies were largely focused on layered compounds. In this class of materials, the dispersion of the energy bands as a function of $k_\perp$ is expected to be small. Thus, the observed energy dispersion is attributable to $\mathbf{k}_\parallel$ which can be precisely measured.

The procedure for mapping out $E(\mathbf{k})$ depends on measuring the energy dependence of peaks that appear in the energy-distribution curves (EDCs), obtained for a series of angles. In general, it is useful to have data acquired at small increments of θ so that peaks can be identified with confidence. This point is illustrated in Fig. 12 which shows ARPES data for the layered material InSe (Larsen *et al.*, 1977). These are EDCs taken with 18-eV photons at 5° intervals in θ. We see that the EDC structure depends strongly on θ and rapid changes occur even for small variation in θ. Intensities of peaks are also observed to be strongly dependent on emission angles and on photon energy. However, to date, the intensity information in ARPES remains virtually unexamined.

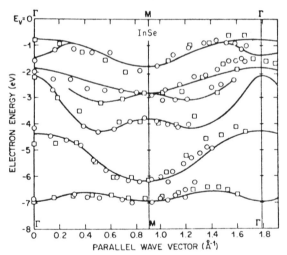

Fig. 13. Two-dimensional band structure for InSe along $\Gamma M \Gamma$ direction. Open circles represent data for $h\nu = 18$ eV; open squares represent data for $h\nu = 24$ eV. Smooth curves with mirror symmetry about the point M have been drawn to pass near experimental points. (After Larsen *et al.*, 1977.)

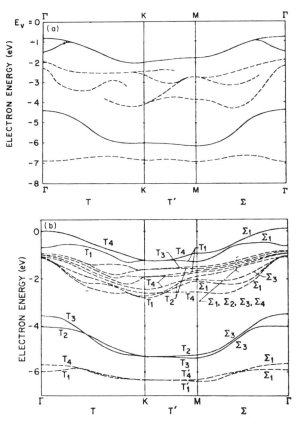

Fig. 14. (a) Experimental results for the two-dimensional energy bands of InSe compared with (b) the pseudopotential band calculations on the related material GaSe.

Figure 13 shows a portion of the band structure of InSe along the $\Gamma M \Gamma$ direction (in the repeated zone scheme). Data here were taken using 18-eV (open circles) and 24-eV (squares) photons with appropriate variation in the polar angle to provide a scan in $k_{\parallel}$ between 0 and 1.8 Å^{-1}. $k_{\parallel}$ is trivially calculated from expression (6). Structure in the EDCs is assumed to be associated with the initial state. A very satisfying and dramatic conclusion of this work, nicely illustrated in Fig. 13, is that the measured E vs $k_{\parallel}$ is symmetrical about the point M as θ is varied to permit $k_{\parallel}$ to span the region between 0 and 1.8 Å^{-1}. Figure 14 summarizes the available measurements for InSe (Larsen *et al.*, 1977) compared with a pseudopotential calculation for the similar compound GaSe, reported by Schlüter *et al.* (No calculation is available for InSe.) Except for a $\sim$ 1-eV relative shift, the experimental and theoretical band structures are very similar.

For three-dimensional materials (as opposed to layered compounds), a number of other approaches have been used to exploit ARPES to study energy-dispersion curves (Smith, 1978). We will briefly consider one of those approaches, e.g., normal emission. In this case, only those electrons emitted normal to the crystal surface are collected. Then $K_{\parallel} = 0$ and we sample bulk electronic states that have $k_{\perp}$ along a specific line in **k** space. Now Eq. (7) is appropriate and information is obtained from EDCs acquired at different $h\nu$. Stöhr et al. (1976b, 1978) have reported such EDCs for different orientations of single-crystal copper. Curves for (001), (110), and (111) emission, taken for $32 \leq h\nu \leq 160$ eV, are shown in Fig. 15. Spectra for the different orientations are quite dissimilar, particularly the (110) spectrum in the vicinity of $h\nu = 45$ eV. For these spectra, photon energies are high enough so that final-state effects should be negligible and experimental peak positions will depend only on the valence-band dispersion $E(\mathbf{k})$. By applying Eq. (7) to calculate $k_{\perp}$ (assuming m* is the free-electron mass and $V_0 = 13.8$ eV), one obtains the $E(\mathbf{k})$ curves (bars) shown in Fig. 16 (Shirley et al., 1977). The photon energy scale on the bottom of the $E(\mathbf{k})$ plots gives the approximate correspondence between the measured $h\nu$ and the displayed $k_{\perp}$ as determined by Eq. (7). The bars represent measurements that are superimposed on the calculated $E(\mathbf{k})$ curves of Burdick (1963). In general, very good correspondence between theory and experi-

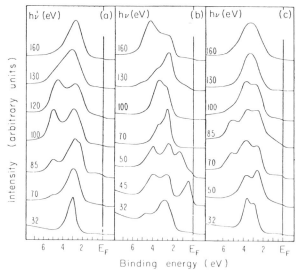

Fig. 15. Photoemission spectra obtained from the three-low index crystal faces of Cu using angle-resolved normal photoemission. (a) (001), (b) (110), and (c) (111). The angular resolution was ∼ ±5° and the energy resolution was ≤ 0.2 eV. (After Shirley *et at.*, 1977.)

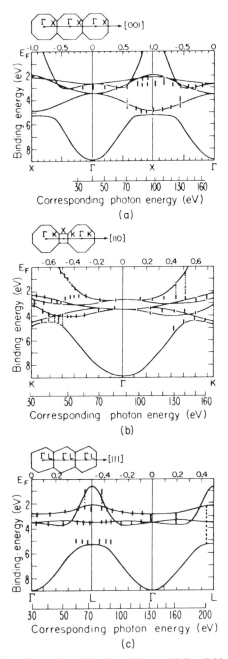

Fig. 16. Comparison of experimental peak positions with Burdick's band structure of Cu for the three directions investigated. The inset at the top of each panel shows the propagation direction of the photoelectron in the extended zone scheme. The illustrations also show the correspondence between photon energy and the part of the Brillouin zone sampled. (a) Cu[001], (b) Cu[110], and (c) Cu[111]. Note difference scales. (After Shirley *et al.*, 1977.)

ment is again obtained. Particularly impressive is the detailed mapping of the portion of $E(\mathbf{k})$ that cuts through E_F between Γ and K in Fig. 16b. In a very small region of k space along ΓK, this band moves from the d-band complex (~ 3 eV—Fig. 16b) to the Fermi level. It is this effect that is so strikingly apparent in the spectra of Fig. 15b near $h\nu = 45$ eV.

This Fermi level cutoff effect has also been seen by Knapp *et al.* (1979) in normal emission at lower $h\nu(\sim 10$ eV). The work of Knapp *et al.* (1979) also emphasizes the ability of the ARPES technique to provide $E(\mathbf{k})$ information for both initial- and final-state energy bands. For this complex situation, the step between measured EDCs and $E(\mathbf{k})$ curves is made with the help of a detailed band calculation.

E. Spin-Polarized Photoemission

1. DESCRIPTION

The electron spin polarization (ESP) for photoemitted electrons is another measurable parameter that yields valuable electronic structure information, particularly for magnetic materials. For magnetically ordered materials, photoemitted electrons have a characteristic spin polarization which reflects the electron spin orientation that occurs in the sample prior to the optical excitation (Busch *et al.*, 1970). In principle, then, a direct measurement of the photoemission density of states and the corresponding net ESP can be obtained as a function of energy relative to E_F. In practice, however, the polarization analysis (using Mott scattering from a gold foil) severely attenuates the photoemitted electron flux so that serious intensity limitations are encountered. Currently available techniques utilize low-energy monochromatic photons ($h\nu < 11$ eV) with the photoelectron flux monitored in an energy-integrated mode. That is, at a given $h\nu$, all photoemitted electrons are analyzed to determine their polarization without regard to their kinetic energy. When the ESP moment is aligned parallel (antiparallel) to the total sample magnetization, the spin polarization is designated as positive (negative). I_+ (I_-) then designates the positively (negatively) polarized electron current and the polarization fraction is expressed as $P = (I_+ - I_-)/(I_+ + I_-)$. Experimentally, P is usually measured as a function of photon excitation energy. The ESP measurement gives information on the net spin of those electrons within $h\nu - \phi$ of E_F (ϕ is the work function; see Fig. 1). By scanning $h\nu$ (typically between 4 and 11 eV), the valence-electron spin polarization can be studied at E_F and in a window ~ 5 eV below E_F. Because an integrated photoelectron current is monitored, measurements just above threshold tend to be the most meaningful.

2. ITINERANT AND LOCAL ELECTRON STATES

In itinerant ferromagnets, the magnetic moment results from spin-polarized electron subbands that are shifted in energy relative to one another. The process of magnetic ordering splits a strong peak at E_F into two subband peaks, one of which may lie predominantly above E_F, the other below. This results in a substantial spin polarization at energies near E_F, as seen in the schematic diagram of Fig. 17. Consequently, one would expect a very strong positively polarized electron fraction at photothreshold ($h\nu \simeq \phi$) where electrons at E_F are sampled. For increased $h\nu$, the polarized fraction should drop since essentially unpolarized electrons are added to the integrated photoelectron current. We observe in Fig. 18 that for ferromagnetic iron, P is indeed large and positive at photothreshold ($h\nu \simeq 4.7$ eV) and falls rapidly with increasing $h\nu$ (Eib and Reihl, 1978). These results are consistent with the itinerant model of ferromagnetism (Fig. 17) and furthermore, as discussed below, appear to be inconsistent with the localized electron model of ferromagnetism.

A case of local electron ferromagnetism can be illustrated with ESP data for magnetite (Fe_3O_4). Magnetite contains both Fe^{2+} and Fe^{3+} ions, which

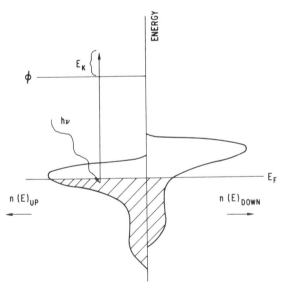

Fig. 17. Schematic representation of the electron density of states for an itinerant ferromagnet. Electrons with opposite spin are separately represented. For photoemission near threshold ($h\nu \gtrsim \phi$, the work function), emission from spin-up electrons is dominant. The integrated polarization fraction decreases as $h\nu$ increases (see text). E_F is the Fermi level and E_k is the electron kinetic energy.

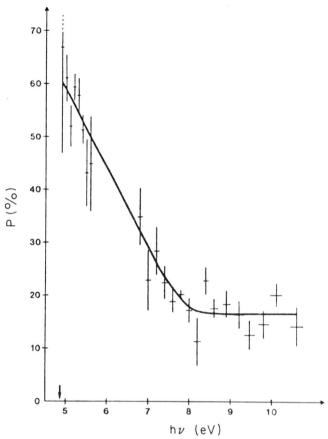

Fig. 18. Photoelectron spin polarization in % as function of photon energy in eV, for Fe(111) at 200 K and 11 kG. The arrow indicates the photothreshold ϕ. The polarization is large at the threshold and decreases rapidly with increasing $h\nu$ (see Fig. 17). (After Eib and Reihl, 1978.)

have d^6 and d^5 configurations, respectively. The problem is complicated by the fact that Fe ions appear in two valence states and in two sublattices (for which the signs of the crystal-field splitting are reversed). Figure 19 shows an energy-level scheme for the Fe^{2+} ions which depicts the electron occupancy of the exchange-split crystal-field levels designated T_{2g} and E_g. This is a high-spin configuration in which the least tightly bound electron is aligned opposite the direction of magnetization (spin down). At photothreshold, we expect electron emission to occur first from the partially occupied, spin-down state. Thus, photothreshold ESP should be negative, and with increasing $h\nu$, ESP should rapidly change to a large positive value as emission from the spin-up states begins to contribute to the photoemitted

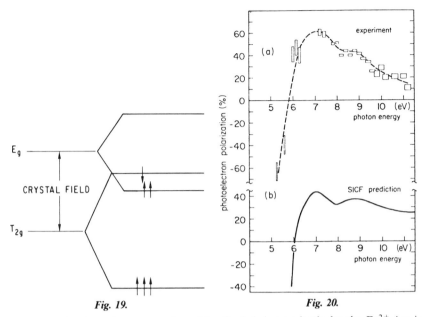

Fig. 19.

Fig. 20.

Fig. 19. A schematic representation of localized d-electron levels for the Fe^{2+} ion in Fe_3O_4. The Eg and T_{2g} crystal-field levels are split by the exchange interaction. For the levels shown here, the integrated spin polarization is negative at threshold and, with increasing $h\nu$, rapidly turns positive (see Fig. 20).

Fig. 20. (a) Electron spin polarization (ESP) as a function of $h\nu$ for Fe_3O_4 at 200 K compared with (b) a model prediction. As indicated in Fig. 19, the ESP is negative at threshold and rapidly turns positive. (After Alvarado *et al.*, 1976.)

flux. Figure 20 shows the ESP for Fe_3O_4, which confirms these expectations (Alvarado *et al.*, 1976). Emission from Fe^{3+} ion states appears at higher $h\nu$ (i.e., above photothreshold) and with positive ESP since the configuration is d^5 (see Fig. 19).

The iron study also led the authors to conclude that ferromagnetism in iron could not be attributable to localized 3d electrons. If iron were in a high-spin state occupying a d^7 or d^8 configuration, then one would expect a negative photothreshold ESP, as seen in Fig. 19 for Fe_3O_4.

3. CONDUCTION-ELECTRON POLARIZATION

Because the energy of the exciting radiation is so low, information about f-electron states is not directly available in ESP studies. Recall that the photoionization cross section for a given orbital angular momentum (l) state depends on the energy of the exciting radiation and that, at small $h\nu$, transition probabilities are very small for electrons with large *l* (see Section II, B). This insensitivity can be used to advantage in studying conduction-electron polarization in lanthanide or actinide compounds. For these materials, the magnetization is primarily determined by the occupied 4f (lanthanide) or 5f (actinide) electrons. Since the photoyield of the f electrons is so small relative to s-, p-, or d-electron yields, the observed ESP must result from conduction electrons that are polarized by electrons in the partially filled f shell.

Conduction-electron polarization is induced by the total *spin* moment **S** of localized f electrons, whereas the magnetization is determined by the total angular momentum **J** (Jena *et al.*, 1978). If the f electron ground state can be specified by Hund's rules, then $\mathbf{J} = \mathbf{L} - \mathbf{S}$ for shells less than half full ($n < 7$) and $\mathbf{J} = \mathbf{L} + \mathbf{S}$ if $n \geq 7$. Thus, **S** and **J** may be either parallel or antiparallel in alignment, and correspondingly, the conduction ESP may be either positive or negative. Studies of several lanthanide compounds with $4f^7$ configurations show the expected positive ESP (Alvarado *et al.*, 1978).

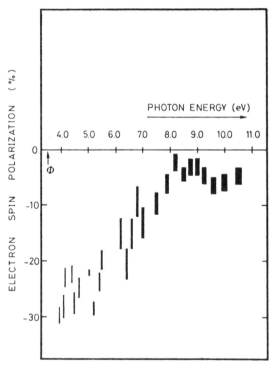

Fig. 21. Spectrum of the spin polarization of USe ($T < 20$ K, $H = 8.4$ kG). Note that the polarization is negative for all *hν*. (After Erbudak *et al.*, 1979.)

In contrast, ESP that is negative for all $h\nu$ is observed for a number of uranium intermetallic compounds (Greuter et al., 1979). These uranium compounds contain 5f electrons in less-than-half-filled 5f shells. If the 5f's are localized and obey Hund's rules, then the conduction-electron polarization should be antiparallel to the magnetization and hence negative for all $h\nu$. The results for USe, shown in Fig. 21, are consistent with this picture. [We note, however, that differences of opinion exist with regard to the nature of the 5f's in the uranium compounds. Both localized and itinerant models have been invoked (Freeman and Darby, 1974; Greuter et al., 1979).]

We have seen that both photo ESP and ARPES measurements hold bright promise as tools for materials characterization. However, to date, the techniques are in highly developmental stages with regard to both instrumentation and analysis. Thus, in the subsequent discussion, we shall be content to focus on the somewhat more mature (and perhaps more mundane) phototemission spectroscopic capabilities.

III. Valence-Electron Studies—XPS

A. Itinerant States—Simple Concepts Illustrated

1. Evolution of Core States from Valence States

We have noted that, for itinerant electrons, XPS measures the valence-band density of states with possible distortions resulting from matrix-element effects. Frequently, the distortions are small and the simple XPS valence-band measurement is well reproduced in a (band) calculation of $n(E)$. The valence-band XPS measurement is, therefore, a graphic, readily interpretable one that can be obtained for a wide range of materials. Systematic effects can be readily and instructively examined. To illustrate these XPS capabilities, we shall, in this section, discuss some simple but fundamental concepts of band theory which can be nicely demonstrated.

The outermost electrons of atoms are, of course, the electrons most directly responsible for chemical behavior and are responsible for essentially all of the common physical properties of materials. In solids, outer electrons of adjacent atoms overlap and hybridize to form bands which are increasingly filled with increasing atomic number Z. The degree of occupation of the bands and the character of the electrons occupying the bands determine the physical properties of the solid. As in free atoms, bands are filled in accordance with the rules of the Pauli exclusion principle so that, in the absence of hybridization, filled bands are specified by the usual designation

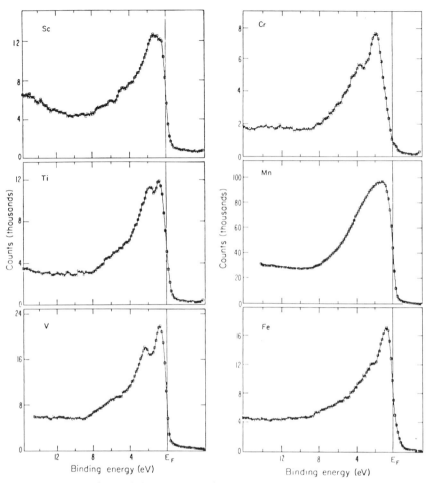

Fig. 22. X-ray photoemission spectra for the valence bands of the 3d transition metals Sc, Ti, V, Cr, Mn, and Fe. (After Ley *et al.*, 1977.)

for atomic shells, i.e., $1s^2$, $2s^2$, $2p^6$, $3s^2$, $3p^6$, $3d^{10}$, etc. An unfilled band will, of course, be pinned to the Fermi level (E_F) of a solid. The band density of states at the Fermi level, $n(E_F)$, will determine or strongly influence most of the transport properties of the material. Figures 22 and 23 show XPS spectra for the 3d transition-metal series from scandium through copper (Ley *et al.*, 1977; Hüfner and Wertheim, 1974). We see that for scandium through nickel the spectra are very similar. The addition of electrons with increasing Z produces little increase of bandwidth and no retreat from E_F. Once the 3d band is filled, however, it rapidly retreats from E_F. This is dramatically demonstrated in the Cu spectrum of Fig. 23 where the 3d band

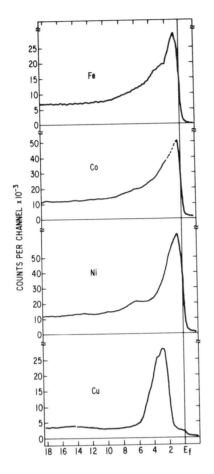

Fig. 23. X-ray photoemission spectra for the valence bands of the 3d elements Fe, Co., Ni, and Cu. (After Hüfner and Wertheim, 1974.)

is found to be centered about 3 eV below E_F, overlapping a now partially filled 4s band.

Let us now switch our attention to the next row of the periodic table and recall that Pd is a 4d element lying below Ni. Pd, like Ni, is a high density of states material whose d band is "pinned" to E_F. XPS spectra for the elements palladium through tellurium are shown in Fig. 24 (Pollak *et al.*, 1972). We see in Fig. 24 that the d band of Ag, like Cu, has filled and retreated from E_F. However, the d band overlaps the 4s band and some s–d hybridization occurs. Also, the d electron wave functions are sufficiently extended so that considerable d–d overlap occurs (Christiansen, 1972; Segall, 1962). Thus, the 4d bands of Ag (and Cu) retain substantial width. In Cd, the 4d band narrows dramatically and continues the rapid retreat to core-level status. This process continues from Cd through Te and a new s–p valence band emerges to define the materials properties of these heavier

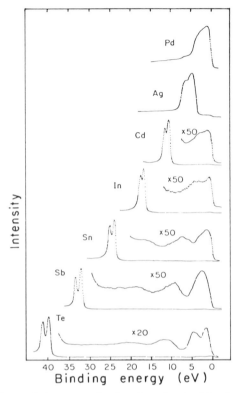

Fig. 24. X-ray photoemission spectra for the valence bands of the elements Pd through Te. For the transition metals, the d-band edge is pinned to E_F (see Figs. 22 and 23). When the d band is filled, it rapidly retreats to core-level status as electrons are added (increasing the atomic number), and a new s–p band emerges at E_F. (After Pollak *et al.*, 1972.)

elements. These s–p bands are much broader than the d bands and the materials have much lower densities of states. The d bands split into their spin–orbit components, the $4d_{3/2}$ and $4d_{5/2}$ levels, which become increasingly well resolved with higher Z. Meanwhile the s band begins to emerge from the hybridized s–p band so that, by Sb and Te, a peak has emerged (near 12 eV binding energy) that is predominately s in character. This level, like the d's, will continue to narrow and retreat from E_F as electrons are added.

2. Local Density of States

This section illustrates another important concept in band theory, namely that of a local density of states. Clearly, for the heavier elements in Fig. 24, the d electrons are *atomiclike* with rather limited radial extent. The wave functions do not have significant overlap with neighboring atoms. In this case, the 3d charge density of, for example, Te is by far the greatest in close proximity to the Te nucleus. Thus, a "local $n(E)$" becomes a meaningful concept, since the states are densely concentrated in a local region of space. However, as the core level advances toward E_F (with decreasing Z), the wave function of the level increasingly delocalizes and hybridization begins. The wave function begins to occupy much more of the interstitial region between atoms. Nonetheless, this delocalization is a gradual process and some local character is nearly always preserved in $n(E)$.

The importance of a local $n(E)$ is also underscored in the study of alloy properties. The subsequent discussion of Ag–Pd and Cu–Ni (Section V, B) shows that a local d-band density of states from the Pd (or Ni) is preserved when the transition metal is dissolved into the noble metal.

3. Hybridization

Hybridization results from the overlap (and interaction) of the wave functions on different atoms. In silver, for example, the 4d-electron band overlaps the 5s band (Fig. 24) but, more importantly, the 4d's are strongly hybridized with d electrons on adjacent atoms (Christiansen, 1972). Figure 25 shows the 4d radial wave function for atomic silver (Herman and Skillman, 1963). In silver metal, the near-neighbor interatomic distance is 2.89 Å (Pearson, 1958). The region of greatest overlap occurs at 1.45 Å, the

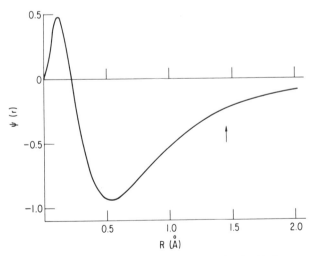

Fig. 25. The radial wave function $\Psi(r)$ versus distance R for the 4d electrons of atomic Ag. The arrow indicates the intermediate position between near-neighbor Ag atoms. These results indicate that considerable overlap of 4d-electron wave functions of adjacent atoms will occur.

intermediate position between adjacent atoms (indicated by the arrow in Fig. 25). Separation of the Ag atoms in the lattice should act to reduce the hybridization and make the d electrons more *atomiclike*. This effect can be nicely observed in the comparison between Ag metal and the ternary (pseudo–II–VI) compound $AgInSe_2$ (Rife *et al.*, 1977). Since In and Se do not contain d electrons near their valence bands, these elements serve to dramatically dilute the d-electron concentration in the valence-band region. The Ag atoms in $AgInSe_2$ are crystallographically ordered so that d–d overlap is always small. Figure 26 shows the Ag metal valence-band region compared with that of the ternary compound. The d bands in the ternary have dramatically narrowed. The level is now even narrower than the nearby In 4d core level (which shows a small spin–orbit splitting). It is unclear whether the spin–orbit splitting of the Ag 4d level in the ternary is obscured by residual hybridization or whether the XPS instrument has inadequate resolution to detect the splitting. [For the free atom, the splitting

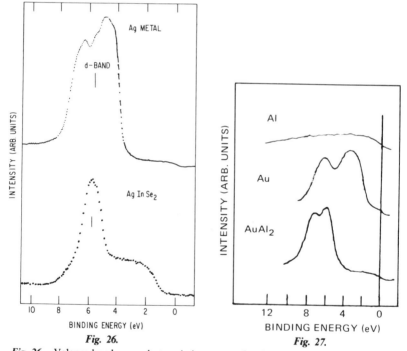

Fig. 26.   Valence-band x-ray photoemission spectra for Ag metal and for the semiconductor $AgInSe_2$. The intense band at ~ 6 eV corresponds to Ag 4d electrons. The 4d bands of the compound are less hybridized and are narrower than the 4d bands in Ag metal.

Fig. 27.   Valence-band x-ray photoemission spectra for Al, Au, and $AuAl_2$. Note that, as Au is diluted in $AuAl_2$, the Au d band appreciably narrows. (After Fuggle *et al.*, 1973.)

is expected to be about 0.65 eV (Herman and Skillman, 1963).] Clearly, the 4d level in the ternary has become very much like the free-atom level.

A similarly dramatic result is seen in the comparison of the valence-band spectra of Au and $AuAl_2$ (Fig. 27). Here we see the measured 5d band narrow from 5.3 to 3.2 eV (Fuggle et al., 1973). In the compound, the residual spin–orbit splitting is approximately the same as the free-atom value of 1.83 eV (Herman and Skillman, 1963).

B. Localized Electrons in Unfilled Shells

1. MULTIPLETS

The wave functions of *itinerant* valence electrons in a solid always experience substantial hybridization with neighboring wave functions located on the same or neighboring atoms. (It is the hybridization that is responsible for the itinerant character.) At the same time, electrons that occupy a partially filled shell may be spatially localized very close to an atomic site and experience very little hybridization with neighboring electrons. These localized electrons do not participate directly in chemical or metallic bonding but behave more like the deep atomic core levels. A profound difference from the core states, however, is that the shells are only partially filled, a feature responsible for many important materials properties in solids that contain such states. Notable examples of materials containing unfilled shells of localized electrons are lanthanide (Ln) elements and their compounds.

Unlike XPS spectra for *itinerant* electron states, XPS spectra for *local* states in unfilled shells do not closely correspond to the ground state $n(E)$. The contrast between local and itinerant states is a manifestation of the relative importance of Coulomb correlation (the tendency of electrons to repel one another and consequently to correlate their relative orbital motions) and effective bandwidth (a consequence of orbital overlap and hybridization) (see Chapter 1, this volume). In the itinerant case, Koopman's theorem (Koopman, 1933) is valid, measured ionization energies are representative of ground-state energies, and the electron states may be treated within a one-electron or band-theory formalism. However, in a correlated electron system, a perturbation of one electron in the system will be strongly felt by its neighboring electrons. The shock of a perturbation that ejects an electron from a local system (e.g., photoemission) will be shared by all correlated participants. In these systems, photoemission observes a final-state multiplet structure (Wertheim, 1978). That is, a photon impinging on a localized n-electron system will photoeject an electron to the vacuum but will also excite the remaining n − 1 electrons into allowed final-state multiplets. The energy absorbed by the n − 1 system in producing the various excitations will be reflected in the measured energy of the photo-ejected electron. Thus, the spectra are a direct measure of the multiplets in the n − 1 system that can be excited in the photoemission process. Since crystalline field effects are small in the lanthanides, photoemission can be understood within the framework of conventional atomic physics. As pointed out by Cox (1975), a theoretical basis for predicting allowed *multiplet*

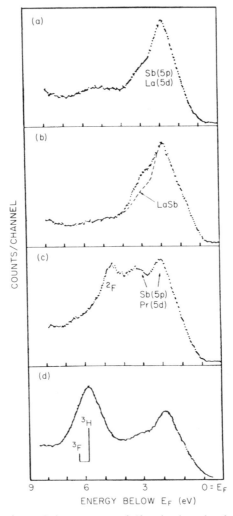

Fig. 28. X-ray photoemission spectrum of 4f and valence-band region of (a) LaSb, (b) CeSb, (c) PrSb, and (d) NdSb. (After Campagna *et al.*, 1974.)

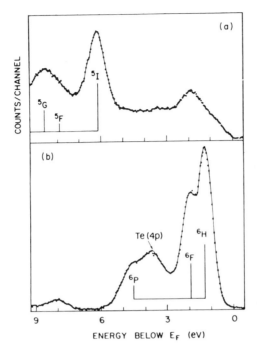

Fig. 29. X-ray photoemission spectrum of 4f and valence-band region of (a) SmSb and (b) SmTe. (After Campagna *et al.*, 1974.)

excitations and for computing their *relative intensities* was available in Racah's fractional-parentage formalism. For the lanthanides (Wertheim, 1978), multiplet *energies* were assumed to be the same (with small corrections) as the known intra-4f optical transitions. Since XPS measures the $n - 1$ multiplet structure, a measured multiplet for an atom with nuclear charge Z was associated with the $Z - 1$ intra-4f optical spectrum.

Lanthanide elements are chemically very active and the preparation and maintenance of suitably clean sample surfaces is a demanding problem. Experimental problems were somewhat eased by examining the NaCl-type binary lanthanide-antimony compounds (Campagna *et al.*, 1974). These are metallic, can be prepared in large single-crystal form, and can be cleaved in situ. Figure 28 shows XPS data for binaries of the lighter lanthanides. In addition to the unfilled 4f shell, lanthanides also have an s–p–d electron population that permits chemical or metallic bonding. Only the itinerant (band) electrons (including some from Sb) are observed in LaSb, whereas a superposition of band spectra and simple multiplet spectra (associated with small 4f populations) are seen in the other compounds of Fig. 28. For these compounds, the 4f population is small relative to the Ln–Sb valence-electron population. However, the 4f intensities at XPS energies are much greater

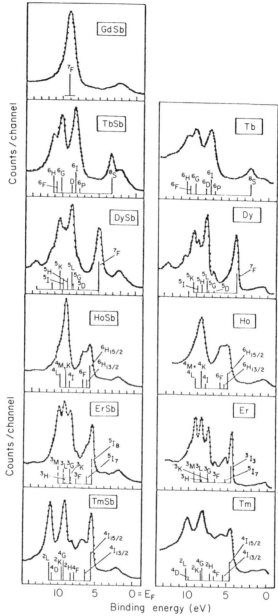

Fig. 30. Valence-band and 4f x-ray photoemission spectra for a series of rare-earth elements and antimonides. The spectra are dominated by 4f emission. Note the strong similarity between the adjacent metal and compound spectra. These adjacent pairs have the same 4f occupation. Calculated final-state multiplet spectra are also shown. Antimonide data from Campagna *et al.* (1974); metal data from Baer and Busch (1974).

than are band-electron intensities. Thus, in general, the 4f electrons can be
easily seen. (Note, for example, in NdSb, where there are three occupied 4f
states and eight bonding electrons, that the 4f- and bonding-electron
intensities are of comparable magnitude.) For the Ce compound, the 4f
electron is not clearly distinguished but, by comparison with LaSb, is
assumed to be a single peak near 3-eV binding energy. This assignment is
further reinforced by examining systematics of other cerium compounds
(Sato, 1976; Platau and Karlsson, 1978).

In Fig. 29, the two compounds SmSb and SmTe are shown (Campagna *et
al.*, 1974). Since Sb and Te occupy groups V and VI, respectively, of the
periodic table, the Sb (or Te) binary compounds leave the Sm ion in the $+3$
(or $+2$) ionization state. The ground-state f configuration is then $4f^5$ in
SmSb and $4f^6$ in SmTe. The corresponding multiplets produce the dramatically different spectra shown in Fig. 29. The SmSb spectra are, however,
remarkably well reproduced by the spectrum of pure Sm (Baer and Busch,
1974), which has the same $4f^5$ ground state.

Figure 30 shows XPS data for the pure lanthanide elements Gd through
Yb and for corresponding antimonides (Campagna *et al.*, 1974; Baer and
Busch, 1974). As seen in Figs. 28 and 29, the predicted multiplets and their
intensities are designated with usual spectroscopic notations. In each case,
the Hund's rule ground state (of the n-1 electron system) is the energy level
with the lowest ionization energy. The remarkably strong similarity between
the pure elements and the compounds (with corresponding 4f occupation)
dramatically emphasizes the atomic nature of the 4f levels. The transition
between the complex multiplet spectrum of Tm metal and the simple
spin–orbit spectrum of the adjacent Yb metal (with a filled 4f shell) further
dramatizes the *atomiclike* character. Adding an additional electron (to

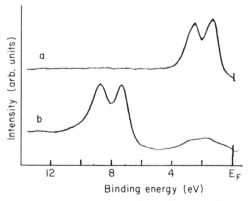

Fig. 31. X-ray photoemission spectra for the valence bands of (a) ytterbium and
(b) lutetium. (After Lang *et al.*, 1974.)

produce Lu) causes the newly formed 4f core level to rapidly retreat from E_F, in the manner described in Section III, A for the 4d transition metals. However, this effect, shown in Fig. 31 (Lang *et al.*, 1974) is unlike that observed for the 4d elements in that no significant hybridization is observable at any time.

2. CONFIGURATION FLUCTUATIONS

The understanding of static multiplet structure provided the framework for investigating homogeneous mixed-valence compounds (Campagna and Wertheim, 1976). These compounds contrast with inhomogeneous mixed-valence compounds, in which a given element might appear in different valence states but these states would occupy inequivalent crystallographic sites. (For example, Fe_3O_4 contains both Fe^{2+} and Fe^{3+} ions in inequivalent sites). The homogeneous compounds are often referred to as valence- or configuration-fluctuation systems. The implication in this designation is that the system fluctuates, in time, between well-defined configurations and may even do so at $T = 0$ K (Varma, 1976). However, all crystallographic sites are (time-averaged) equivalent.

The investigation of these systems is important for improving the understanding of the electronic structure of systems that exhibit intermediate degrees of itinerancy. Well-developed formalism exists to treat both the extreme local and the extreme itinerant systems. The intermediate region is rich in phenomenological effects but no single theoretical framework will suitably account for all phenomena observed. Generally, the 4f's in the lanthanides show localized electron behavior but the fluctuation phenomenon contains the first signs of bridging the intermediate region; in this case, from the local extreme. An itinerant system may be viewed as an extreme example of a configuration-fluctuation system in which many of the possible configurations simultaneously appear. [Many examples of the intermediate case exist in actinide compounds. The actinides contain an unfilled shell of 5f electrons which have a radial extent significantly larger than the 4f's. The 5f electrons have a greater tendency to hybridize and the distinctly *atomiclike* character may be lost (Freeman and Darby, 1974).]

A number of homogeneous mixed-valence *lanthanide* compounds are known (Campagna and Wertheim, 1976). These compounds are particularly attractive for XPS studies since 4f-electron cross sections are large and the well-understood static multiplet spectra are so profoundly dependent on 4f occupation. XPS thus contains the potential for directly displaying the 4f configurations involved.

A striking example of the success of this type of study is demonstrated with the homogeneous mixed-valence compound SmB_6 (Chazalviel *et al.*,

1976). In this compound, Sm ions appear in both 2 + and 3 + valence states so that both f^6 and f^5 configurations appear. The compound SmB_6 also contains s–p valence electrons which overlap the multiplet structure. The s–p electron intensity should be comparable to the valence-electron intensity in the related compound LaB_6, which has no occupied 4f states. As shown in Fig. 32, the s–p intensity is small relative to the f-electron intensity. The upper curve of Fig. 33 shows the measured SmB_6 spectrum. Here, the LaB_6 s–p structure has been subtracted, leaving only 4f spectra. The Sm^{2+} ($4f^6$ ground-state) multiplet structure appears near E_F. Intense Sm^{3+} multiplet structure also appears in Fig. 33 but at higher binding energies, well separated from the 2 + structure. The lower curve of Fig. 33 shows a theoretical calculation of the Sm^{2+} and Sm^{3+} multiplets. Expected multiplets and their relative intensities (within a multiplet) were computed, energies were obtained from optical data or calculated, and relative multiplet intensities were scaled for best fit to the data. The multiplets were appropriately broadened using an energy-dependent broadening function to simulate experiment (Sm^{2+}) or to account for lifetime effects (Sm^{3+}). The theoretical curve is a remarkably good representation of the experimental spectrum, thus providing direct and dramatic evidence of configuration fluctuations. Furthermore, the multiplet intensity ratios provide a measure of the average valence (2.7), and the multiplet positions measure the energy shift (7 eV) between the ground levels (^6H and ^5I) of the two final-state configurations.

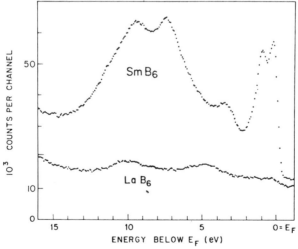

Fig. 32. XPS spectra of the valence-band and 4f region of mixed-valence SmB_6 and of metallic trivalent LaB_6.

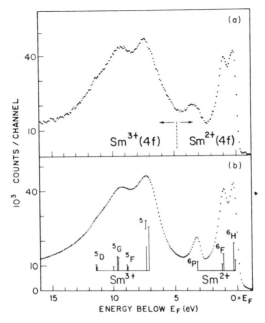

Fig. 33. (a) The Sm contribution to the valence band of SmB_6, obtained as the difference curve of the data of Fig. 32. (b) Theoretical curve calculated using the intensity of final-state multiplets.

IV. Core-Electron Studies

The early development of XPS concentrated largely on the measurements of absolute binding energies of electron core levels (Siegbahn *et al.*, 1967). It was observed that the absolute binding energies of the *atomiclike* electron core levels are dependent on the chemical state (or more precisely, on the local charge environment) of the atom under study. Furthermore, intensity measurements of the core lines provide the capability of performing quantitative chemical analyses. These observations led to the widespread use of XPS for basic and applied chemistry studies. Accordingly, the x-ray photoemission technique was called ESCA, for electron spectroscopy for chemical analysis.

Core-level studies are essentially the domain of XPS as opposed to UPS, since very few core levels are accessible to UPS with its low photon excitation energies. UPS has a higher resolution capability which could, in some cases, be used to advantage in core-level studies. However, it is frequently observed that linewidths are primarily determined not by instrument resolution limitations but rather by broadening mechanisms intrinsic to the solid. The wider linewidths are commonly found in insulating materials. Consequently, the resolution capability of commercial XPS instruments ($\sim$ 0.5 to 1.5 eV) is adequate for most chemistry applications.

For elemental analysis, core-level studies are not highly sensitive. Sensitivities exceed 1 at. % only in those favorable cases where the element under study has levels with high subshell photoemission cross sections (SPC). The heavier elements (5d series or heavier) are particularly favorable since they have high SPC 4f core levels. For quantitative chemical analysis, techniques other than PES are generally to be preferred unless analytical information in the near-surface (within $\sim$ 30 Å) is desired. SPC's for elements Li through U have been tabulated by Cardona and Ley (1978).

An extensive literature documents the substantial effort devoted to chemical-shift studies of chemical compounds (see literature reviews cited in

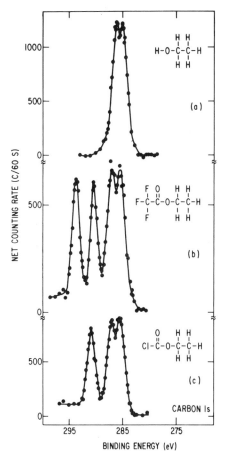

Fig. 34. Carbon 1s photoelectron spectra from (a) ethanol, (b) ethyl trifluoroacetate, and (c) ethyl chloroformate. (After Gelius *et al.*, 1970.)

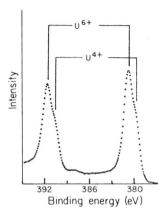

Fig. 35. XPS spectra for the spin–orbit split U 4f levels in U_3O_8. Chemically shifted peaks corresponding to U^{4+} and U^{6+} ions are observed. (After Verbist *et al.*, 1974.)

Section I). In small molecules, chemical shifts of as high as 10 eV may be observed between atoms in extreme oxidation states. Figure 34 shows C 1s core-level spectra for ethanol, ethyl trifluoracetate, and ethyl chloroformate. The different carbon coordinations in these compounds are shown by the structural formulas in Fig. 34. Different degrees of electron transfer (from the carbon site) are expected with the different ligand coordinations. These different charge distributions associated with the inequivalent sites result in the shifted XPS C 1s lines observed in Fig. 34. More typically, however, significantly smaller chemical shifts are observed when the oxidation state of an atom is changed even by several formal elementary charges. The 2p level of aluminum, for example, is shifted by 2.7 eV between Al metal and the Al^{3+} compound Al_2O_3.

In the compound U_3O_8, the uranium atom appears in crystallographically inequivalent sites where it apparently resides in $4+$ and $6+$ ionic configurations. If U_3O_8 is written as $UO_2 \cdot 2UO_3$, then the hexavalent form of uranium should appear with twice the intensity of the tetravalent form and at the higher binding energy. Figure 35 shows spin–orbit split U $4f_{5/2}$ and U $4f_{7/2}$ levels (separated by 10.8 eV), which display an additional small (1.2-eV) splitting corresponding to the chemically shifted U^{6+} and U^{4+} sites. The $6+$ and $4+$ sites appear with the appropriate $2:1$ ratio (Verbist *et al.*, 1974).

In the U_3O_8 study, the uranium charge state was identified by comparing the observed shifts with binding energies of the reference compounds UO_2 and UO_3. This kind of "fingerprinting" procedure is often used for identifying chemical environments. It has proved to be of considerable value in studies involving a wide range of applications including industrial catalysts, polymers, devices, and aerosols (Carlson, 1978; Grant and Housley, 1979).

Although chemical-shift information is clearly useful, the thorough understanding of all factors affecting chemical shifts constitutes a formidable problem. Both ground-state properties (charge distributions) and effects associated with the excitation process must be considered. [See Fadley (1978) for a discussion of interpretive procedures and relevant literature citations.] The ground-state energy of an electron in a tightly bound core level is determined by the attractive potential of the nuclei and the repulsive Coulomb interaction with all other electrons. Charge rearrangement associated, for example, with valence change clearly affects this energy. As valence charge is transferred from an atom, the electron–electron repulsion is decreased and the electronic level moves to lower total energy (higher binding energy). Conceptually, one might prefer to view this charge-transfer process as a reduction of the nuclear screening charge so that the nucleus exerts a stronger attractive force on the core electron state, pulling it to higher binding energy. However, the lattice Madelung energy partially compensates for the level shift associated with charge transfer.

In addition to energy shifts associated with different ground states, shifts that are associated with the photoelectron excitation process can also occur. Upon photoejection of an electron, a negative charge flows toward the hole in order to screen the suddenly appearing local positive charge. The screening thus diminishes the local charge of the (frozen orbital) hole state and reduces the measured binding energy as well. This "relaxation energy" contains both intra-atomic and extra-atomic contributions. The former contribution involves rearrangement of only those electrons within an atom; the latter contribution refers to charge readjustment involving neighboring atoms. Chemical-shift measurements involve the comparison of two measured binding energies which contain relaxation effects that may differ. The relaxation shifts are difficult to estimate precisely (Shirley, 1978). Furthermore, observed chemical shifts are often disappointingly small and, in rare cases (Kim et al., 1973), have the wrong sign. These interpretive difficulties, coupled with the experimental problem of sample charging in insulating samples (Johansson et al., 1973; Urch and Webber, 1974), present drawbacks for core-level XPS studies in chemistry applications.

The determination of charge transfer in alloys appears, at first, to be a problem well suited to a core-level chemical-shift analysis. In practice, however, interpretation of such chemical shifts can be quite complex. Watson and Perlman (1975) have considered the case of chemical shifts in Au–Sn alloys. Gold is the more electronegative element so that, in the compounds, one would expect Au to act as an electron acceptor. (Mössbauer-effect studies do indicate electron transfer to the Au site.) Correspondingly, the Au core levels in the alloys should appear at a lower binding energy than in the elemental metal. However, the Au core levels are observed at a higher (by as much as 1.1 eV) binding energy. It is argued that this shift, occurring with added Sn, is largely a consequence of decreased d-electron character at the Au site.

Another aspect of XPS core-level spectroscopy is the variety of many-electron effects that can be observed in the photoejection process. Observation of these effects is of interest for elucidating the physics of the excitation processes. In addition, the many-electron processes can serve as probes of the electronic structure. (The multiplet structure discussed in Section III is a many-electron process in the unfilled 4f shell.) Commonly, many-electron effects produce satellite peaks that appear on the high-binding-energy side of the main peak in an XPS spectrum. Sometimes the satellites are interpreted as one-electron–like excitations in the n-1 system. That is, a second electron (in the n-1 system) may be excited "concurrently" with the primary emission in response to sudden changes in the local atomic charge that accompany the primary electron ejection. When the final state of the second excited electron is bound, a "shake-up" process is said to be involved; when it is unbound, "shake-off" has occurred. These secondary low-energy electrons are excited to energetically well-defined unoccupied levels in the vicinity of E_F, and hence can probe outer electron states. Charge transfer in insulators or excitation across a band gap in semiconductors (Vernon *et al.*, 1976; Rosencwaig *et al.*, 1971) are examples. Other many-electron phenomena associated with core-level photoemission include plasmons (Ley *et al.*, 1975), Fermi surface excitations in metals (Wertheim and Citrin, 1978), multiplet splittings (Shirley, 1978), and correlation satellites (Kowalczyk *et al.*, 1975). Sometimes many-electron effects produce "satellite" structure so complicated that one cannot identify a main XPS line.

Clearly, core-level PES including studies of the main line and substructure, provides many opportunities to obtain new electronic structure information. Sometimes, however, the proper interpretation of the spectra requires a rather complex analysis.

V. Some Illustrative Studies

A. Elemental Metals

We have indicated earlier that XPS often provides an accurate direct measure of the valence-band density of states $n(E)$. However, it was cautioned that photoemission cross sections can vary with the initial wavefunction character of the emitted electrons. In this section, we shall present XPS valence-band spectra for elemental metals and show some comparisons with calculated $n(E)$ curves. Recall that these comparisons are appropriate only for materials with bandlike (itinerant) electrons and thus must exclude the 4f electrons of the lanthanide elements and perhaps the 5f's in the actinide series (Freeman and Darby, 1974).

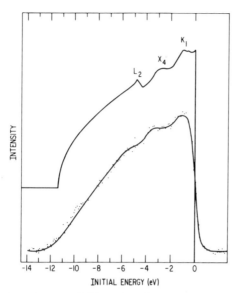

Fig. 36. Comparison of corrected XPS valence-band spectrum (lower curve) with calculated density of states. (After Baer and Busch, 1973.)

Results for aluminum, a nearly free-electron metal, are shown in Fig. 36 (Baer and Busch, 1973). The upper curve, showing very weak structural features, is a calculated $n(E)$ (Rooke, 1968). The lower curve is the XPS spectrum (corrected for inelastic electrons). The subtle predicted features are observed experimentally and the predicted bandwidth is nicely confirmed.

The more localized d electrons in the transition elements produce a higher and generally more structured valence band $n(E)$ than is observed in the simpler metals. In Figs. 22 and 23, we saw XPS spectra for the series of 3d transition elements scandium through copper. As noted earlier, the d bands are pinned to E_F until the band is filled. With the addition of more electrons, the filled d band retreats rapidly from E_F as seen in the Cu spectrum. In Cu, the small $n(E_F)$ is determined by the partially filled Cu 4s band. The valence bands of the 3d elements are relatively narrow and show generally similar spectral features grossly consistent with a "rigid-band" concept. That is, the (paramagnetic) bands of the 3d elements are similar to one another except for increased band filling for the higher-Z elements.

Figure 37 shows valence-band spectra for eight elements from the 3d, 4d, and 5d series compared with recent theoretical band structure calculations (Smith *et al.*, 1974; Hüfner *et al.*, 1973). Again, the data contain a correction for the inelastic electron background. The graphic nature of these measurements again illustrates expected band structure trends. The radial extent of

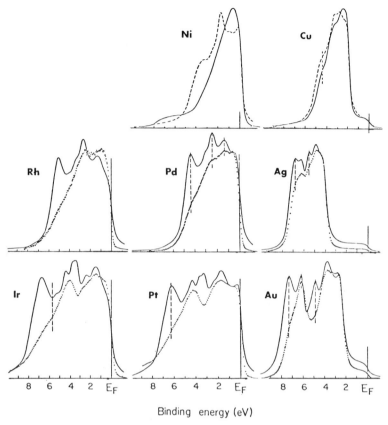

Fig. 37. XPS data for the 3d, 4d, and 5d transition metals Ni and Cu; Rh, Pd, and Ag; and Ir, Pt, and Au. For Ni and Cu (after Hüfner *et al.*, 1973) the solid lines are experimental, the dashed lines theoretical. For all other elements (after Smith *et al.*, 1974), the dotted spectra are experimental results, the solid lines theoretical.

the d-electron wave functions increases as one moves down a column of the periodic table (Herman and Skillman, 1963). The greater radial extent produces an increased d-electron bandwidth, a result clearly seen in the columns of Fig. 37. Furthermore, the spin–orbit interaction, which increases with atomic number Z, contributes to the band splitting observed in the 5d series.

The theoretical $n(E)$ curves in Fig. 37 show generally good agreement with experiment. Most predicted structural features are accounted for in the measured spectra. Furthermore, bandwidths are essentially correct. However, one notable and persistent discrepancy between theory and experiment is the tendency, at the bottom of the d bands, for the experimental results to

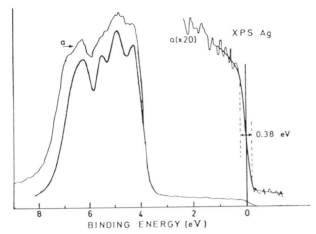

BINDING ENERGY (eV)

Fig. 38. Curve a is the experimental XPS spectrum of Ag. The lower curve is a theoretical spectrum which includes an energy-dependent broadening function. (After Barrie and Christiansen, 1976.)

fall beneath the calculated $n(E)$, apparently as the consequence of an energy-dependent broadening mechanism. At E_F, the resolution is limited by the instrument capability, but features are less well resolved at higher energies. Barrie and Christiansen (1976) have considered the Ag d bands, convoluting their theoretical $n(E)$ with an E-dependent Lorentzian, plus the instrument resolution function. [This effort was stimulated by the observation of line-shape distortions of core spectra resulting from coupled collective excitations at the Fermi surface (see Wertheim and Citrin, 1978).] Figure 38 shows their corrected Ag results compared with experiment; agreement is now remarkably good. Application of this broadening function to the calculated $n(E)$ for the other elements of Fig. 37 would also result in significantly better agreement with experiment.

Of the transition metals shown in Fig. 37, those located toward the right of the periodic table are chemically less active than those toward the left of the table. Because the less active elements are easier to deal with experimentally, a greater effort has been directed toward them. However, for the transition elements to the left of the table that have been carefully studied, XPS data also correspond well with calculated state densities. Figure 39 shows the comparison for elemental niobium (Höchst *et al.*, 1976). In the upper portion of this figure, experimental data are compared with the calculated $n(E)$. [The theoretical $n(E)$ was given an energy-dependent *intensity* correction to compensate for the effect, treated above as an energy-dependent *broadening*.] In the lower half of Fig. 39, an effort is made to effectively reduce the instrument-induced broadening by using data-

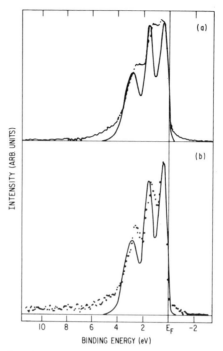

Fig. 39. (a) XPS valence-band spectrum of Nb corrected for inelastic background; the solid curve gives a theoretical density of states (DOS) smoothed with the 0.55-eV-wide resolution function and lifetime broadened with a 0.4-eV-wide Lorentzian. The dots refer to experiment. (b) XPS valence-band spectrum of Nb, deconvoluted with a 0.55-eV-wide resolution function. The solid curve gives the theoretical DOS smoothed with a 0.3-eV-wide resolution function and a 0.4-eV-wide Lorentzian. The dots refer to experiment. (After Höchst *et al.*, 1976.)

deconvolution techniques. Very similar theoretical and experimental curves, both showing pronounced three-peak structures, are observed.

The examples presented in this section and in Section III, A illustrate the great success of band theory for calculating electronic structures over an extended energy range in a wide range of itinerant systems and demonstrates the value of XPS for providing simple, direct tests of the band calculations. The large number of tests that have been successfully undertaken reinforces one's confidence both in the computational procedures and in the interpretation of the XPS data.

B. Alloys

Early investigations of electronic properties in alloys made extensive use of the rigid band theory (Jones, 1937). According to the theory, varying the relative concentrations of the two components in a binary continuous solid

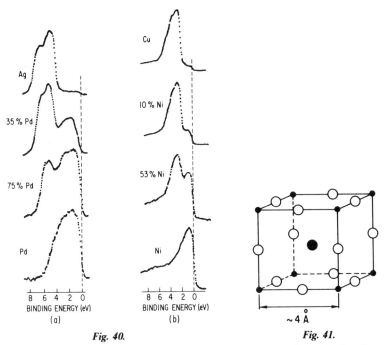

Fig. 40.   ***Fig. 41.***

Fig. 40. XPS valence-band data for (a) Ag, Pd, and Ag–Pd alloys; (b) Cu, Ni, and Cu–Ni alloys. (After Hüfner *et al.*, 1973.)

Fig. 41. Unit cell for the perovskite structure (general formula ABO₃). •, B; ●, A; ○, O²⁻

solution would merely result in a Fermi level shift. The same $n(E)$ curve was assumed to be appropriate for either element if the Fermi level was appropriately shifted to accommodate all electrons. Numerous tests of the theory were devised (Hume-Rothery and Raynor, 1962; Friedel, 1954), most extensively involving noble metals and their adjacent neighbors in the periodic table. The model accounted adequately for many experimental observations, but increasingly came to be regarded as an extreme oversimplification (Seib and Spicer, 1970). For example, the "coherent potential" approximation assumes that, in first approximation, separate d bands are associated with the individual atomic species. In second approximation, the d bands might to some extent be hybridized with one another (Kirkpatrick *et al.*, 1970). Prior to the development of the photoelectron spectroscopies, the definitive experiment needed to test the conflicting theories remained elusive. With the publication of photoemission results for Ag–Pd and Cu–Ni alloys, however, failure of the rigid-band concept was dramatically demonstrated (Hüfner *et al.*, 1973; Seib and Spicer, 1970).

Figure 40a shows XPS measurements for Ag, Pd, and two alloys (Hüfner *et al.*, 1973). One does not observe (as predicted by the rigid-band model) a single d band that shifts from E_F to several eV below E_F as one progresses through the series from Pd to Ag. Rather, the preservation, in the alloy, of the d-band character appropriate to the individual elements is clearly seen. Figure 40b shows XPS results for the Cu–Ni alloy series; again, the d-band identities of the Cu and Ni constituents are preserved.

These results emphasize the local character of the d-band density of states, a concept briefly discussed in Section III,A. A dilution effect is also seen here. As one constituent of the alloy series becomes more dilute and the distance between its atoms increases, its d band narrows. This dilution effect is readily seen in the d-band spectrum of Ag (Fig. 40).

C. Oxides

PES studies of oxides provide a valuable extension of the electronic properties studies of elemental metals or alloys. Oxidation represents a different kind of bonding than that encountered in simple metal systems. Electronegativities of the metal and ligand are very different and, with bonding, appreciable charge transfer from the metal to the ligand site (ionicity) is expected. If multiple oxidation states of the metal can be obtained, systematics of the variable metal valence can be monitored. The role of d or f electrons in the formation of molecular orbitals can be studied. Any localized, nonbonding orbitals that are occupied can likewise be observed and, for itinerant systems, band structure effects monitored.

XPS data for trioxides of the adjacent 5d elements Ta, W, and Re (groups V, VI, and VII, respectively) provide a band structure study that emphasizes the rigid-band character of these oxides. For ReO_3, the electron concentration can be increased by adding Na to the binary oxide without introducing important crystallographic structure changes. The structure can be described with reference to the unit cell for the perovskite lattice, shown in Fig. 41. This structure is appropriate to the ternaries $NaTaO_3$ or $NaWO_3$. If, however, Na is removed, then the transition metal and oxygen still maintain their relative positions in the lattice. The only change is that Na, quite distant from both the transition-metal and oxygen elements, is removed. In this lattice, Na should primarily function as a supplier of electrons without otherwise seriously altering the band structure.

As seen in Fig. 42, the band structure of WO_3 and the isoelectronic $NaTaO_3$ are, indeed, very similar. Now if the electron concentration is increased, the Fermi level should move into the conduction band (left unfilled in the compounds of Fig. 42) without altering the occupied bands. The effect of adding Na to WO_3 is seen in the upper curve of Fig. 43. Except for the new peak that has appeared near E_F, the spectrum is essentially identical to the WO_3 spectrum in Fig. 42. We also see that this

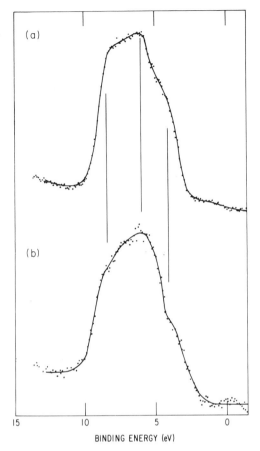

Fig. 42. XPS valence-band data for (a) $NaTaO_3$ and (b) WO_3.

spectrum is much like that of the (approximately) isoelectronic compound ReO_3 (Campagna *et al.*, 1975).

Band-theory results (e.g., Mattheiss, 1972) indicate that the occupied electronic states are composed of hybridized O 2p and transition-metal 5d electrons. The calculations also show that the peak appearing near E_F (Fig. 43) is a partially filled, predominantly d band that is separated by a small gap from a filled valence band (Fig. 42). This prediction is consistent with the XPS data and with the observation that ReO_3 and Na_xWO_3 (sodium–tungsten–bronzes) are excellent electrical conductors whereas WO_3 and $NaTaO_3$ are insulating. The XPS data thus indicate that for ABO_3 or BO_3-type oxides of the 5d elements in groups V, VI, and VII, the electronic structure can be nicely described within a simple rigid-band framework. Furthermore, a large Fermi level shift, appropriate to one electron per

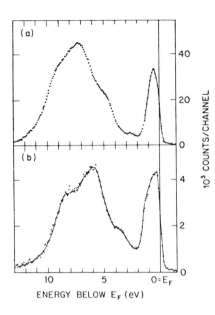

Fig. 43. XPS valence-band data for (a) Na_xWO_3 ($x = 0.805$) and (b) ReO_3. (After Campagna *et al.*, 1975.)

formula unit, can be observed. Without introducing important band structure changes, this shift changes the compound from an insulator to a highly conducting metal.

Binary oxides of uranium provide the opportunity to observe localized 5f electrons and to monitor their response to a valence change at the uranium ion. A number of different binary uranium oxides are known (Katz and Rabinowitz, 1951). This is a consequence of the fact that uranium ions can appear in different oxidation states. By an appropriate choice of oxides, the number of uranium valence electrons that participate in metal-ligand bonding can be varied. Figure 44 shows UO_2 spectra for the first 45 eV below E_F, compared with XPS data for ThO_2 (Veal and Lam, 1974a). Of special interest in Fig. 44 is the spectral comparison of UO_2 and ThO_2 near the Fermi level E_F. The broad peak labeled O 2p consists of overlapping oxygen and metal orbitals which make up the covalent (or ionic) bond of these oxides. All four electrons of thorium (that are outside the radon core) participate in the metal–oxide chemical bond. For UO_2, on the other hand, only four of the six available electrons participate in the bond. The remaining two uranium electrons are localized 5f electrons that do not appreciably hybridize with O 2p's. These localized levels are seen in the sharp, very intense peak near E_F in the valence-electron spectrum of UO_2. Except for the existence of the 5f-electron spectrum in UO_2, the ThO_2 and UO_2 valence-band spectra are very similar. Thus, the most profound differences (notably the conductivities, color, and magnetism) are directly related to the 5f electrons present in UO_2 and missing in ThO_2.

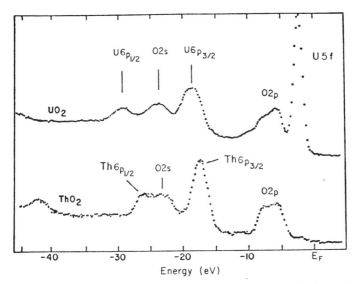

Fig. 44. XPS data for UO_2 and ThO_2. The narrow peak in UO_2 near E_F is associated with U 5f electrons. (After Veal and Lam, 1974a.)

As seen in Fig. 45, the 5f-electron peak intensity exhibits a systematic dependence on the degree of oxidation of the uranium atom (Veal and Lam, to be published). As the uranium oxidation state increases, the 5f occupancy decreases. Thus, in going from UO_2 to UO_3, the 5f peak systematically shrinks (if all spectra are normalized to a uranium core level). As the oxidation state of uranium in the binary oxides is increased, electrons are transferred from the localized 5f states into the "bonding" molecular orbitals which are predominately O 2p in character. For UO_3, containing hexavalent uranium, the 5f peak has entirely disappeared. Of course, the uranium core-level binding energies also increase with increasing uranium valence (see Fig. 35 and Section IV).

The systematics of the uranium oxides appear in contrast to the results for the 5d transition-metal oxides. For the U oxides, addition of (electronegative) oxygen depletes excess localized electrons while, for the 5d oxides, addition of (electropositive) Na adds electrons to an itinerant d band.

D. Hydrides

The electronic structure of transition-metal hydrides has long been a subject of considerable interest and controversy (for references, see Veal *et al.*, 1979). Historically, rather simple (and contradictory) models were advocated to account for the observed electronic properties of the metallic hydrides. On the one hand, it was proposed that with the addition of

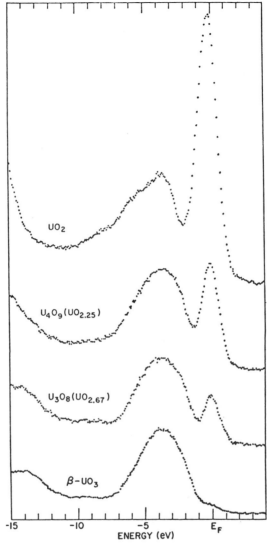

Fig. 45. XPS valence-band data for several uranium oxides. The U 5f peak near E_F is attenuated with increasing uranium oxidation. (From Veal and Lam, to be published.)

hydrogen to a transition metal, electrons migrate from the hydrogen to the valence band of the metal ("protonic" model). The dominant effect in this model is to produce a shift in the position of the Fermi level and attendant effects associated with filling of the conduction-band states. An alternative anionic model assumed that hydrogen tends to act as an electron acceptor with the net result that electrons should shift from the vicinity of metal-atom sites toward the hydrogen sites. A third model assumed that the metal and hydrogen atoms are covalently bonded. Recently, band calculations have been reported for several transition-metal hydrides (e.g., Switendick, 1972). These one-electron–type calculations provide results that contradict the rigid-band ideas often invoked, where band filling (or emptying) was expected to occur with varying hydrogen concentration while otherwise leaving the band structure essentially unchanged. The recent band calculations show the appearance of a new intense peak several electron volts below the Fermi level E_F in the calculated density of states $n(E)$ for the hydride. This dominant feature, first reported by Switendick (1970), results from predominantly s-type states that are associated with hydrogen–metal and hydrogen–hydrogen bonding.

Recent PES studies (Westlake *et al.*, 1978; Veal *et al.*, 1979) provide new information about charge transfer and lend supporting evidence to the validity of the band-theory predictions. An example of this work is an XPS study of the Zr–H system (Veal *et al.*, 1979). The valence-band results for Zr metal and the hydride, compared with appropriate $n(E)$ calculations, are shown in Figs. 46 and 47. The hydride spectra are very different from the valence-band spectra of zirconium metal, a result that clearly rules out use of a rigid-band–type model where E_F is assumed to shift with hydrogen concentration to compensate for an altered electron concentration. Figure 47 shows the $ZrH_{1.65}$ valence-band data compared with the interpolated $n(E)$ vs E curve computed for TiH_2 (Switendick, 1972). This comparison is

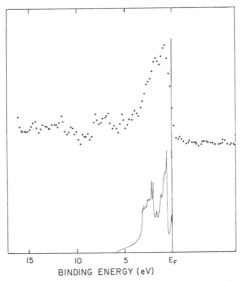

BINDING ENERGY (eV)

Fig. 46. Zr metal valence-band XPS spectrum (dots) of Veal *et al.* (1979) compared with the density of states calculation of Jepsen *et al.* (1975).

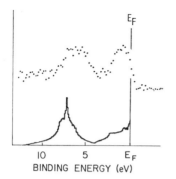

Fig. 47. ZrH$_{1.65}$ XPS valence-band spectrum (dots) of Veal *et al.* (1979) compared with density of states for TiH$_2$ calculated by Switendick (1972).

made because no such calculations exist for zirconium hydride and because one would expect the titanium and zirconium hydride systems (like the pure metals) to be very similar. The rather featureless theoretical density of states curve is remarkably well confirmed by experiment, especially with respect to the location of the s-type bonding peak ~ 7 eV below E_F and the high intensity of this peak relative to the peak near the Fermi energy.

Binding energies of Zr core levels were also monitored. The levels in the hydride were shifted, by ~ 1 eV, to higher binding energies. If relaxation energy differences are small, then the hydride core-level shift to higher binding energy (relative to fixed E_F) indicates that charge is transferred away from the vicinity of the zirconium atom with the addition of hydrogen. The system must, of course, be electrically neutral, so that significant charge redistribution must be occurring in the hydride stimulated by the presence of the added hydrogen. Presumably, enough of the available charge is "tied up" in a hydrogen–metal bond so that a net reduction in electron density at the zirconium site is realized.

Using a simplified cluster model, Jena (see Chapter 6, this volume) theoretically addressed this question of charge transfer. Results of the calculations are shown in Fig. 48. The dashed curve is the electron charge contained in a sphere of radius R about the free hydrogen atom. This integrated charge will approach the value 1 as R becomes infinite. The solid curve shows the integrated charge for a cluster of Zr$_4$H about the tetrahedral (hydrogen) site when the ambient zirconium metal background has been subtracted out. For $R < 3$ Å, this integrated charge is everywhere greater than the charge associated with the free hydrogen atom. This result is also evident in plots of (spherically averaged) electron density for the two cases discussed above (shown in the inset of Fig. 48).

The integrated charge $Z(R)$ near the proton site exceeds the value of 1, the maximum for free hydrogen, and exhibits oscillatory behavior indicating

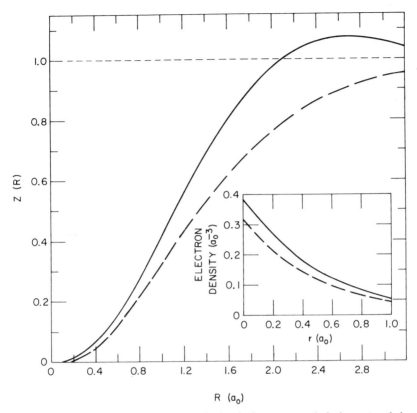

Fig. 48. Integrated charge density about the free hydrogen atom (dashed curve) and about the hydrogen site in Zr–H (solid curve), the latter calculated for a Zr_4H cluster with the contribution from the ambient Zr metal subtracted out. The larger magnitude for the hydride suggests that charge is transferred from metal to hydrogen sites when the hydride is formed. The inset shows the corresponding charge densities about the hydrogen site. (After Veal *et al.*, 1979.)

that regions of electron depletion (relative to zirconium metal) must occur near the zirconium site. Relative depletion occurs for $R \gtrsim 2.6$ Å. The substantial increase in electron density about the proton site (relative to free hydrogen) means that charge is transferred from the vicinity of the zirconium atom, leaving a net positive charge near the zirconium site. This positively charged environment will be sensed by the highly localized atomic core electrons of the zirconium atom and will be reflected as a shift in the XPS core-level energies relative to the Fermi level of the material. Thus, for zirconium hydride, hydrogen appears to act as an electron acceptor rather than an electron donor.

E. Silicate Glasses

XPS appears to be a valuable probe for studying electronic properties of bulk silicate glasses as well as the leaching properties of glass surfaces. An illustration of the utility of the technique for bulk glasses is available from a study of the O 1s line in a series of sodium silicate glasses (Veal and Lam, 1978). In the amorphous SiO_2 network, the silicon atom resides at the center of a tetrahedron of oxygen atoms. Silicons in adjacent tetrahedra are "bridged" by oxygens:

$$-\overset{|}{\underset{|}{Si}}-O-\overset{|}{\underset{|}{Si}}-\,.$$

When Na_2O is added to SiO_2, it is expected that bridging oxygen (O_I) linkages will be destroyed and new "nonbridging" oxygen (O_{II}) sites of the form

$$-\overset{|}{\underset{|}{Si}}-O\overset{Na}{\diagup}\qquad\overset{Na}{\diagdown}O-\overset{|}{\underset{|}{Si}}-$$

will appear (Warren and Briscoe, 1938). In this case, the addition of a *molecule* of Na_2O should produce two nonbridging oxygen sites. Thus, as Na_2O is added to the SiO_2 matrix, the number of nonbridging oxygen *atoms* should appear in direct proportion to the number of sodium atoms added. Since the bonding of the O_I and O_{II} atoms is markedly different, the local charge distributions and hence the XPS binding energies of the O 1s core levels of the bridging and nonbridging oxygen atoms should be different.

Figure 49 shows the O 1s spectra for amorphous SiO_2 and for several $(Na_2O)_x(SiO_2)_{1-x}$ glasses. Nearby sodium Auger lines are also shown. With the addition of Na_2O to SiO_2, a new peak appears on the low-binding-energy side of the O 1s peak. This peak scales with sodium concentration, as would be expected if the Na_2O were entering the lattice of SiO_2 in the manner suggested by Warren and Briscoe (1938).

The O_I and O_{II} intensity ratios (for spectra normalized to a constant Si line) are displayed in Fig. 50 for a series of $(Na_2O)_x(SiO_2)_{1-x}$ samples. At $x = 0$, the O/Si ratio is 2.0 and all oxygens occupy the O_I site. For increasing x, the O_{II} concentration should increase at twice the rate of decrease of O_I concentration until, at $x = 0.67$, the process is terminated with all oxygen atoms in the O_{II} bonding configuration. The model results are shown as the solid lines in Fig. 50. For small values of x, the model predictions are nicely confirmed by the XPS data. For large values of x, however, the experimental results deviate significantly from the predicted behavior. The high-x samples are more difficult to prepare and are very

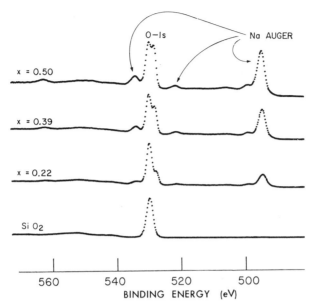

Fig. 49. The O 1s XPS spectra for SiO$_2$ and several (Na$_2$O)$_x$(SiO$_2$)$_{1-x}$ glasses. The low-binding-energy component of the O 1s peak, which scales with sodium concentration, is a measure of the nonbridging oxygen. (After Veal and Lam, 1978.)

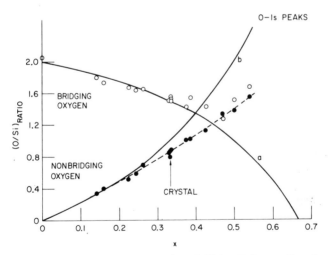

Fig. 50. The intensities of the bridging and nonbridging O 1s core lines for a series of (Na$_2$O)$_x$(SiO$_2$)$_{1-x}$ glasses, plotted as a function of Na$_2$O concentration. (After Veal and Lam, 1978.)

hygroscopic, so that experimental errors could be encountered in those data. If, however, the deviations at high x from the model results are confirmed, then it must be that Na_2O in high concentrations is incorporated into the glass in some unknown manner.

The simple model (Warren and Briscoe, 1938) is clearly inappropriate for the region $x > 0.67$, since all possible O_{II} sites have been filled. Any additional Na_2O in the high-x region must then be dissolved in the glass in some other way (perhaps incorporating H_2O). The dissolution mechanism appropriate for the high-x region may also be operable at $x < 0.67$, causing a reduction in the expected number of O_{II} sites. These XPS data clearly demonstrate the effect of bridging-bond disruption when Na_2O is added to silica glass, and enable the effect to be quantified. It may also be that deviations from the expected simple model will provide new and unexpected insights into the mechanisms of glass modification by alkali oxides.

The utility of the XPS technique is further demonstrated in an extension of the sodium silicate glass study, in which a series of sodium disilicate glasses containing varying amounts of Fe_2O_3 were systematically examined (Lam *et al.*, 1979). In this $(Na_2O \cdot 2SiO_2)_{1-x}(Fe_2O_3)_x$ series, it was observed that when Fe_2O_3 is dissolved in sodium disilicate glass, matrix oxygen bonds are disrupted to permit the Fe^{3+} ion to bond. In Fig. 51, curve a shows the measured O 1s line for an iron glass and curve b shows the O 1s levels for the sodium disilicate glass. (In these spectra, Na Auger structure has been subtracted out.) Subtracting the sodium disilicate contribution from the iron glass spectrum, we obtain the difference or residual spectrum shown in Fig. 51c. To visualize the significance of the residual result, it is useful to consider a simple mixture of Fe_2O_3 and $Na_2O \cdot 2SiO_2$. If Fig. 51a were the O 1s spectrum of the mixture, then the residual, Fig. 51c, should be a simple, nearly Gaussian line corresponding to the O 1s line of Fe_2O_3. However, in Fe_2O_3, only 1.5 oxygens are associated, on the average, with each iron atom. If Fe chooses to bond to more than 1.5 oxygens, then the associated oxygen must be obtained from the $Na_2O \cdot 2SiO_2$ matrix to which Fe_2O_3 was added. In this case, if one is to subtract out the $Na_2O \cdot 2SiO_2$ XPS contribution from the total $(Na_2O \cdot 2SiO_2)_{0.7}(Fe_2O_3)_{0.3}$ XPS spectrum to recover the O 1s associated with the Fe atoms, then the $Na_2O \cdot 2SiO_2$ part must be scaled to account for the loss of the matrix oxygens, i.e., those oxygens which now bond to the Fe's. Without the scaling, too much $Na_2O \cdot 2SiO_2$ will be subtracted out and one will obtain an "overshoot" spectrum such as is seen in Fig. 51c. These results clearly indicate that with the addition of Fe_2O_3 to $Na_2O \cdot 2SiO_2$, some O 1s bonds of the starting matrix are destroyed and new sites associated with Fe–O bonds are produced. To obtain a simple, nearly Gaussian line shape (Fig. 51d) by subtraction of curve b from curve a, curve b must be scaled by a factor whose magnitude implies that for each added molecule of Fe_2O_3, three oxygen bonds of the disilicate matrix are broken and replaced by bonds to the iron atoms.

This simple interpretation of the XPS results implies that Fe in sodium disilicate glass is coordinated with three near-neighbor oxygen atoms. This result has not been verified with independent experimental evidence (for references to other models, see Frischat and Tomandl, 1969; Lam *et al.*, 1979), and could represent an oversimplified interpretation of the XPS data. In any case, the XPS results provide a new line of pursuit for attacking the difficult problem of elucidating structure and bonding properties in amorphous materials.

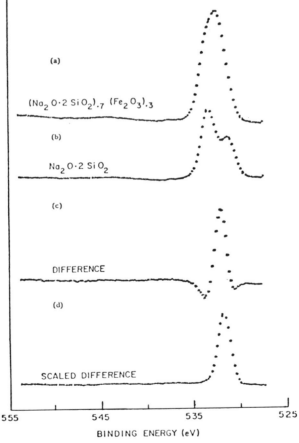

Fig. 51. XPS O 1s lines for (a) sodium disilicate containing Fe_2O_3 and (b) for sodium disilicate. The simple difference spectrum (c) indicates that O 1s bonds of the disilicate matrix are broken by the added Fe_2O_3. The scaled difference (d) indicates that three bonds of the matrix are broken for each added molecule of Fe_2O_3.

VI. Concluding Remarks

A brief description of the current state of the art of photoelectron spectroscopy has been presented here. The various specializations within PES provide complementary electronic-properties information that can be profitably exploited by the materials scientist. The PES field is young but has grown rapidly and has found application in a large number of specialized fields. The techniques have been used in problems related to physical, inorganic, and organic chemistry; biochemistry; solid-state physics; surface chemistry and physics; industrial chemistry; and environmental science. Yet some branches of PES are surely "infant" spectroscopies with significantly improved understanding and instrumentation, and widespread exploitation in materials science, still to be realized. For example, serious problems still remain in the interpretation of angular resolved UPS data. Also, virtually no use has yet been made of angular-dependent *intensity* information. No doubt, significant gains will be forthcoming in instrumentation for spin-polarized photoemission so that the spin dependence of energy distribution curves can be obtained. At the same time, understanding of phenomena observed with the older, angle-integrated spectroscopies will be improved. Photoemission cross sections and various collective electron phenomena are in need of improved understanding and quantification. For many purposes, higher instrument resolution and photon fluxes would be desirable. Important advances relevant to these problems are currently being made by utilizing synchrotron radiation. The promise of this intense radiation source for studies of the solid state has been recognized with the result that new and improved synchrotron radiation sources are now planned or under construction. New advances in instrumentation, the level of understanding, and the range of new applications should, in coming years, make the field of PES an exciting one for the practicing spectroscopist and a fruitful one for the materials scientist.

Acknowledgments

The author wishes to acknowledge helpful suggestions contributed by D. J. Lam, A. P. Paulikas, and A. J. Arko.

References

Alvarado, S. F., Erbudak, M., and Munz, P. (1976). *Phys. Rev. B* **14**, 2740.
Baer, Y., and Busch, G. (1973). *Phys. Rev. Lett.* **30**, 280.
Baer, Y., and Busch, G. (1974). *J. Electron Spectrosc. Relat. Phenom.* **5**, 611.

Baker, A. D., Brisk, M., and Liotta, D. (1978). *Anal. Chem.* **50**, 328R.

Barrie, A., and Christiansen, N. E. (1976). *Phys. Rev. B* **14**, 2442.

Brundle, C. R. (1974). *J. Vac. Sci. Technol.* **11**, 212.

Brundle, C. R., and Baker, A. D., eds. (1977–1979). "Electron Spectroscopy: Theory, Techniques, and Applications," Vols. 1–3, Academic Press, New York.

Buck, O., Tien, J., and Marcus, H., eds. (1979). "Electron and Positron Spectroscopies in Material Science and Engineering." Academic Press, New York.

Burdick, G. A. (1963). *Phys. Rev.* **129**, 138.

Busch, G., Campagna, M., and Siegmann, H. C. (1970). *J. Appl. Phys.* **41**, 1044.

Campagna, M., and Wertheim, G. K. (1976). *Struct. Bonding (Berlin)* **30**, 99.

Campagna, M., Bucher, E., Wertheim, G. K., Buchanan, D. N. E., and Longinotti, L. D. (1974). *Proc. Rare Earth Res. Conf. 11th, Traverse City, Mich.* p. 810.

Campagna, M., Wertheim, G. K., Shanks, H. R., Zumsteg, F., and Banks, E. (1975). *Phys. Rev. Lett.* **34**, 738.

Cardona, M., and Ley, L., eds. (1978). "Photoemission in Solids I," Topics in Applied Physics, Vol. 26. Springer-Verlag, Berlin and New York.

Carlson, T. A. (1975). "Photoelectron and Auger Spectroscopy." Plenum, New York.

Carlson, T. A., ed. (1978). "X-ray Photoelectron Spectroscopy," Benchmark Papers in Physical Chemistry and Chemical Physics, Vol. 2. Dowden, Hutchinson & Ross, Stroundsburg, Pennsylvania.

Caudano, R., and Verbist, J., eds. (1974). "Electron Spectroscopy: Progress in Research and Applications." Amer. Elsevier, New York.

Chazalviel, J. N., Campagna, M., Wertheim, G. K., and Schmidt, P. H. (1976). *Phys. Rev. B* **14**, 4586.

Christiansen, N. E. (1972). *Phys. Status Solidi B* **54**, 551.

Cox, P. A. (1975). *Struct. Bonding (Berlin)* **24**, 59.

Craseman, B., ed. (1975). "Atomic Inner-Shell Processes." Vol. 2 Academic Press, New York.

Dekeyser, W., Fiermans, L., Vanderkelen, G., and Vennik, J., eds. (1973). "Electron Emission Spectroscopy," Reidel Publ., Boston, Massachusetts.

Eastman, D. E. (1972). In "Techniques of Metals Research" (E. Passaglia, ed.), Vol. 6, p. 413. Wiley (Interscience), New York.

Eastman, D. E. (1974). In "Vacuum Ultraviolet Radiation Physics" (E. E. Koch, R. Haensel, and C. Kunz, eds.), p. 417. Pergamon, Oxford.

Eastman, D. E., and Grobman, W. D. (1972). *Phys. Rev. Lett.* **28**, 1327.

Eastman, D., and Kuznietz, M. (1971). *J. Appl. Phys.* **42**, 1396.

Eib, W., and Reihl, B. (1978). *Phys. Rev. Lett.* **40**, 1674.

Einstein, A. (1905). *Ann. Phys. (Leipzig)* **17**, 132.

Erbudak, M., Greuter, F., Meier, F., Reihl, B., and Vogt, O. (1979). *J. Appl. Phys.* **50**, 2099.

Evans, S. (1977). *J.C.S. Faraday II* **73**, 1341.

Fabian, D. J., and Watson, L. M., eds. (1973). "Band Structure Spectroscopy of Metals and Alloys." Academic Press, New York.

Fadley, C. S. (1978). In "Electron Spectroscopy: Theory, Techniques and Applications" (A. D. Baker and C. R. Brundle, eds.), Vol. 2, p. 2. Academic Press, New York.

Feibelman, P. J., and Eastman, D. E. (1974). *Phys. Rev. B* **10**, 4932.

Feuerbacher, B., and Willis, R. F. (1976). *J. Phys. C* **9**, 169.

Freeman, A. J., and Darby, J. B., eds. (1974). "The Actinides: Electronic Structure and Related Properties," Vols. 1 and 2. Academic Press, New York.

Freeouf, J., Erbudak, M., and Eastman, D. E. (1973). *Solid State Commun.* **13**, 771.

Friedel, J. (1954). *Adv. Phys.* **3**, 446.

Frischat, G. H., and Tomandl, G. (1969). *Glastech. Ber.* **42**, 182.

Fuggle, J. C., Watson, L. M., Fabian, D. J., and Norris, P. R. (1973). *Solid State Commun.* **13**, 507.

Gelius, U., Hedén, P. F., Hedman, J., Lindberg, B. J., Manne, R., Nordberg, R., Nordling, C., and Seigbahn, K. (1970). *Phys. Scr.* **2**, 70.

Grandke, T., Ley, L., And Cardona, M. (1978). *Phys. Rev. B* **18**, 3847.

Grant, R. W., and Housley, R. M. (1979). In "Electron and Positron Spectroscopies in Materials Science and Engineering" (O. Buck, J. Tien, and H. Marcus, eds.), p. 219. Academic Press, New York.

Greuter, F., Eib, W., Erbudak, M., Meier, F., and Reihl, B. (1979). *J. Appl. Phys.* **50**, 2099.

Grobman, W. D., Eastman, D. E., Freeouf, J. L., and Shaw, J. (1974)). *Proc. Int. Conf. Semicond., 12th, Stuttgart, 1974* p. 1275. Teubner, Stuttgart.

Grobman, W. D., Eastman, D. E., and Freeouf, J. L. (1975). *Phys. Rev. B* **12**, 4405.

Hercules, D. M. (1972). *Anal. Chem.* **44**, 106R.

Hercules, D. M. (1976). *Anal. Chem.* **48**, 294R.

Hercules, D. M., and Carver, J. C. (1974). *Anal. Chem.* **46**, 133R.

Herman, F., and Skillman, S. (1963). "Atomic Structure Calculations." Prentice-Hall, Englewood Cliffs, New Jersey.

Hertz, H. (1887). *Ann. Phys. Chem.* **31**, 983.

Höchst, H., Hüfner, S., and Goldman, A. (1976). *Solid State Commun.* **19**, 899.

Hüfner, S., and Wertheim, G. K. (1974). *Phys. Lett. A* **47**, 349.

Hüfner, S., Wertheim, G. K., and Wernick, J. H. (1973). *Phys. Rev. B* **8**, 4511.

Hume-Rothery, W., and Raynor, G. V. (1962). "The Structure of Metals and Alloys," 4th ed. Inst. Met., London.

Janak, J. R., Williams, A. R., and Moruzzi, V. L. (1975). *Phys. Rev. B* **11**, 1552.

Jena, P., Emmons, R., Lam, D. J., and Ray, D. K. (1978). *Phys. Rev. B* **18**, 3562.

Jepsen, O., Andersen, O. K., and Mackintosh, A. (1975). *Phys. Rev. B* **12**, 3084.

Jorgensen, C. K. (1975). *Struct. Bonding (Berlin)* **22**, 49.

Johansson, G., Hedman, J., Berndtsson, A., Klasson, M., and Nelsson, R. (1973). *J. Electron Spectrosc. Relat. Phenom.* **2**, 295.

Jones, H. (1937). *Proc. Phys. Soc. London* **47**, 250.

Kane, P. F., and Larrabee, G. B. (1974). "Characterization of Solid Surfaces." Plenum, New York.

Katz, J. J., and Rabinowitz, E. (1951). "The Chemistry of Uranium," National Nuclear Energy Series, Vol. VIII-5. McGraw-Hill, New York.

Kim, K. S., O'Leary, T. J., and Winograd, N. (1973). *Anal. Chem.* **45**, 2213.

Kirkpatrick, S., Velicky, B., and Ehrenrich, H. (1970). *Phys. Rev. B* **1**, 3250.

Knapp, J. A., Himpsel, F. J., and Eastman, D. E. (1979). *Phys. Rev. B* **19**, 4952.

Koch, E. E., Haensel, R., and Kunz, C., eds. (1974). *Vac. Ultraviolet Radiat. Phys. Proc. Int. Conf. Vac. Ultraviolet Radiat. Phys. 4th, Hamburg*.

Koopman, T. (1933). *Physica (Utrecht)* **1**, 104.

Kowalczyk, S. P., Ley, L., Martin, R. L., McFeely, F. R., and Shirley, D. A. (1975). *Faraday Discuss. Chem. Soc.* **60**, 7.

Lam, D. J., Veal, B. W., Chen, H., and Knapp, G. S. (1979). In "Science Underlying Radioactive Waste Management" (G. J. McCarthy, ed.), p. 97. Plenum, New York.

Lang, W. C., Padalia, B. D., Fabian, D. J., and Watson, L. M. (1974). *J. Electron Spectrosc. Relat. Phenom.* **5**, 207.

Larsen, P. K., Chiang, S., and Smith, N. V. (1977). *Phys. Rev. B* **15**, 3200.

Ley, L., and Cardona, M., eds. (1980). "Photoemission in Solids II," Topics in Applied Physics, Vol. 27. Springer-Verlag, Berlin and New York. To be published.

Ley, L., McFeely, F. R., Kowalczyk, S. P., Jenkins, J. G., and Shirley, D. A. (1975). *Phys. Rev. B* **11**, 600.

Ley, L., Dabbousi, O. B., Kowalczyk, S. P., McFeely, F. R., and Shirley, D. A. (1977). *Phys. Rev. B* **16**, 5372.

Mahan, G. D. (1970). *Phys. Rev. B* **2**, 4334.

Mattheiss, L. F. (1972). *Phys. Rev. B* **6**, 4718.

Pearson, W. B. (1978). "Handbook of Lattice Spacings and Structures of Metals." Pergamon, New York.

Platau, A., and Karlsson, S. E. (1978). *Phys. Rev. B* **18**, 3820.

Pollack, R. A., Kowalczyk, S., Ley, L., and Shirley, D. A. (1972). *Phys. Rev. Lett.* **29**, 274.

Rife, J. C., Dexter, R. N., Bridenbaugh, P. M., and Veal, B. W. (1977). *Phys. Rev. B* **16**, 4491.

Rooke, G. A. (1968). *J. Phys. C* **1**, 767.

Rosencwaig, A., Wertheim, G. K., and Guggenheim, H. J. (1971). *Phys. Rev. Lett.* **27**, 479.

Sato, S. (1976). *J. Phys. Soc. Jpn.* **41**, 1913.

Schwartz, W. E. (1973). *Anal. Chem.* **45**, 789A.

Segall, B. (1962). *Phys. Rev.* **125**, 109.

Seib, D. H., and Spicer, W. E. (1970). *Phys. Rev. B* **2**, 1676.

Shirley, D., ed. (1972). "Electron Spectroscopy." North Holland, Amsterdam.

Shirley, D. (1978). In "Photoemission in Solids I" (M. Cardona and L. Ley, eds.), Topics in Applied Physics, Vol. 26, p. 165. Springer-Verlag, Berlin and New York.

Shirley, D. A., Stöhr, J., Wehner, P. S., Williams, R. S., and Apai, G. (1977). *Phys. Scr.* **16**, 398.

Siegbahn, K., Nordling, C., Fahlman, A., Nordberg, R., Hamrin, K., Hedman, J., Johansson, G., Bergmark, T., Karlsson, S. -E., Lindgren, I., Lindgren, B. (1967). "ESCA: Atomic, Molecular, and Solid State Structure Studied by Means of Electron Spectroscopy," Nova Acta Regiae Soc. Sci. Upsaliensis, Ser. IV, Vol. 20. Almquist and Wiksells, Stockholm.

Smith, N. V. (1971). *Phys. Rev. B* **3**, 1862.

Smith, N. V. (1978). In "Photoemission in Solids I" (M. Cardona and L. Ley, eds.), Topics in Applied Physics, Vol. 26, p. 237. Springer-Verlag, Berlin and New York.

Smith, N. V., Wertheim, G. K., Hüfner, S., and Traum, M. T. (1974). *Phys. Rev. B* **10**, 3197.

Spicer, W. E. (1967). *Phys. Rev.* **154**, 385.

Spicer, W. E. (1972). In "Optical Properties of Solids" (F. Abeles, ed.), p. 755. North-Holland Publ., Amsterdam.

Spicer, W. E. (1976). In "Optical Properties of Solids—New Developments" (B. O. Seraphin, ed.), p. 631. North-Holland Publ., Amsterdam.

Spicer, W. E., Yu, K. Y., Lindau, I., Pianetta, P., and Collins, D. M. (1976). In "Surface and Defect Properties in Solids" (J. M. Thomas and M. W. Roberts, eds.), Vol. 5, p.103. Chem. Soc., London.

Stöhr, J., McFeely, F. R., Apai, G., Wehner, P. S., and Shirley, D. A. (1976a). *Phys. Rev. B* **14**, 4431.

Stöhr, J., Apai, G., Wehner, P. S., McFeely, F. R., Williams, R. S., and Shirley, D. A. (1976b). *Phys. Rev. B* **14**, 5144.

Stöhr, J., Wehner, P. S., Williams, R. S., Apai, G., and Shirley, D. A. (1978). *Phys. Rev. B* **17**, 587.

Switendick, A. C. (1970). *Solid State Commun.* **8**, 1463.

Switendick, A. C. (1972). *Ber. Bunsenges. Phys. Chem.* **76**, 535.

Urch, D. S., and Webber, M. (1974). *J. Electron Spectrosc. Relat. Phenom.* **5**, 791.

Varma, C. (1976). *Rev. Mod. Phys.* **48**, 219.

Veal, B. W., and Lam, D. J. (1974a). *Phys. Rev. B* **10**, 4902.

Veal, B. W., and Lam, D. J. (1974b). *Phys. Lett. A* **49**, 466.

Veal, B. W., and Lam, D. J. (1978). In "The Physics of SiO_2 and Its Interfaces" (S. Pantelides, ed.), p. 299. Pergamon, New York.

Veal, B. W., and Lam, D. J. (1980). In "Uran" (C. Keller, ed.), Gmelin-Handbuch der Anorganischen Chemie, Chap. 10.7. Springer-Verlag, Berlin and New York. To be published.

Veal, B. W., Lam, D. J., and Westlake, D. G. (1979). *Phys. Rev. B* **19**, 2856.

Verbist, J., Riga, J., Pireaux, J. J., and Caudano, R. (1974). *J. Electron Spectrosc. Relat. Phenom.* **5**, 193.

Vernon, G. A., Stucky, G., and Carlson, T. A. (1976). *Inorg. Chem.* **15**, 278.

Warren, B. E., and Briscoe, J. (1938). *J. Am. Ceram. Soc.* **21**, 259.

Watson, R. E., and Perlman, M. L. (1975). *Struct. Bonding (Berlin)* **24**, 83.

Wertheim, G. K. (1978). In "Electron Spectroscopy: Theory, Techniques and Applications" (A. D. Baker and C. R. Brundle, eds.), Vol. 2, p. 259. Academic Press, New York.

Wertheim, G. K. and Citrin, P. H. (1978). In "Photoemission in Solids I" (M. Cardona and L. Ley, eds.), Topics in Applied Physics, Vol. 26, p. 197. Springer-Verlag, Berlin and New York.

Westlake, D. G., Satterthwaite, B., and Weaver, J. H. (1978). *Phys. Today* **31**, 32.

3

Electronic Structure and the Electron — Phonon Interaction†

W. H. BUTLER

Metals and Ceramics Division
Oak Ridge National Laboratory
Oak Ridge, Tennessee

I. Introduction to the Formalism of the Electron–Phonon Interaction

A. *The Many Faces of the Electron–Phonon Interaction*

The interaction between the electrons and the phonons in a metal manifests itself in numerous ways. Its most obvious effect is the scattering of electrons off of the distortions of the lattice caused by the phonons. This scattering process degrades the energy and momentum which the electron may have acquired from an applied field or thermal gradient and thereby contributes to the electrical and thermal resistivity. Electron–phonon scattering is the dominant contributor to the resistivity for reasonably pure

†Research sponsored by the Material Sciences Division, U.S. Department of Energy under contract W-7405-eng-26 with the Union Carbide Corporation.

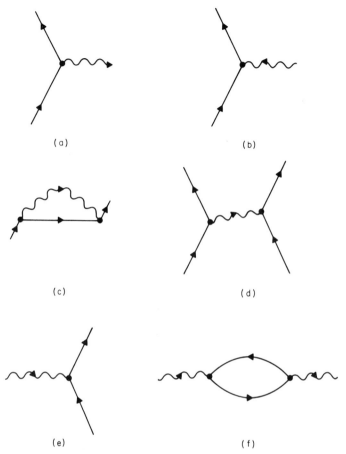

Fig. 1. Electron–phonon processes in a metal. Straight lines coming into a vertex represent electrons. Wavy lines represent phonons. The dot at a vertex where two straight lines and a wavy line join represents the electron–phonon interaction. A straight line with its arrow pointing away from a vertex can represent either an electron leaving the vertex or a hole entering the vertex. (a) Electron scattering with phonon emission, (b) electron scattering with phonon absorption, (c) phonon emission and reabsorption, (d) electron–electron interaction via phonon exchange, (e) phonon decay into an electron–hole pair, and (f) phonon emission and reabsorption of an electron–hole pair.

metals except at very low temperatures where impurity scattering and (possibly) electron–electron scattering are important. The electron–phonon scattering process can be viewed as the emission or absorption of a phonon by the electron and is depicted in Figs. 1a (emission) and 1b (absorption).

A less obvious effect of the electron–phonon interaction is that it slows down the electrons near the Fermi energy. This effect should not be confused with the resistivity discussed above because it does not cause energy or momentum dissipation. This reduction in the electronic velocities is sometimes referred to as the *electron–phonon mass enhancement* because for a metal with a free-electron dispersion relation the Fermi velocity is $v_F = \hbar k_F / m$ where $\hbar k_F$ is the Fermi momentum and m is the effective mass in the absence of electron–phonon interactions. The Fermi momentum is not affected by the electron–phonon interaction so that the velocity reduction $v_F \rightarrow v_F^* = v_F / (1 + \lambda)$ is equivalent to a mass enhancement $m \rightarrow m^* = m(1 + \lambda)$ where λ is the electron–phonon mass enhancement parameter. λ varies from approximately 0.1 for the alkali metals to 1.5 for Pb. The process leading to the mass enhancement is the emission and reabsorption of "virtual" phonons by the electrons (Fig. 1c). The word "virtual" implies that energy need not be conserved during the short time after emission and before reabsorption of the phonon. Physically, one may picture the electronic mass enhancement as arising from the fact that an electron polarizes the lattice (i.e., causes small displacements of nearby ions) and must move more slowly due to this encumbering polarization cloud.

Another surprising aspect of the electron–phonon interaction is that it leads to an attractive interaction between the electrons which for most metals is strong enough to overcome the Coulomb repulsion and at low enough temperatures causes a transition to a new phase with remarkable superconducting properties. The mechanism responsible for the interaction between the electrons is the exchange of phonons (Fig. 1d). A simplified physical picture of the process is shown in Fig. 2. Imagine a lattice of positively charged ions immersed in a sea of electrons. Concentrate on a single electron near the Fermi energy as it moves through the lattice at the Fermi velocity $v_F \approx 10^8$ cm/s. The electron will attract the nearby positively charged ions and they will respond by moving toward the electron, but due to their large inertia, their maximum response will occur at a time $\tau \approx 1/\omega_{ph} \approx 10^{-13}$ s after the electron has passed (ω_{ph} is a typical phonon frequency). Thus, the electron will leave a wake of positively charged lattice polarization which has its maximum strength a distance $l \approx v_F \tau$ (several hundred lattice spacings) behind the electron. A second electron may then be attracted by this region of positive polarization. This amounts to an effective attractive interaction between the electrons. The Coulomb repulsion between the electrons is much reduced by the slow response of the ions which allows the interacting electrons to be well separated in space.

The electron–phonon interaction also affects the phonons. The process depicted in Fig. 1b can also be considered from the point of view of the phonon (Fig. 1e) which finds itself susceptible to decay by turning into an "electron–hole pair." A hole coming out of a vertex is of course equivalent to an electron coming in and we can also describe this process for phonon decay as the absorption of the phonon by an electron in the Fermi sea, the electron being promoted by the energy of the phonon out of the Fermi sea.

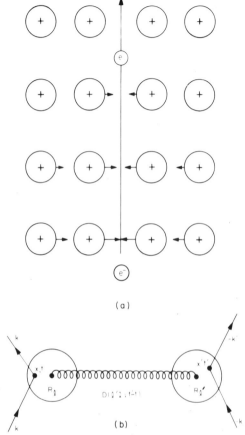

Fig. 2. (a) An electron leaves a wake of positive polarization which can attract a second electron. (b) Mathematical representation of 2a. An electron exerts a force on the atom at R_l, the electron momentum changes from k to k', and the force exerted on the atom causes displacements [via $D(l'l; t' - t)$] of the other atoms. Finally, a second electron is scattered by the displaced atom.

This process limits the phonon lifetime to a finite value τ_{ph} and causes it to acquire a corresponding energy width $\gamma = \hbar/\tau_{ph}$ which can be observed by inelastic neutron scattering experiments.

The phonon frequencies are also modified by the electron–phonon interaction. The "bare" phonon frequencies, i.e., those calculated by neglecting the electron–phonon interaction, are reduced by the emission and reabsorption of virtual electron–hole pairs (Fig. 1f). The physical process involved in the reduction of the phonon frequencies is the motion of the electrons to screen out part of the interionic forces.

It should be clear from consideration of the diagrams of Fig. 1 that resistivity, electronic mass enhancement, superconductivity, phonon lifetimes, and phonon frequency renormalization are all closely related phenomena. It should not be surprising that, for example, one can learn much about the tendency of metals to become superconductors from a knowledge of their high-temperature resistivity and vice versa. Figure 1 also illustrates a second important point. Namely that only three things are necessary for a quantitative first-principles treatment of all of the physical phenomena described above:

1. The electronic structure of the metal which is needed to construct the electron propagators (the straight lines of Fig. 1).

2. The phonon frequencies and polarizations which are used to construct the phonon propagators (wavy lines of Fig. 1).

3. The electron–phonon matrix elements (dots where two electron lines and one phonon line intersect).

These three ingredients are now available for most of the cubic elements and will likely be available for most crystalline metals in the near future. The past 15 years have witnessed remarkable progress and success for the one-electron band theory of solids as described in Chapter 1. Particularly noteworthy are recent accurate calculations entirely from first principles of the cohesive energy, bulk modulus, and lattice constants of the elements (Moruzzi *et al.*, 1978). The electronic structure (wave functions and energy bands) of ordered crystalline solids can now be calculated reliably and routinely. There has also been much progress in our ability to understand and to calculate the electronic structure of solid solution alloys (Gyorffy and Stocks, 1979). At present, phonon frequencies can be calculated accurately from first principles only for the free-electron–like metals (see, e.g., Dagens *et al.*, 1975) although much progress has been made for transition metals (Sinha and Harmon, 1976; Cooke, 1979; Varma and Weber, 1979). Fortunately, there is accurate experimental data mostly from inelastic neutron scattering (Larose and Vanderwall, 1974) that allows us to reconstruct the phonon frequencies for most elements and many compounds. The third ingredient, the electron–phonon matrix element, can in principle be calculated from first principles but, at present, we have only prescriptions and approximations. It appears, however, that some of these approximations are sufficiently accurate and reliable that detailed comparisons between theory and experiment are now in order.

The remainder of this section is devoted to a brief review of the formal expressions for the resistivity, the electronic mass enhancement, the superconducting transition temperature, the phonon lifetime, the phonon self-energy, and the lattice thermal conductivity. Section II discusses the electron–phonon matrix elements with particular emphasis on the variation of the strength of the electron–phonon interaction among the elements and in transition-metal compounds.

B. Electronic Transport and the Electron–Phonon Interaction

The Boltzmann equation which describes the transport of electrons in a metal with a uniform electric field $\mathbf{E}$ and a uniform thermal gradient ∇T is (see Chapter 5, this volume; see also Ziman, 1960)

$$\left(e\mathbf{E} \cdot \mathbf{v}_k + (\varepsilon_k/T)\nabla T \cdot \mathbf{v}_k\right)\partial f_k/\partial \varepsilon_k = \sum_{k'} Q_{kk'}\phi_{k'}, \tag{1}$$

where $\mathbf{v}_k$ is the electronic group velocity $\mathbf{v}_k = \hbar \nabla_k \varepsilon_k$, f_k is the equilibrium Fermi–Dirac distribution function, and ϕ_k is the deviation function which describes the departure of the electronic distribution function F_k from equilibrium:

$$F_k = f_k - (\partial f_k/\partial \varepsilon_k)\phi_k. \tag{2}$$

It will be assumed in the following that the phonons remain in equilibrium. This assumption is justified for not too low temperatures, especially for metals with complicated Fermi surfaces.

The scattering operator has both "scattering out" and "scattering in" terms which describe respectively the loss and gain of electrons in state k due to scattering between the other states

$$Q_{kk'} = Q_{kk'}^{\text{out}} - Q_{kk'}^{\text{in}}, \tag{3}$$

$$Q_{kk'}^{\text{out}} = \frac{\delta_{kk'}}{k_B T} \sum_{k''} P_{kk''}, \tag{4}$$

$$Q_{kk'}^{\text{in}} = \frac{1}{k_B T} P_{kk'}, \tag{5}$$

where $P_{kk'}$ is the equilibrium probability for an electron to scatter from state k to state k'. The electrical or thermal current density can be calculated from the deviation function ϕ_k:

$$\mathbf{J}_e = \frac{-2e}{V_a} \sum_k \mathbf{v}_k \phi_k (-\partial f/\partial \varepsilon_k), \tag{6}$$

$$\mathbf{J}_Q = \frac{2}{V_a} \sum_k \varepsilon_k \mathbf{v}_k \phi_k (-\partial f/\partial \varepsilon_k), \tag{7}$$

where V_a is the volume per atom. ϕ_k is in principle obtained by solving the Boltzmann equation

$$\phi_k = \sum_{k'} [Q^{-1}]_{kk'}\left[e\mathbf{E} \cdot \mathbf{v}_k + (\varepsilon_k/T)\nabla T \cdot \mathbf{v}_k\right]\partial f/\partial \varepsilon_k. \tag{8}$$

but the inversion of the scattering operator is a difficult task. Fortunately, there is a variational principle for the Boltzmann equation (Kohler, 1948) which allows one to perform a relatively good calculation of the conductivity by using a rather crude trial function for ϕ_k.

In particular, for the case of an applied electric field which we take to be in the x direction the simplest sensible trial function is

$$\phi_k = -eE_x v_x(k)\tau. \tag{9}$$

I shall call this approximation the *lowest-order variational approximation* (LOVA) for the resistivity. In the LOVA, it is assumed that ϕ_k is proportional to $v_x(k)$ for all values of k and it will be seen that τ plays the role of a relaxation time. The conductivity is obtained in terms of τ by substitution of Eq. (9) into Eq. (6):

$$\sigma = \frac{2e^2 N(0)\langle v_x^2\rangle \tau}{V_a} = e^2(n/m)_{\text{eff}}\tau, \tag{10}$$

where $N(0)$ is the (single-spin) Fermi energy density of states per atom and $\langle v_x^2\rangle$ is defined by

$$2N(0)\langle v_x^2\rangle = 2\sum_k v_x^2(k)(-\partial f/\partial \varepsilon_k), \tag{11}$$

where $2N(0)\langle v_x^2\rangle$ plays the role of $(n/m)_{\text{eff}}$ in the classical formula. At moderate temperature for most metals the electronic structure does not vary on a scale of $4k_B T$ which is the approximate width of $(-\partial f/\partial \varepsilon)$ so that in the sum over k of Eq. (11) $(-\partial f/\partial \varepsilon)$ behaves like a delta function $[-\partial f/\partial \varepsilon \approx \delta(\varepsilon)]$ and $\langle v_x^2\rangle$ is simply the average over the Fermi surface of the square of the x component of the Fermi velocity.

The relaxation time τ can be obtained by substituting Eq. (9) into Eq. (1) and operating on both sides with $\sum_k v_x(k)$:

$$N(0)\langle v_x^2\rangle = \tau \sum_{kk'} Q_{kk'} v_x(k) v_x(k'). \tag{12}$$

Using Eqs. (3)–(5) and making use of the symmetry of the transition probability $P_{kk'} = P_{k'k}$, we have

$$2k_B T N(0)\langle v_x^2\rangle /\tau = \sum_{kk'} P_{kk'}[v_x(k) - v_x(k')]^2. \tag{13}$$

Further progress requires explicit consideration of the fact that the transition probability arises from electron–phonon interactions. Its form can be deduced from the golden rule of perturbation theory. The probability of an electron initially in state k making a transition to state k' by emitting a phonon of momentum $q = k - k'$ and energy $\hbar\omega = \varepsilon_k - \varepsilon_{k'}$ as in Fig. 1a is

$$P_{kk'}^e = (2\pi/\hbar)\sum_j |g_{kk'}^j|^2 f_k(1 - f_{k'})\big[n\big(\omega_{k-k'}^j\big) + 1\big]\delta\big(\varepsilon_k - \varepsilon_{k'} - \hbar\omega_{k-k'}^j\big),$$

(14)

where $g_{kk'}^j$ is the electron–phonon matrix element for transitions between electronic states k and k' caused by a phonon of polarization index j, f_k is the Fermi–Dirac function $f(\varepsilon_k) = [\exp(\varepsilon_k/k_BT) + 1]^{-1}$, and $n(\omega)$ is the Bose–Einstein function $n(\omega) = [\exp(\hbar\omega/k_BT) - 1]^{-1}$. The factor $f_k(1 - f_{k'})$ enters because the initial electronic state must be occupied and the final state empty. The two terms in the phonon factor $n(\omega) + 1$ can be thought of as arising from the sum of the transition probabilities due to stimulated $[n(\omega)]$ and spontaneous [1] emission in analogy with the theory of radiation (Schiff, 1955). Similarly, the probability for the absorption of a phonon of energy $\hbar\omega = \varepsilon_{k'} - \varepsilon_k$ by an electron in state k (Fig. 1b) is

$$P_{kk'}^a = (2\pi/\hbar)\sum_j |g_{kk'}^j|^2 f_k(1 - f_{k'})n\big(\omega_{k'-k}^j\big)\delta\big(\varepsilon_k - \varepsilon_{k'} + \hbar\omega_{k'-k}^j\big) \quad (15)$$

and the total transition probability is $P_{kk'} = P_{kk'}^e + P_{kk'}^a$, which when used in Eq. (1) gives an expression for $1/\tau$ in this approximation. Equations (13)–(15) can be greatly simplified by making use of the fact that the matrix elements $g_{kk'}^j$ (which will be discussed in detail in Section II) and the group velocities $v_x(k)$ change with energy on the scale of eV[†] whereas the temperature and phonon frequencies are of the order of meV. This difference in energy scales allows a separation of the **k** dependence of the sums in Eq. (13) into an energy dependence and a variation over the Fermi surface. In order to achieve this separation the sums over k and k' are replaced by integrals over constant energy surfaces

$$\sum_k = \int d\varepsilon \sum_k \delta(\varepsilon - \varepsilon_k) = \int d\varepsilon \frac{V_a}{(2\pi)^3}\int_\varepsilon \frac{dS_k}{\hbar v_k}, \quad (16)$$

where V_a is the volume per atom and dS_k is an element of the constant energy surface at energy ε. It is also helpful to insert an integral over ω. Thus, Eq. (13) becomes

$$\frac{k_BTN(0)\langle v_x^2\rangle}{\tau\pi} = \int d\varepsilon \int d\varepsilon' \int d\omega \left\{ \frac{V_a^2}{(2\pi)^6}\int_\varepsilon \frac{dS_k}{\hbar v_k}\int_{\varepsilon'} \frac{dS_{k'}}{\hbar v_{k'}}\sum_j |g_{kk'}^j|^2 \right.$$

$$\left. \times [v_x(k) - v_x(k')]^2 \delta\big(\hbar\omega - \hbar\omega_q^j\big)\right\}$$

$$\times f(\varepsilon)[1 - f(\varepsilon')]\{[n(\omega) + 1]$$

$$\times \delta(\varepsilon - \varepsilon' - \hbar\omega) + n(\omega)\delta(\varepsilon - \varepsilon' + \hbar\omega)\}. \quad (17)$$

[†]A few metals with extremely narrow bands near the Fermi energy such as some of the high T_c A-15 materials may require special consideration.

This clumsy expression is greatly simplified by evaluating the surface integrals at the Fermi surface. This approximation makes only a slight error for nonpathological materials except at high temperatures and it allows us to replace the expression within the first set of curly brackets by a simple function of ω,

$$\frac{[V_a^2/(2\pi)^6]\int_{FS}(dS_k/\hbar v_k)\int_{FS}(dS_{k'}/\hbar v_{k'})\sum_j|g_{kk'}^j|^2 \times [v_x(k) - v_x(k')]^2\delta(\hbar\omega - \hbar\omega_q^j)}{2N(0)\langle v_x^2\rangle} = \alpha_{tr}^2(\omega)F(\omega). \quad (18)$$

$\alpha_{tr}^2(\omega)F(\omega)$ is a dimensionless function of ω introduced by Allen (1971a) which tends to be similar in shape to the phonon density of states $F(\omega)$ (hence, the strange notation). Figure 3 shows $F(\omega)$ and $\alpha_{tr}^2(\omega)F(\omega)$ for Nb and Pd. Also shown is a very similar function $\alpha^2(\omega)F(\omega)$ which occurs in the theory of the electron–phonon mass enhancement and in the theory of the superconducting transition temperature [see Eqs. (36)–(38)]. Equation (17) may now be written as

$$\hbar/\tau = (2\pi k_B T)2\int(d\omega/\omega)\alpha_{tr}^2(\omega)F(\omega)I(\hbar\omega/2k_B T), \quad (19)$$

where the function $I(\hbar\omega/2k_B T)$ defined by

$$I(\hbar\omega/2k_B T) = (\hbar\omega/2k_B^2 T^2)\int d\varepsilon\int d\varepsilon' f(\varepsilon)[1 - f(\varepsilon')]$$

$$\times\{[n(\omega) + 1]\delta(\varepsilon - \varepsilon' - \hbar\omega) + n(\omega)\delta(\varepsilon - \varepsilon' + \hbar\omega)\} \quad (20)$$

can be shown to be simply

$$I(x) = (x/\sinh x)^2. \quad (21)$$

At temperatures sufficiently high that $kT \gtrsim \hbar\omega_{max}$ where ω_{max} is the maximum phonon frequency, $I(x)$ tends to unity and the electrical resistivity becomes linear in temperature in this approximation

$$\rho = V_a 2\pi k_B T\lambda_{tr}/2e^2 N(0)\langle v_x^2\rangle\hbar \quad (kT \gtrsim \hbar\omega_{max}), \quad (22)$$

$$\lambda_{tr} = 2\int(d\omega/\omega)\alpha_{tr}^2(\omega)F(\omega). \quad (23)$$

The quantity λ_{tr} defined in Eq. (23) is quite similar to the electron–phonon mass enhancement λ (see Eq. 38). The LOVA expression for the resistivity

$$\rho = \frac{V_a 2\pi k_B T}{e^2 N(0)\langle v_x^2\rangle\hbar}\int_0^\infty \frac{d\omega}{\omega}\alpha_{tr}^2(\omega)F(\omega)I(x) \quad (24)$$

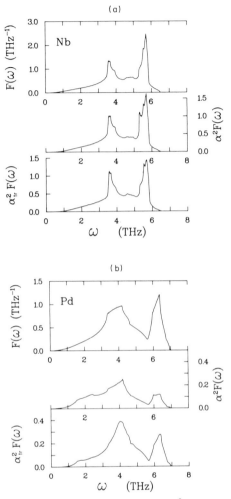

Fig. 3. Phonon density of states and spectral functions $\alpha^2(\omega)F(\omega)$ and $\alpha_{tr}^2(\omega)F(\omega)$ for (a) Nb and (b) Pd. The spectral functions were calculated using the "rigid muffin-tin" approximation. (Pinski *et al.*, 1978b, 1980; Butler *et al.*, 1979.)

can be thought of as a modern version of the Bloch–Grüneisen theory (Allen and Butler, 1978). In fact, the Bloch–Grüneisen formula (Bloch, 1928) can be obtained by replacing the function $\alpha_{tr}^2(\omega)F(\omega)$ in Eq. (24) by $C\omega^4$, where C is an adjustable constant, and cutting off the integral at ω_D (also an adjustable constant). It can be shown that $\alpha_{tr}^2(\omega)F(\omega)$ *does* vary as ω^4 as $\omega \to 0$ so that Eq. (24) yields a T^5 law for the resistivity at sufficiently low temperature. This temperature may be so low, however, that the LOVA

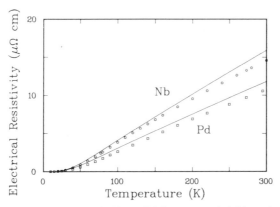

Fig. 4.   Resistivity of Nb and Pd. The solid lines are resistivities calculated from the $\alpha_{tr}^2(\omega)F(\omega)$'s of Fig. 3. The dots are the experimental results of Laubitz and Matsumura (1972) for Pd and of Abraham and Deviot (1972) for Nb.

expression is no longer reliable. Figure 4 shows calculations of the resistivity of Nb and Pd together with the corresponding experimental data (Pinski *et al.*, 1981). These calculations actually go beyond the LOVA since they use an energy and *k*-dependent distribution function obtained from an approximate inversion of the scattering operator [Eq. (8)]. These improvements to the LOVA are very important below about 50 K but they would only be barely discernable on Fig. 4. The agreement with experiment is quite gratifying since the calculations use *no* adjustable parameters.

C.   *Electron–Phonon Renormalization of the Electronic Mass*

The electrons in a metal interact with each other and with the phonons, yet most of us persist in picturing them as independent particles. This picture is not really correct but it has proved to be surprisingly useful. Indeed, only a few observable effects arising from the many-body interactions which qualitatively violate this simple concept have been discovered. The most spectacular of these effects is, of course, superconductivity, but there are many-body corrections to some normal-state properties as well. One important normal-state effect is the enhancement of the electronic mass in which electrons (more properly called quasielectrons) at the Fermi energy have their velocities reduced by a factor $1 + \lambda$ due to their interaction with the phonons and with the other electrons. It is this decreased velocity which is observed in de Haas–van Alphen and cyclotron resonance experiments. The low-temperature electronic specific heat is enhanced by this effect because it is proportional to the Fermi energy density of states which is itself inversely proportional to the electron velocity [cf. Eq. (16)],

$$N(\varepsilon) = \left[V_a / (2\pi)^3 \right] \int_\varepsilon dS_k / \hbar v_k. \qquad (25)$$

Surprisingly, many properties are unaffected by this velocity decrease. The size and shape of the Fermi surface is unaffected as is the dc electrical conductivity. The conductivity [Eq. (10)] is unaffected because the decrease of $N(0)\langle v_x^2 \rangle$ by the factor $1 + \lambda$ is precisely canceled by an increase in τ by the same factor. The other dc transport coefficients are also unaffected except perhaps for the thermopower. The question of whether or not the thermopower is enhanced by a factor (or factors) of $1 + \lambda$ is unsettled at present (Lyo, 1978; Vilenkin and Taylor, 1979). Grimvall (1976) gives a list of the experimental properties which are not affected and a discussion of those that are.

Suppose that one has calculated the band structure of a metal by solving the one-electron Schrödinger equation

$$\left[(-\hbar^2 / 2m) \nabla^2 + \Sigma_n v_{eff}(\mathbf{r} - \mathbf{R}_n) - \varepsilon_k \right] \psi_k(r) = 0, \qquad (26)$$

where $v_{eff}(\mathbf{r} - \mathbf{R}_n)$ is an effective one-electron potential describing the self-consistent field including exchange and correlation which an electron sees due to the ions and all of the other electrons. v_{eff} may be nonlocal but we require it to be real and energy independent. The quasiparticle spectrum of the interacting system is not given by the one-electron eigenvalues ε_k (no matter how adroitly one chooses v_{eff}). It must be obtained by considering the change in the total energy of the system caused by the addition of an extra electron of momentum k. This energy is ε_k plus an additional term,

$$E_{N+1}(k) - E_N = \varepsilon_k + \Sigma(k,E). \qquad (27)$$

The additional term $\Sigma(k,E)$ is called the self-energy. One contribution to the self-energy arises from that part of the electron–electron interaction that has not been adequately treated by v_{eff}. This part of the self-energy and especially its variation with energy are not known very well except perhaps for the alkalis (Hedin and Lundqvist, 1969). This contribution to $\Sigma(k,E)$ will not be discussed further. It can probably be made rather small for electrons near the Fermi energy by judicious choice of v_{eff} except perhaps for strongly paramagnetic metals such as Pd.

The electron–phonon contribution to the self-energy is very important because of its rapid energy variation near the Fermi energy. At low temperature, the electron–phonon self-energy is given by (Scalapino, 1969)

$$\Sigma_{ep}(k,E) = \sum_{k'j} |g_{kk'}^j|^2 \frac{(1 - f_{k'})}{E - \varepsilon_{k'} - \hbar\omega_{k-k'}^j} - \sum_{k'j} |g_{kk'}^j|^2 \frac{f_{k'}}{\varepsilon_{k'} - E - \hbar\omega_{k-k'}^j}. \qquad (28)$$

The first term in Eq. (28) is an energy shift which arises in second-order Brillouin–Wigner perturbation theory from transitions between the state in which the added electron has energy E and the state in which the added electron has emitted a phonon and is in band state $\varepsilon_{k'}$. Note that this transition can only occur if the state k' is unoccupied, hence, the factor $1 - f_{k'}$. The second term arises because previous to the introduction of the extra electron, the energy of the system was reduced by transitions in which an electron within the Fermi sea emits a phonon and goes to state k. These transitions are now blocked because of the added electron in this state. This blocking energy is lost to the system when the extra electron is added.

It is helpful in evaluating the electron–phonon self-energy to define an auxiliary quantity called the (k-dependent) electron–phonon spectral function

$$\alpha_k^2(\omega)F(\omega) = \frac{V_a}{(2\pi)^3\hbar} \int_{FS} \frac{dS_{k'}}{\hbar v_{k'}} \sum_j |g_{kk'}^j|^2 \delta(\omega - \omega_{k-k'}^j). \tag{29}$$

This k-dependent spectral function is quite similar to the transport spectral function defined in Eq. (18). Equation (28) can be simplified in a manner analogous to the treatment of Eq. (13). The sum over k' can be replaced by an integral over energy and an integration over constant energy surfaces as in Eq. (16); thus,

$$\Sigma_{ep}(k,E) = \int d\omega \int d\varepsilon' \frac{V_a}{(2\pi)^3} \int_{\varepsilon'} \frac{dS_{k'}}{\hbar v_{k'}} \sum_j |g_{kk'}^j|^2 \delta(\omega - \omega_{k-k'}^j)$$

$$\times \left[\frac{1 - f(\varepsilon')}{E - \varepsilon' - \hbar\omega_{k-k'}^j} + \frac{f(\varepsilon')}{E - \varepsilon' + \hbar\omega_{k-k'}^j} \right]. \tag{30}$$

The major contributions to the ε' integral occurs for $\varepsilon' = E \pm \hbar\omega$ where ω is less than the maximum phonon frequency ω_{max}. If we restrict our interest to energies E within a few times ω_{max} of the Fermi energy, the surface integral over the constant energy surface ε' might as well be evaluated at the Fermi surface since the matrix elements, Fermi surface, and Fermi velocity are essentially constant over an energy range of this magnitude.[†] This yields

$$\Sigma_{ep}(k,E) = \hbar \int d\omega\, \alpha_k^2(\omega)F(\omega) \int d\varepsilon' \left[\frac{1 - f(\varepsilon')}{E - \varepsilon' - \hbar\omega} + \frac{f(\varepsilon')}{E - \varepsilon' + \hbar\omega} \right], \tag{31}$$

which may be integrated to yield

$$\Sigma_{ep}(k,E) = \hbar \int d\omega\, \alpha_k^2(\omega)F(\omega) \log\left| \frac{E - \hbar\omega}{E + \hbar\omega} \right|. \tag{32}$$

[†] The same caveats concerning exceptional metals with rapidly varying electronic structure near the Fermi energy mentioned in Section I,B apply here as well.

This result is valid at $T = 0$. Expressions valid at finite temperature have been derived by Allen (1971b) and by Grimvall (1969).

The dispersion relation is given by the solution of $E - \varepsilon_k - \Sigma_{ep}(k, E) = 0$. Note that $\Sigma_{ep}(k, 0) = 0$. This implies that the size and shape of the Fermi surface are not affected by the electron–phonon interaction. The Fermi velocity *is* affected, however. It is obtained by taking the gradient with respect to k of the quasiparticle energy

$$\hbar v_k^* = \nabla_k E = \nabla_k \varepsilon_k + \nabla_k \Sigma_{ep}(k, E). \tag{33}$$

Now the explicit k derivative of $\Sigma_{ep}(k, E)$ vanishes at $E = 0$ according to Eq. (32), but there remains an implicit dependence on k arising from the energy dependence. Solving Eq. (33) for $\nabla_k E$ at $E = 0$ we have

$$\hbar v_k^* = \nabla_k \varepsilon_k / (1 + \lambda_k) = \hbar v_k / (1 + \lambda_k), \tag{34}$$

where

$$\lambda_k = -\left[\partial \Sigma_{ep}(k, E) / \partial E \right]_{E=0}. \tag{35}$$

Equation (35) can be evaluated easily using Eq. (32):

$$\lambda_k = 2 \int (d\omega/\omega) \alpha_k^2(\omega) F(\omega)$$

$$= \left[2V_a / (2\pi)^3 \right] \int_{FS} dS_k / \hbar v_k \sum_j |g_{kk'}^j|^2 / \hbar \omega_{k-k'}^j. \tag{36}$$

The Fermi surface average of λ_k, denoted by λ, is an important quantity because it determines the enhanced Fermi energy density of states $N^*(0)$ which is observed in low-temperature specific heat measurements. Recall that $N(0)$, the band-theory density of states, may be written as an integral over the Fermi surface of the inverse of the Fermi velocity. $N^*(0)$, the quasiparticle density of states, may be written the same way,

$$N^*(0) = \sum_k \delta[E(k)] = \left[V_a / (2\pi)^3 \right] \int_{FS} dS_k / \hbar v_{k^*} = N(0)(1 + \lambda). \tag{37}$$

The Fermi surface averaged mass enhancement can be written in terms of the (Fermi surface averaged) spectral function $\alpha^2 F(\omega)$:

$$\lambda = 2 \int (d\omega/\omega) \alpha^2(\omega) F(\omega); \tag{38}$$

as will be seen in the next section it is primarily $\alpha^2(\omega) F(\omega)$ which determines the superconducting transition temperature.

D. *The Superconducting Transition Temperature*

In this section, I shall sketch only a "bare-bones" outline of the theory of T_c, but shall attempt to emphasize the physics of the electron–phonon coupling which in most elementary treatments is represented solely by an uninteresting coupling constant g^2. A very thorough treatment of the formal aspects of this problem including the effects of gap anisotropy and impurities (both magnetic and nonmagnetic) has been given recently by Allen (1980a).

The normal to superconducting transition occurs due to the formation of Cooper pairs (Cooper, 1956; Bardeen *et al.*, 1957) which are bound states formed from an electron in band state k with spin up and an electron in the time reversed band state $(-k,$ spin down). The total energy of the pair E_p is less than the sum of the energies of the states from which it was formed $(E_p < \varepsilon_{k\uparrow} + \varepsilon_{-k\downarrow} = 2\varepsilon_k)$.

The pair binding energy comes from the attractive interaction due to the exchange of phonons (Fig. 1d). It can be derived in terms of a simple physical picture (Fig. 2b). Suppose an electron is at position x in the crystal at time t. It will experience a force which arises from the potential fields of the ions whose positions we denote by $\{R_l\}$. If the total potential field at point x is $V(x, \{R_l\})$, then by translational invariance we can write $V(x + \delta, \{R_l\}) = V(x, \{R_l - \delta\})$. This result can be used to express the total force exerted by the electron on the crystal as a sum of contributions from each lattice site. If F_α^T is the total force exerted on the crystal by the electron in direction α we have

$$F_\alpha^T = \nabla_\alpha V(x, \{R_l\}) = -\sum_l \delta V(x, \{R_l\})/\delta R_{l\alpha}, \qquad (39)$$

and the force on an atom at R_l in direction α due to an electron at (x, t), i.e., $F_{l\alpha}(t)$ can be identified with $-\delta V/\delta R_{l\alpha}$.

A force exerted on an ion at lattice site l and time t will propagate through the lattice causing displacements of the other ions. The calculation of this effect for a harmonic lattice is a standard problem in the theory of lattice dynamics (see, e.g., Reissland, 1973). The function $D_{\beta\alpha}(l'l; t' - t)$ which gives the displacement in direction β of the ion at $R_{l'}$ at time t' due to a force exerted on the ion at R_l in the direction α at time t is called the lattice Green's function or propagator, i.e.,

$$u_{l'\beta}(t') = \sum_{l\alpha} \int dt\, D_{\beta\alpha}(l', l; t' - t) F_{l\alpha}(t). \qquad (40)$$

The propagator can be calculated from the phonon frequencies ω_q^j and polarization vectors $\hat{\varepsilon}^j(q)$

$$D_{\beta\alpha}(l'l; t' - t) = -\theta(t' - t) \sum_{qj} \frac{\varepsilon_\beta^j(q)\varepsilon_\alpha^j(q)}{M\omega_q^j} \sin\left[\omega_q^j(t' - t)\right] e^{iq \cdot (R_l - R_{l'})}.$$

(41)

Given the displacement of the atom at l', we can calculate the effect on the electrons in its vicinity. The change in the energy of an electron at (x', t') due to a displacement $\mathbf{u}_{l'}(t')$ is

$$\delta V = \sum_{l'\beta} \left[\delta V\left(x', \{R_l\}\right)/\delta R_{l'\beta}\right] u_{l'\beta}(t').$$

(42)

Thus, the change in energy of an electron at (x', t') due to the existence of an electron at (x, t) for a short time Δt is

$$v(x't'; x, t) = - \sum_{l'\beta, l\alpha} \frac{\delta V(x')}{\delta R_{l'\beta}} D_{\beta\alpha}(l'l; t' - t) \frac{\delta V(x)}{\delta R_{l\alpha}} \Delta t.$$

(43)

This interaction can be converted into a function of energy and momentum by noting that an electron with wave function $\psi_k(x)e^{i\varepsilon_k t/\hbar}$ is annihilated at the vertex at (x, t) (see Fig. 2b) while an electron with wave function $\psi_{k'}(x)e^{i\varepsilon_{k'} t/\hbar}$ is created there and similarly at (x', t'), except that the states are time reversed ($k \to -k$, $t \to -t$, etc.). The electron spins may be ignored in this case since they are not affected by the electron–phonon interaction. Thus, the interaction is

$$v(k, k'; \varepsilon_k - \varepsilon_{k'})$$

$$= (-1/V_a^2)\int d^3x' \int d^3x \int d\tau \sum_{l'l\beta\alpha} \psi_{-k'}^*(x') \frac{\delta V(x')}{\delta R_{l'\beta}} \psi_{-k}(x') D_{\beta\alpha}(l', l; \tau)$$

$$\times \psi_{k'}^*(x) \frac{\delta V(x)}{\delta R_{l\alpha}} \psi_k(x) e^{i(\varepsilon_k - \varepsilon_{k'})\tau}.$$

(44)

Using the Bloch property of the wave functions $\psi_k(x + R_l) = e^{ik \cdot R_l}\psi_k(x)$, time reversal symmetry $\psi_k(x) = \psi_{-k}^*(x)$, and the translational symmetry of $V(x, \{R_l\})$, we have

$$\int d^3x \psi_{k'}^*(x)\left[\delta V(x)/\delta R_{l\alpha}\right]\psi_k(x) = e^{i(k-k') \cdot R_l} V_a I_{k'k}^\alpha,$$

(45)

where $I_{k'k}^\alpha$ is the electronic part of the electron–phonon matrix element and $V_a = V/N_a$ is the volume per atom

$$I_{k'k}^\alpha = (1/V_a)\int d^3x \psi_{k'}^*(x)\left[\delta V(x)/\delta R_{0\alpha}\right]\psi_k(x).$$

(46)

Note that the region of integration in Eq. (46) is confined to the neighborhood of the atomic cell at the origin, i.e., to a volume of approximately V_a. Using Eq. (41) and Eqs. (45) and (46) in Eq. (44) we obtain

$$v(k',k;\varepsilon_{k'} - \varepsilon_k) = \sum_j |g^j_{kk'}|^2 \frac{2\hbar\omega^j_{k'-k}}{\left(\varepsilon_{k'} - \varepsilon_k\right)^2 - \left(\hbar\omega^j_{k'-k}\right)^2} \tag{47}$$

where

$$|g^j_{kk'}|^2 = \hbar \sum_{\alpha\beta} \frac{\varepsilon^j_\alpha \varepsilon^j_\beta I^\alpha_{k'k} I^\beta_{kk'}}{2M\omega^j_{k'-k}} \tag{48}$$

is the square of the electron–phonon matrix element. It will be shown in Section II that this is the same matrix element that appeared in Eqs. (17) and (30) in the expressions for the resistivity and for the mass enhancement.

The derivation of the effective attractive interaction is only part of the solution to the problem of calculating T_c because the bound state is obtained by considering repeated phonon exchange as shown in Fig. 5. Let the function $\Gamma(k',k;\varepsilon_{k'} - \varepsilon_k)$ denote the total scattering amplitude for electrons in states k and $-k$ to interact and be scattered into states k' and $-k'$. Figure 5 shows how an infinite series of repeated interactions can be summed to yield an integral equation for Γ. In a simplified notation, the first row of diagrams in Fig. 5 may be written algebraically as

$$\Gamma = v + vKv + vKvKv + \cdots, \tag{49}$$

where Γ represents the scattering amplitude or "vertex function" (box), v represents the phonon mediated electron–electron interaction (wavy phonon line with attached dots), and K represents the lowest-order approximation to the two-particle propagator (pair of straight lines connecting one

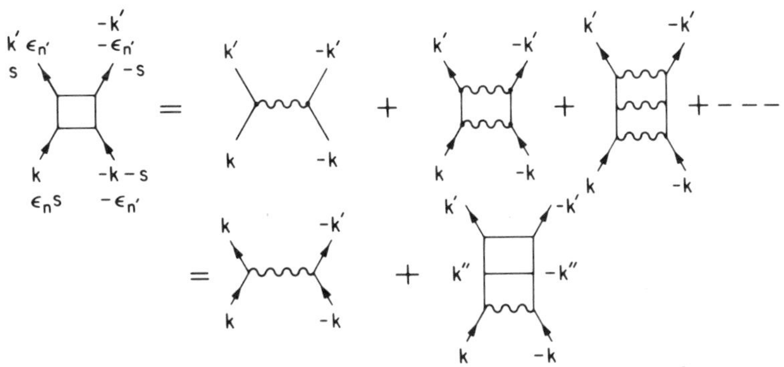

Fig. 5. Repeated phonon exchange leads to superconductivity. The infinite sum of terms each involving one more phonon exchange than the last can be summed as shown in the second row.

interaction with the next). This interaction may be summed formally to yield (second row of Fig. 5) $\Gamma = v + vK\Gamma$. The formation of a bound state between interacting particles is signaled by a divergence in the scattering amplitude. For the case under consideration here, the normal to superconducting transition will occur when the temperature becomes sufficiently low that Γ diverges or equivalently when the equation $\Gamma = vK\Gamma$ has a nontrivial solution. Inserting all of the necessary indices into this equation yields

$$\Gamma(k',k;\varepsilon',\varepsilon) = \sum_{k''} \int d\varepsilon'' v(k',k'';\varepsilon' - \varepsilon'') K(k'',\varepsilon'') \Gamma(k'',k;\varepsilon'',\varepsilon). \quad (50)$$

In order to solve this equation for the transition temperature it is necessary to introduce the lowest-order pair propagator

$$K(k,\varepsilon) = \frac{1}{(\varepsilon - \varepsilon_k)} \frac{1}{(-\varepsilon - \varepsilon_{-k})} \quad (51)$$

and to convert this zero-temperature formalism into a finite-temperature formalism by means of the trick devised by Matsubara (1955) of replacing energies by imaginary integral quantities $\varepsilon \to i\varepsilon_n = i(2n + 1)\pi k_B T$; thus,

$$\Gamma(k,k';\varepsilon_n,\varepsilon_{n'}) = k_B T \sum_{n''} \sum_{k''j} |g_{kk''}^j|^2 \frac{2\hbar\omega_{k-k''}^j}{(\varepsilon_n - \varepsilon_{n''})^2 + (\hbar\omega_{k-k''}^j)^2}$$

$$\times \frac{1}{\varepsilon_{n''}^2 + \varepsilon_{k''}^2} \Gamma(k'',k';\varepsilon_{n''},\varepsilon_{n'}). \quad (52)$$

In a manner similar to the previous treatment of the self-energy, the sum over k'' can be replaced by an integral over $\varepsilon_{k''}$ and an integral over constant energy surfaces. The k' and $\varepsilon_{n'}$ arguments of Γ may be suppressed and it will again be assumed that the band structure does not change significantly over an energy interval about the Fermi energy equal to a typical phonon energy. This assumption allows the surface integral to be placed at the Fermi energy and the ε'' integral to be performed analytically. Defining a set of temperature-dependent coupling functions

$$\lambda_{kk'}^{(m)} = N(0) \sum_j |g_{kk'}^j|^2 \frac{2\hbar\omega_{k-k'}^j}{(2\pi m k_B T)^2 + (\hbar\omega_{k-k'}^j)^2}, \quad (53)$$

we have

$$\Gamma(k,n) = \frac{N(0)^{-1} V_a}{(2\pi)^3} \int \frac{dS_{k'}}{\hbar v_{k'}} \sum_{n'} \frac{\lambda_{kk'}(n - n')}{|2n' + 1|} \Gamma(k',n'). \quad (54)$$

$\Gamma(k,n)$ is sometimes referred to as the gap function and Eq. (54) may be called the linearized gap equation. It is only valid very near T_c; below T_c a nonlinear equation must be solved to determine $\Gamma(k,n)$.

The variation of $\Gamma(k,n)$ with k from point to point on the Fermi surface is known as "gap anisotropy." For many materials, this anisotropy is relatively weak, i.e., $\Gamma(k,n) = \bar{\Gamma}(n) + \gamma_k(n)$, where the Fermi surface average of $\gamma_k(n)$ vanishes and $\gamma_k(n) \ll \bar{\Gamma}(n)$. If we neglect γ_k entirely, Eq. (54) may be simplified to yield a simple equation for T_c,

$$\bar{\Gamma}(n) = \sum_{n'=-\infty}^{\infty} \left[\lambda(n-n')/|2n'+1| \right] \bar{\Gamma}(n'), \tag{55}$$

where $\lambda(m)$ is a temperature-dependent generalization of the isotropic mass enhancement parameter

$$\lambda(m) = \langle\langle \lambda_{kk'} \rangle_k \rangle_{k'} = 2 \int \frac{\alpha^2(\omega) F(\omega) \omega \, d\omega}{\omega^2 + (2\pi m k_B T)^2}. \tag{56}$$

The brackets $\langle \ \rangle_k$ in Eq. (56) indicate a Fermi surface average.

The highest temperature for which Eq. (55) has a nontrivial solution is T_c. Since T_c does not depend critically on the details of the m dependence of $\lambda(m)$ but on its overall magnitude let us attempt to find an approximate solution to Eq. (55) by assuming $\lambda(n-n')$ to have the (rather unlikely) form

$$\lambda(n-n') = \lambda\Theta(N-|n|)\Theta(N-|n'|), \tag{57}$$

where

$$\Theta(N-|n|) = \begin{cases} 0, & |n| \geq N \\ 1, & |n| < N \end{cases} \tag{58}$$

and the cutoff Matsubara frequency is set by the scale of the phonon frequencies $(2N+1)\pi k_B T = \hbar\omega_0$. With this assumption Eq. (55) is satisfied by $\bar{\Gamma}(n) = \bar{\Gamma}\Theta(N-|n|)$ and T_c is determined by

$$1/\lambda = 2 \sum_{n=0}^{N-1} \frac{1}{2n+1} = \gamma + 2\ln 2 + \psi(N+\tfrac{1}{2}), \tag{59}$$

where $\gamma = 0.5772$ is Euler's constant and $\psi(z)$ is the Digamma function (Abramowitz and Stegun, 1964), which for large values of its argument may be approximated by $\psi(z) \approx \ln z$ so that the T_c expression is

$$k_B T_c = (2e^\gamma/\pi)\hbar\omega_0 e^{-1/\lambda} = 1.13\hbar\omega_0 e^{-1/\lambda}. \tag{60}$$

This expression is identical to the result obtained by Bardeen et al. (1957). It turns out that although Eq. (57) is not a realistic model for the variation of $\lambda(m)$ the result (Eq. 60) has essentially the correct form if $\alpha^2(\omega)F(\omega)$ is a physically reasonable function.

Two important effects are not included in Eq. (55). One is the effect of the electron–phonon mass enhancement which should have been included in the two-particle propagator K. Its effect is essentially to replace the exponent $1/\lambda$ in Eq. (60) by $(1 + \lambda)/\lambda$. The second effect is the Coulomb repulsion between the electrons. The magnitude of this effect is difficult to calculate precisely, but it is usually argued (Morel and Anderson, 1962) that since the Coulomb repulsion acts instantaneously on the scale of a lattice relaxation time, it may be replaced by a frequency-independent pseudo-potential μ^* over the frequency range where the electron–phonon interaction is attractive. The approximate effect of the Coulomb repulsion is to replace the exponent of Eq. (60) $(1 + \lambda)/\lambda$ including the mass enhancement effect by $(1 + \lambda)/(\lambda - \mu^*)$. Superconducting tunneling experiments can be analyzed to obtain $\alpha^2(\omega)F(\omega)$ and μ^* (McMillan and Rowell, 1969). The values of μ^* obtained in this way for simple metals are ≈ 0.1. The small value is due to the short range of the screened Coulomb interaction in a metal and to the slow response of the lattice which allows an attractive interaction between electrons which are well separated in space.

Allen and Dynes (1975) have solved the equation analogous to Eq. (55) which includes mass renormalization and Coulomb repulsion effects using several realistic forms for $\lambda(m)$. The results of their calculations can be summarized accurately by an analytic expression for T_c [†],

$$T_c = \frac{\langle \omega \rangle_{\log}}{1.2} \exp\left[-\frac{1.04(1 + \lambda)}{\lambda - \mu^* - 0.62\lambda\mu^*} \right],$$ (61)

where

$$\langle \omega \rangle_{\log} = \exp\left[\frac{2}{\lambda} \int (d\omega/\omega)\alpha^2 F(\omega) \ln \omega \right].$$ (62)

As mentioned previously, Eqs. (55) and (61) pertain to a material with an isotropic gap function. This is sometimes referred to as the "dirty limit" because scattering due to impurities tends to eliminate the anisotropy of $\Gamma(k,n)$. If the material is very pure the applicable equations are analogous to Eq. (54) and retain the anisotropy in $\lambda_{kk'}$. It is not difficult to show that in the weak coupling–weak anisotropy approximation, T_c is enhanced by the anisotropy by an amount

$$\delta T_c/T_c = \langle \lambda_k^2 - \lambda^2 \rangle_k / \lambda^2,$$ (63)

where $\langle \ \rangle_k$ indicates a Fermi surface average. A more detailed treatment of the effect of gap anisotropy on T_c can be found in Butler and Allen (1976), Peter et al. (1977), and Allen (1980a).

[†] This equation is a slight revision of an equation proposed earlier by McMillian (1968). Allen and Dynes also showed that Eq. (61) needs further revision for $\lambda \gtrsim 1.5$.

E. Phonon Lifetimes and Energy Shifts

In Section I, B the scattering of the conduction electrons by emission and absorption of phonons was discussed. These same processes when seen from the point of view of a phonon mode appear to be the annihilation and creation of phonons by means of emission or recombination of electron–hole pairs (Figs. 1e and 1a). The transition rate which governs the number of phonons n_{qj} in the phonon mode of momentum q and polarization j, [i.e., $\tau_j^{-1}(q) = -n_{qj}^{-1} dn_{qj}/dt$], is given by the Fermi "golden-rule" formula (Allen, 1972)

$$\tau_j^{-1}(q) = (2\pi/\hbar) \sum_{kk'} |g_{kk'}^j|^2 \delta\left(\varepsilon_k - \varepsilon_{k'} + \hbar\omega_q^j\right)$$
$$\times \left[f_k(1 - f_{k'}) - f_{k'}(1 - f_k) \right] \delta(k - k' + q). \tag{64}$$

The first and second terms in Eq. (64) correspond to Figs. 1e and 1a, respectively. Since the phonon and thermal energies are small on an energy band scale, $f_k - f_{k'}$ may be approximated by $\delta(\varepsilon_k)\hbar\omega_q^j$ and Eq. (64) may be approximated by

$$\hbar/\tau_j(q) = 2\pi\hbar\omega_q^j \frac{V_a^2}{(2\pi)^6} \int_{FS} \frac{dS_k}{\hbar v_k} \int_{FS} \frac{dS_{k'}}{\hbar v_{k'}} |g_{kk'}^j|^2 \delta(k - k' + q). \tag{65}$$

The lifetimes defined by Eq. (65) may be used to calculate the electron–phonon component of the lattice thermal resistivity (Butler and Williams, 1978)

$$\left(W_{ep}^p\right)^{-1} = \frac{1}{V_a} \sum_{qj} C_j(q,T) u_{jx}^2(q)\tau_j(q), \tag{66}$$

where $C_j(q,T) = (x/\sinh x)^2$ where $x = \hbar\omega_q^j/2k_B T$ is the specific heat of phonon mode (q,j), $u_{jx}(q)$ is the x component of the phonon group velocity, and V_a is the atomic volume. Perhaps more importantly it can be used to define a phonon linewidth

$$\gamma_j(q) = \tfrac{1}{2}\tau_j^{-1}(q) \tag{67}$$

which is experimentally observable in careful inelastic neutron scattering experiments. The factor $\tfrac{1}{2}$ in Eq. (67) arises because $1/\tau_j(q)$ as defined in Eq. (64) is a decay rate for the phonon *probability* which decays twice as fast as the phonon *amplitude*.

The study of the phonon linewidth is important because it yields a very direct and detailed measure of the electron–phonon interaction. The mass enhancement λ and the spectral function $\alpha^2(\omega)F(\omega)$ can both be written as simple averages over the phonon linewidth,

$$\lambda = \left(\hbar/\pi N(0)\right) \sum_{qj} \gamma_j(q) / \left(\hbar\omega_q^j\right)^2, \tag{68}$$

$$\alpha^2(\omega)F(\omega) = \left(1/2\pi N(0)\hbar\omega\right) \sum_{qj} \gamma_j(q)\delta\left(\omega - \omega_q^j\right). \tag{69}$$

Since λ is a measure of the overall strength of the phonon mediated electron–electron coupling it can be seen from Eq. (68) that $\gamma_j(q)/(\omega_q^j)^2$ is a measure of the amount of electron–electron coupling contributed by phonon mode (q,j). Figure 6 shows calculated and experimental linewidths in Nb and Pd (Butler *et al.*, 1977, 1979; Pinski and Butler, 1979; Youngblood *et al.*, 1979). There is much structure and variation in $\gamma_j(q)$ which can be

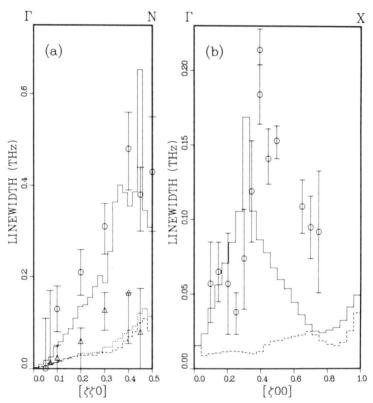

Fig. 6. Phonon linewidths due to the electron–phonon interaction for (a) Nb and (b) Pd. Histograms are calculated values—solid for longitudinal modes and dashed or dotted for transverse modes. Circles (triangles) are experimental values for longitudinal (transverse) modes from inelastic neutron scattering.

quantitatively related via first-principles calculations to the underlying electronic structure. These linewidths were used to calculate the spectral functions of Fig. 3 and the resistivities shown in Fig. 4.

Closely related to the phonon linewidths are the phonon energy shifts (Allen, 1980b). Just as the electronic dispersion relation is modified by the emission and reabsorption of virtual phonons so also is the phonon dispersion relation modified by the emission and reabsorption of virtual electron–hole pairs. The phonons acquire a complex self-energy so that the modified dispersion relation is given by

$$\left(\omega_q^j\right)^2 - 2i\gamma_j(q)\omega_q^j = \left(\Omega_q^j\right)^2 + 2\Omega_q^j\Pi_j\left(q,\omega_q^j\right), \tag{70}$$

where $\Pi_j(q,\omega)$ is the phonon self-energy and Ω_{qj} is the original or "bare" phonon frequency. The self-energy has both a real and an imaginary part

$$\Pi_j(q,\omega) = \Delta_j(q,\omega) - i\Gamma_j(q,\omega). \tag{71}$$

The real part causes the shift of the phonon frequencies while the imaginary part causes a phonon wave function to decay exponentially with time, $\phi_j(q) \propto \exp(i\omega_q^j t)\exp(-\gamma_j(q)t)$, where the damping factor $\gamma_j(q) = \Gamma_j(q,\omega_q^j)\Omega_q^j/\omega_q^j$ is the linewidth discussed previously in this section.

Unlike the linewidth, the frequency shift $\Delta_j(q)$ is not unique but depends upon one's choice of the bare frequencies Ω_q^j. The physical phonon frequencies ω_q^j are completely determined by the interatomic force constants. The force constant $\Phi_{\alpha\beta}(l,l')$ is defined to be the negative of the force on atom l in direction α per unit displacement of the atom at l' in the direction β, all other atoms being held fixed. The nonuniqueness of $\Omega_j(q)$ and $\Delta_j(q)$ results from the different ways that $\Phi_{\alpha\beta}(l,l')$ may be separated into "electronic" and "ionic" or "bare" contributions.

A rigorous starting point for a discussion of the interatomic force constants is a theorem of Hellmann (1937) and Feynman (1939) which states that the force on an atomic nucleus in a molecule or solid is simply the classical force on the nucleus due to the electrostatic field of the other nuclei and the electrons. Using this theorem, the force on nucleus l in direction α can be written in terms of the nuclear potential energy V_{NN}, the electron–nucleus potential energy V_{eN}, and the electronic density ρ:

$$F_{l\alpha} = -\frac{\delta V_{NN}\{R_l\}}{\delta R_{l\alpha}} - \int d^3r \frac{\delta V_{eN}\left(r,\{R_l\}\right)}{\delta R_{l\alpha}}\rho(r), \tag{72}$$

$$V_{NN} = \frac{1}{2}\sum_{ll'}' \frac{Z^2 e^2}{|R_l - R_{l'}|}, \tag{73}$$

$$V_{eN}(r) = \sum_l \frac{-Ze}{|r - R_l|}. \tag{74}$$

The force constant can then be obtained from the gradient of $F_{l\alpha}$ with respect to $R_{l'\beta}$ (assuming $l \neq l'$):

$$\Phi_{\alpha\beta}(l,l') = \frac{\delta^2 V_{NN}\{R_l\}}{\delta R_{l\alpha}\delta R_{l'\beta}} + \int dr \frac{\delta V_{eN}(r,\{R_l\})}{\delta R_{l\alpha}} \frac{\delta\rho(r)}{\delta R_{l'\beta}}. \qquad (75)$$

The charge density perturbation $\delta\rho$ in the second term or Eq. (75) is related to the crystal potential perturbation through the "bare" or lowest-order susceptibility

$$\delta\rho(r) = \int \chi_0(r,r')\delta V(r') dr', \qquad (76)$$

where $\chi_0(r,r')$ is given by

$$\chi_0(r,r') = \frac{e}{V_a^2} \sum_{kk'} \frac{\psi_k^*(r)\psi_{k'}(r)\psi_k(r')\psi_{k'}^*(r')}{\varepsilon_k - \varepsilon_{k'}}(f_k - f_k'). \qquad (77)$$

Substitution of Eqs. (77) and (76) into the second term of Eq. (75) allows the electronic part of $\Phi_{\alpha\beta}$ to be written as

$$\Phi_{\alpha\beta}^e(l,l') = \sum_{kk'j} \frac{I_{kk'}^{\alpha 0} I_{k'k}^{\beta}(f_k - f_{k'})}{\varepsilon_k - \varepsilon_{k'}} e^{i(k'-k)\cdot(R_l - R_{l'})} \qquad (78)$$

where $I_{kk'}^{\beta}$ is the electronic part of the electron–phonon matrix element used previously and $I_{kk'}^{\alpha 0}$ is the matrix element of the bare gradient of the nuclear potential.

There is a serious difficulty with Eq. (75). The charge density perturbation $\delta\rho/\delta R_{l'\beta}$ is difficult to calculate and it must be calculated very accurately because there is a huge cancellation between the two terms in this equation. At present, there is considerable controversy over the best way to handle this problem and over the physical origin of the anomalies seen in the phonon dispersion curves of metals (especially transition metals).

Pickett and Gyorffy (1976) showed that the force constant matrix can be decomposed into a short-ranged part $\Phi_{\alpha\beta}^S(l,l')$ and a long-ranged part $\Phi_{\alpha\beta}^L(l,l')$ which is given by Eq. (78) except that the bare matrix element $I_{kk'}^{\alpha 0}$ is replaced by $I_{kk'}^{\alpha}$. Presumably, it is $\Phi_{\alpha\beta}^L(l,l')$ which determines the fascinating dips and wiggles of the phonon dispersion curves of metals with strong electron–phonon coupling (see Smith *et al.*, 1976). Varma and Weber (1979) have evaluated $\Phi_{\alpha\beta}^L(l,l')$ using a "modified tight-binding" approximation for the matrix elements $I_{kk'}^{\alpha}$. Their calculated dispersion curves for Nb, Mo and a Nb–Mo alloy are shown in Fig. 7. Even allowing for the adjustable parameters used to fit $\Phi_{\alpha\beta}^S(l,l')$, the agreement between theory and experiment is impressive.

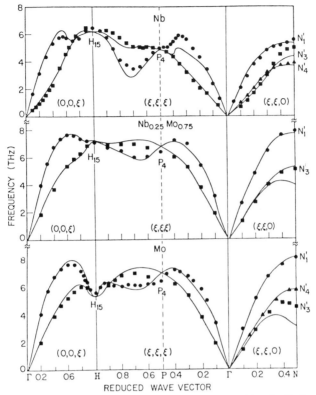

Fig. 7. Phonon dispersion curves for Nb. Circles are observed longitudinal mode frequencies and squares and triangles are observed transverse mode frequencies (Powell *et al.*, 1968). Solid curves are calculated frequencies (Varma and Weber, 1979). (From Weber, 1980.)

F. Summary and Discussion of Formal Results

In the previous parts of this section expressions were derived for

(1) resistivity in lowest-order variational approximation [Eqs. (18) and (24)],

(2) electron–phonon mass enhancement [Eq. (36)],

(3) superconducting transition temperature [Eq. (61)],

(4) phonon lifetimes [Eqs. (65) and (67)],

(5) lattice thermal resistivity [Eq. (66)], and

(6) interatomic force constants [Eqs. (75) and (78)].

The theory has been greatly simplified by the neglect of higher-order terms such as the "vertex correction" to Fig. 1a shown in Fig. 8. Migdal (1958) has

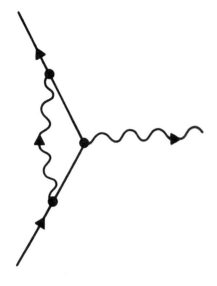

Fig. 8. Electron–phonon vertex correction.

shown that these terms are smaller than the lowest-order term by at least a factor of order $\hbar\omega/E_B$ where ω is a typical phonon frequency of order 10 meV and E_B is a characteristic band structure energy of order eV. The physical reason for this difference in energy scales can be traced back to the small mass of an electron relative to the ions which allows the electrons to follow the ions adiabatically with a low probability for transitions between electronic states. The difference between the energy scales also greatly simplifies the expressions for the resistivity, mass enhancement, transition temperature, and linewidth because it allows these quantities to be expressed in terms of transitions between electronic states at the Fermi energy only.

One important approximation which we make is that of independent electrons whose motion is governed by single-particle–like Schrödinger equations. This procedure is usually justified today in terms of the density functional theory of Hohenberg and Kohn (1964) and Kohn and Sham (1965) but it should be remembered that their theory only provides a rigorous prescription for the calculation of the ground-state energy and electron density. Its use in the calculation of transition rates is an approximation. Numerical calculations by Rasolt and Devlin (1976), however, indicate that it is probably a good approximation for the electron–phonon interaction.

Of the three ingredients necessary to calculate the quantities listed under (1)–(6), i.e., electronic structure, phonons, and electron–phonon matrix elements, the matrix elements present the greatest difficulties and they are the subject of the next section.

II. Electron-Phonon Matrix Elements

A. *Theory of the Electron-Phonon Matrix Element*

In this section we shall be concerned with understanding and calculating the electron-phonon matrix elements $g_{kk'}^j$ which appeared so ubiquitously in the formulae of Section I. In that section, it was shown that the matrix element which enters the attractive phonon mediated electron-electron interaction is given by Eq. (48) and it was asserted that the same matrix elements entered the expressions for the resistivity, mass enhancement, and phonon linewidth. This assertion can be demonstrated by noting that the matrix element needed in each of these instances is

$$g_{kk'}^j = \langle \psi_k | \delta V | \psi_{k'} \rangle, \tag{79}$$

where δV is the change in the electronic Hamiltonian caused by a phonon of mode index j and momentum $q = k' - k$. The change in crystal potential due to a phonon is

$$\delta V = \sum_l \delta V / \delta R_l \cdot \mathbf{u}_l^j, \tag{80}$$

where $\mathbf{u}_l^j$ is the phonon displacement field

$$\mathbf{u}_l^j = \sum_q \hat{\varepsilon}_q^j \sqrt{\hbar / 2MN\omega_q^j} \, e^{iq \cdot R_l}. \tag{81}$$

When Eqs. (80) and (81) are substituted into Eq. (79), the Bloch property of the wave function and the translational symmetry of V can be used to show that

$$g_{k'k}^j = \sum_\alpha \varepsilon_\alpha^j (k' - k) I_{k'k}^\alpha \sqrt{\hbar / 2M\omega_{k'-k}^j}, \tag{82}$$

where $I_{k'k}^\alpha$ is given by Eq. (46). This result is of course equivalent to Eq. (48).

The central problem in calculating the electron-phonon matrix element is the evaluation of $\delta V / \delta R_\alpha$ which appears in $I_{k'k}^\alpha$. As defined in Section I, D, this is the change in total crystal potential due to an infinitesimal displacement of the atom at the origin in the direction α (all other atoms fixed). This is a difficult quantity to calculate for a system of inhomogeneous interacting electrons because as one displaces a nuclear potential the electrons move to compensate the resultant electrostatic fields. Thus, the change in total potential is given by the change in bare potential δV_b plus the compensating induced potential δV_{ind} resulting from the readjustment of the electronic charge density

$$\delta V / \delta R_\alpha = \delta V_b / \delta R_\alpha + \delta V_{ind} / \delta R_\alpha, \tag{83}$$

$$\delta V_{ind} / \delta R_\alpha = \int dr' \, \bar{v}(r, r') \delta \rho(r') / \delta R_\alpha. \tag{84}$$

The kernel $\bar{v}(r,r')$ which converts a charge density perturbation into a potential perturbation has a Hartree term and an exchange-correlation term

$$\bar{v}(r,r') = \frac{e}{|r-r'|} + \frac{e\delta^2 E_{xc}}{\delta\rho(r)\delta\rho(r')V_a}, \qquad (85)$$

where $E_{xc}[\rho]$, the exchange and correlation energy per unit volume of the inhomogeneous electron gas, is often replaced by $E_{xc}^h(\rho)$, the corresponding quantity for the homogeneous electron gas (Kohn and Sham, 1965). The change in bare potential δV_b is rigorously obtained by displacing the nuclear potential [Eq. (74)], but it is often assumed that the core charge density moves rigidly with the nucleus so that V_b can be replaced by a "pseudopotential" which at large distances is the *ionic* potential $Z_i e/|r|$. In this case, $\delta\rho$ in Eq. (84) would be replaced by the valence charge density.

The calculation of the electron–phonon matrix elements is thus seen to hinge on the calculation of $\delta\rho/\delta R_\alpha$ and the difficulties involved in calculating this quantity have been mentioned already. The calculation can be performed formally by introducing the "bare" and "interacting" susceptibilities χ_0 and χ:

$$\delta\rho(r) = \int d^3r \chi_0(r,r')\delta V(r'), \qquad (86)$$

$$\delta\rho(r) = \int d^3r \chi(r,r')\delta V_b(r'). \qquad (87)$$

An expression for $\chi_0(r,r')$ has been given already Eq. (77) and it is easy to see that $\chi(r,r')$ is related to $\chi_0(r,r')$ through

$$\chi = \left(\chi_0^{-1} - \bar{v}\right)^{-1}. \qquad (88)$$

Unfortunately the inversion operations are nontrivial for an inhomogeneous system.

One model in which $\delta\rho/\delta R_\alpha$ *can* be easily obtained is that of a weak pseudopotential. In this case, the susceptibility of the homogeneous electron gas can be used and the result is very simple; the *screened* pseudopotential

$$V_{sc}(r) = V_b(r) + \int dr' \chi(r-r') \int dr'' \bar{v}(r'-r'')V_b(r'') \qquad (89)$$

moves rigidly with its bare pseudopotential. The potential V_{sc} also determines the band structure in this model. $V_{bs}(r) = \Sigma_l V_{sc}(r - R_l)$, and we have the simple result that the potential perturbation which enters the electron–phonon matrix element is simply the rigid displacement of one of the potentials whose sum is the total crystal potential.

This result is consistent with the more general result derived in Section I,D [Eq. (39)] that the sum over all sites of $-\delta V/\delta R_l$ is the force on an electron due to the total crystal potential, i.e., it is the gradient of the band-theory potential

$$\sum_l \delta V(r)/\delta R_{l\alpha} = -\nabla_\alpha V(r). \tag{90}$$

Equation (90) is an important sum rule because most of the quantities of interest can be written in terms of sums of the form $\sum_{ll'} (\delta V/\delta R_l) (\delta V/\delta R_{l'}) F(l, l')$ [cf. Eq. (43)]. The sum rule will be satisfied by any rigid potential approximation in which $V(r)$ is written as a sum of "atomic" potentials which are assumed to move rigidly with the nucleus, and this must be one reason why simple approximations such as the rigid muffin-tin approximation (see Section II, B) appear to work so well. Unfortunately there is more than one way of writing the total potential as a sum of atomic contributions so that the sum rule does not uniquely determine the electron–phonon matrix element.

B. Electron–Phonon Matrix Elements for Simple Metals

The term "simple metal" refers to those elements in which the conduction electrons are derived from loosely bound "s and p" orbitals and have a band structure which roughly approximates a free-electron parabola. The transition metals whose conduction electrons are derived from tightly bound "d" orbitals are certainly excluded from this group and I personally would also exclude the alkaline earths heavier than Mg and the noble metals because their band structures are strongly influenced by d orbitals.

If the electron–ion system can be treated as a weakly perturbed homogeneous electron gas, the many-body analysis is greatly simplified and can rigorously be pushed quite far (Heine *et al.*, 1966; Prange and Sachs, 1967). Unfortunately, this situation never *really* occurs in real metals although it may be a useful approximation especially for Na, K, and Al.

In order to describe the interelement variation of the electron–phonon coupling, it is helpful to focus on the single parameter λ as a measure of the overall strength of the electron–phonon coupling for a given material. McMillan (1968) showed that λ could be factorized into electronic and primarily phonon dependent factors by writing

$$\lambda = N(0)\langle I^2 \rangle / M \langle \omega^2 \rangle, \tag{91}$$

where $N \langle I^2 \rangle$ is independent of the phonon frequencies and is defined by

$$N(0)\langle I^2 \rangle = 2M \int d\omega \, \alpha^2(\omega) F(\omega) \omega. \tag{92}$$

$\langle I^2 \rangle$ may be written with the help of Eqs. (69), (65), and (48) as

$$\langle I^2 \rangle = \int (dS_k/\hbar v_k) \int (dS_{k'}/\hbar v_{k'}) \sum_\alpha |I_{k'k}^\alpha|^2 \Big/ \Big[\int (dS_k/\hbar v_k) \Big]^2. \quad (93)$$

The denominator of Eq. (91) in principle depends upon both the electronic structure and the phonons since it is defined by Eqs. (91), (92), and (38) as

$$\langle \omega^2 \rangle = \int d\omega\, \alpha^2 F(\omega)\omega \Big/ \int d\omega\, \alpha^2 F(\omega)\omega^{-1}. \quad (94)$$

In practice, however, it is found that $\alpha^2 F(\omega)$ is usually similar in shape to $F(\omega)$ (see Fig. 3), and that $\langle \omega^2 \rangle$ can be estimated by "canceling" the factors α^2 in the numerator and the denominator of Eq. (94).

The Fermi surface average of the electronic part of the electron–phonon matrix element $\langle I^2 \rangle$ takes a very simple form in the weak scattering approximation because the wave functions which enter Eq. (46) may be approximated by plane waves so that

$$I_{k'k}^\alpha = iq_\alpha V_{sc}(q), \quad (95)$$

where $q = k' - k$ and $V_{sc}(q)$ is obtained from Eqs. (88) and (89),

$$V_{sc}(q) = [1 - \bar{v}(q)\chi_0(q)]^{-1} V_b(q). \quad (96)$$

This potential may also be used to calculate the band structure. Substitution of Eq. (95) into Eq. (93) yields

$$\langle I^2 \rangle = (1/2k_F^2) \int_0^{2k_F} q^3 V_{sc}^2(q)\,dq. \quad (97)$$

A similar treatment of λ_{tr} [Eq. (23)] yields

$$\lambda_{tr} = N(0)\langle I^2 \rangle_{tr}/M\langle \omega^2 \rangle_{tr} \quad (98)$$

with

$$\langle I^2 \rangle_{tr} = (1/4k_F^4) \int_0^{2k_F} q^5 V_{sc}^2(q)\,dq, \quad (99)$$

and $\langle \omega^2 \rangle_{tr}$ defined similarly to $\langle \omega^2 \rangle$ except that α^2 is replaced by α_{tr}^2.

Table I gives a summary of the variation of λ and its component factors $N(0)\langle I^2 \rangle$ and $M\langle \omega^2 \rangle$. The calculated values of λ are taken from Grimvall's (1976) review and in most cases represent the average of several independent calculations. The empirical values (λ_{emp}) were also taken from Grimvall's compilation except for the alkalis for which the values given are estimates of λ_{tr} obtained from the experimental resistivity assuming an isotropic Fermi surface. Most of the λ_{emp} values are similar to those of McMillan (1968) who obtained them from Eq. (61) using $\theta_D/1.45$ for the preexponential factor and 0.13 for μ^*. Values of the Fermi energy density of

TABLE I

EMPIRICAL QUANTITIES RELATED TO SIMPLE METAL ELECTRON–PHONON INTERACTIONS[a]

	r_s	$N(0)$	$N(0)/N_f$	$\langle I^2 \rangle$	$M\langle \omega^2 \rangle$	$\langle I^2 \rangle/M\langle \omega^2 \rangle$	λ_{emp}	λ_{calc}	T_c
Li	3.24	3.33	1.56	0.00223	0.0181	0.12	(0.31)	0.41	
Na	3.94	3.43	1.09	0.00052	0.0110	0.047	(0.12)	0.16	
K	4.88	5.30	1.09	0.00016	0.0066	0.024	(0.08)	0.13	
Rb	5.22	5.99	1.08	0.00015	0.0058	0.026	(0.12)	0.16	
Cs	5.63	7.94	1.23	0.00009	0.0046	0.020	(0.14)	0.16	
Be	1.87	0.40	0.28	0.073	0.12	0.61	0.24	0.26	0.026
Mg	2.64	2.72	0.96	0.0052	0.040	0.13		0.35	
Zn	2.29	1.32	0.62	0.018	0.056	0.32	0.43	0.42	0.85
Cd	2.57	1.40	0.52	(0.014)	(0.048)	0.30	0.42	0.32	0.52
Hg	2.65	2.04	0.72	0.014	0.0178	0.79	1.60	0.93	4.16
Al	2.07	2.80	1.07	0.011	0.079	0.14	0.39	0.49	1.16
Ga	2.18	1.22	0.42	(0.027)	(0.077)	0.35	0.42	0.24	1.08
In	2.39	2.70	0.77	0.010	0.033	0.30	0.81	0.94	3.40
Tl	2.47	2.36	0.63	0.0104	0.031	0.34	0.80	0.91	2.38
Sn	2.21	2.98	0.75	0.0152	0.063	0.24	0.72	0.80	3.72
Pb	2.29	3.55	0.83	0.0135	0.031	0.44	1.55	1.47	7.19

[a]Units: r_s (Bohr-radii), $N(0)$ (states/Ry-spin), $\langle I^2 \rangle$ [Ry^2/(Bohr-radii)^2], $M\langle \omega^2 \rangle$ [Ry/(Bohr-radii)^2], T_c (K).

states were obtained from the low-temperature specific heat $C_V = (1 + \lambda)N(0)\frac{2}{3}\pi^2 k_B^2 T$ (Hultgren *et al.*, 1973) and seem to be determined reasonably well. The ratio of $N(0)$ to the free-electron Fermi energy density of states N_f gives an idea of the magnitude of the band structure effects. The values of $\langle \omega^2 \rangle$ should be treated with some caution. The best data are for Pb, Sn, Tl, In, and Hg for which tunneling experiments have been performed which yield $\alpha^2(\omega)F(\omega)$. Reasonably good estimates can also be made for the alkalis and Al using an analysis of the moments of their phonon spectra (Martin, 1965; Gilat and Nicklow, 1965). The remaining values of $\langle \omega^2 \rangle$ were estimated from neutron data [Be, Mg, and Zn: (Young and Koppel, 1964)] or from specific heat data [CD: (Craig *et al.*, 1953), Ga: (Adams *et al.*, 1952)]. The estimates for Cd and Ga are quite uncertain. The Debye temperature obtained from the T^3 term in the low-temperature specific heat or elastic constants is *not* a reliable indicator of $\langle \omega^2 \rangle$ because it is only sensitive to the lowest-frequency phonons.

The quantities $\langle I^2 \rangle/M\langle \omega^2 \rangle = \lambda/N(0)$ were obtained by dividing the empirical values of λ by the Fermi energy densities of states (except for the alkalis where calculated values of λ were used). The entries labeled $\langle I^2 \rangle$

were obtained by multiplying $\langle I^2 \rangle / M \langle \omega^2 \rangle$ by $M \langle \omega^2 \rangle$. The column labeled r_s gives a measure of the mean conduction-electron density. A sphere of radius r_s would contain one conduction electron if the conduction electrons formed a homogeneous electron gas.

Hopefully, the uncertainties in Table I have not obscured the important trends. Considering the T_c column, we see that only Be of the simple metals in groups 1a and 2a is a superconductor whereas all of the metals in groups 2b, 3b, and 4b are superconducting with the highest T_c's and λ's occurring among the heavier elements, e.g., Hg and Pb. The variation in λ arises from the variations of its three component factors $N(0)$, $\langle I^2 \rangle$, and $M \langle \omega^2 \rangle$.

The Fermi energy density of states shows a simple variation only for the alkali metals where it increases approximately as r_s^2 as one proceeds down column 1a. This increase is due to the flattening of the conduction band as the electronic kinetic energy is allowed to decrease by the expansion of the lattice in the heavier alkalis. Most of the polyvalent simple metals have Fermi energy state densities which deviate substantially from that expected of a homogeneous electron gas. These deviations are caused by stronger pseudopotentials than in the alkalis. Be, for example, appears to have been made almost into a semiconductor by the electron–ion interaction.

The variation in $\langle I^2 \rangle$, the Fermi surface average of the electronic part of the electron–phonon matrix elements, is complex since it depends on the details of the pseudopotential, and especially on its value near $q = 2k_F$. McMillan (1968) pointed out that $\langle I^2 \rangle$ may be written in terms of a dimensionless average of the pseudopotential $\langle V^2 \rangle$ by taking advantage of the fact that $V_{sc}^2(0) = -\tfrac{2}{3} E_F$ in the weak pseudopotential approximation

$$\langle I^2 \rangle = 2k_F^2 \left(\tfrac{2}{3} E_F\right)^2 \langle V^2 \rangle = \left(12.06/r_s^4\right) \langle V^2 \rangle \tag{100}$$

(in atomic units), where $\langle V^2 \rangle$ is given by

$$\langle V^2 \rangle = \int_0^{2k_F} q^3 V_{sc}^2(q)\, dq \Big/ \int_0^{2k_F} q^3 V_{sc}^2(0)\, dq. \tag{101}$$

Empirically, we find (Fig. 9) that $\langle I^2 \rangle \propto r_s^{-6.25}$ with substantial scatter especially for the elements with $Z \geq 3$.

The average effective force constant $M \langle \omega^2 \rangle$ also correlates with the electron density as shown in Fig. 10. A proportionality of $M \langle \omega^2 \rangle$ to r_s^{-4} might be expected from the "jellium model" (Allen and Cohen, 1969). Empirically, we find a variation closer to r_s^{-3}, but with several anomalously low values especially Hg and Pb.

Combining these two variations, we would predict the highest values of $\langle I^2 \rangle / M \langle \omega^2 \rangle$ (a quantity sometimes denoted V_{ph} in BCS theory) to be the largest for small r_s. This rule may be seen from Table I to be reasonably well obeyed although there are numerous exceptions. The high values of λ for Pb

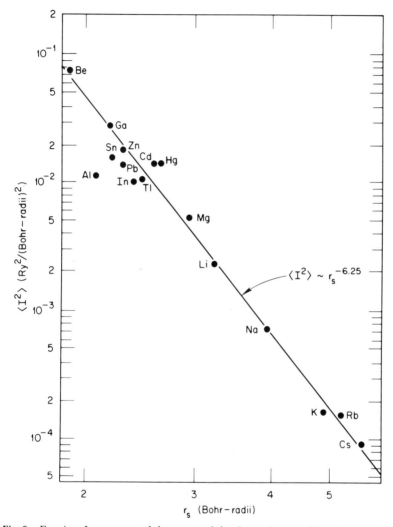

Fig. 9. Fermi surface average of the square of the electronic part of the electron–phonon matrix element for the simple metals.

and Hg result from relatively low values of r_s, anomalously low phonon frequencies, and especially for Pb, a high Fermi energy density of states. The anomalously low value of $\langle I^2 \rangle$ for Al is probably due to $V_{sc}(q)$ being very small in the vicinity of $q = 2k_F$. The second highest value of $\langle I^2 \rangle / M \langle \omega^2 \rangle$ is found for Be whose λ is low mainly because of a low Fermi energy density of states. It is significant in this context that Be fabricated in the form of thin films has been reported to have a T_c of 9.6 K (Glover *et al.*, 1971).

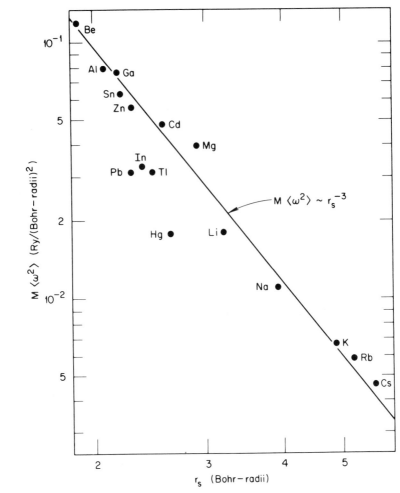

Fig. 10. Effective force constant $M\langle\omega^2\rangle$ for the simple metals versus r_s.

The tendency of $\langle I^2\rangle/M\langle\omega^2\rangle$ to increase as r_s decreases brings to mind the question of metallic hydrogen—whether or not it exists and whether or not it is a high T_c superconductor if it does exist. There has been much speculation and a number of calculations on this material (Ashcroft, 1968; Schneider, 1969; Caron, 1974; Switendick, 1976; Gupta and Sinha, 1976; Papaconstantopoulos and Klein, 1977) with the calculated value of T_c varying between 0.08 and 260 K! The high T_c values were obtained from values of $\langle I^2\rangle$ which lie substantially above the empirical trend of Fig. 9. This does not, of course, mean that they are necessarily incorrect. The calculations of $\langle I^2\rangle$ by Switendick (1976) and by Papaconstantopoulos and

Klein (1977) are especially interesting because they give high values of $\langle I^2 \rangle$ and are based on the "rigid muffin-tin approximation" (to be described in the next section) which usually *underestimates* $\langle I^2 \rangle$ for the simple metals. Both the pseudopotential calculations and the rigid muffin-tin calculations of electron–phonon coupling in metallic H have been severely criticized by Gupta and Sinha (1976).

In recent years the pseudopotential treatments of simple metals have become more sophisticated. Tomlinson and Swihart (1979) and Tomlinson (1979), for example, have calculated the anisotropic mass enhancement and the transport properties of Zn using realistic models for both the phonons and the electronic structure. However, a central problem continues to be the obtaining or devising of an adequate pseudopotential. Usually, the pseudopotential is not obtained by electron gas screening of an ionic potential but is inferred from an experimental quantity such as the phonon dispersion curves or the Fermi surface and even then there are problems of nonuniqueness.

One promising approach is that of Taylor, Rasolt, and collaborators (Rasolt and Taylor, 1975; Dagens *et al.*, 1975; Shukla and Taylor, 1976; Taylor *et al.*, 1976). These authors use a screened ionic pseudopotential but they adjust the parameters of their bare ionic pseudopotential so that it gives the correct charge density in linear response. They have calculated accurate phonon spectra for Na, K, and Al and excellent temperature-dependent resistivities for Na and K with this technique. It would be interesting to calculate both the resistivity and λ for Li using their techniques since most previous pseudopotential calculations appear to have seriously overestimated the strength of its electron–phonon interaction.

C. Electron–Phonon Matrix Elements for Transition Metals

1. TRANSITION METALS

The transition metals form a group of 24 elements which are characterized by partially filled energy bands derived from atomic d orbitals. In the transition metals of the fourth period (Sc through Ni) the 3d and 4s atomic levels have very similar energies with the 3d level decreasing in energy with respect to the 4s level as the atomic number increases. In the fifth and sixth periods similar behavior is observed with the approximately degenerate levels being the 4d and the 5s in the fifth period and the 5d and 6s in the sixth period.

Because the radial wave functions of the d levels have fewer nodes than those of the s levels with which they are degenerate they tend to be concentrated closer to the nucleus. The band states in the solid which are derived from these atomic d levels are not at all like those of free electrons, but it would be a mistake to imagine that they are tightly bound by the individual potential wells in the solid. In fact, the d bands in a transition metal lie above the maximum of the crystal potential.

2. TIGHT-BINDING SCHEMES

There are two different pictures of the d bands in the transition metals which have led to two different views of the electron–phonon interaction. In the simplest version of the first picture which I will call the *tight-binding picture*, one imagines that the solid is built up of atomic potentials V_a which are rigidly brought together to make the crystal potential V_c:

$$V_c = \sum_l V_a(\mathbf{r} - \mathbf{R}_l). \tag{102}$$

Separately, each potential has a fivefold degenerate bound d level with wave function $\phi_\mu(\mathbf{r})$, and it is supposed that the crystal wave function can be constructed as a linear combination of these orbitals

$$\psi_{kn}(\mathbf{r}) = \sum_{l\mu} e^{i\mathbf{k}\cdot\mathbf{R}_l} A_{n\mu}\phi_\mu(\mathbf{r} - \mathbf{R}_l). \tag{103}$$

The energy bands are obtained in principle by minimizing the expectation value of the Hamiltonian

$$E_{kn} = \left\langle \psi_{kn} \middle| T + \sum_l V_a(\mathbf{r} - \mathbf{R}_l) \middle| \psi_{kn} \right\rangle, \tag{104}$$

with respect to variations of the coefficients $A_{n\mu}$.

In the tight-binding picture, it is quite natural when calculating electron–phonon matrix elements to assume that the change in crystal potential due to an atomic displacement is given by the rigid displacement of an atomic potential or perhaps an atomic potential with its long-range tail clipped off or screened out. In fact, very few actual calculations of $\langle \psi_k | \nabla V_a | \psi_{k'} \rangle$ using a tight-binding representation have been reported. Birnboim and Gutfreund (1974) claimed that this procedure gave matrix elements that were much too large and advocated use of the *modified tight-binding* approximation. This latter version of the tight-binding picture was suggested by Fröhlich (1966) and expounded by Mitra (1969) and Barišić (1972a, b). It was their view that one should not calculate the electron–phonon matrix element using the wave functions ψ_k of the unperturbed crystal but one should instead use wave functions made up of a linear combination of *atomic orbitals which move with their atom as it is displaced*. In Fröhlich's *modified tight-binding* picture one no longer calculates the *matrix elements of the potential gradient*,

$$I_{l\mu,l'\mu'}^{TB} = \sum_{l''} \int \phi_\mu(\mathbf{r} - \mathbf{R}_l) \nabla V_a(\mathbf{r} - \mathbf{R}_{l''}) \phi_\mu(\mathbf{r} - \mathbf{R}_{l'}) d^3r, \qquad (105)$$

as in the original Bloch theory, but rather one calculates the *gradient of the tight-binding matrix elements* with respect to interatomic separation

$$I_{l\mu,l'\mu'}^{MTB} = \nabla_{R_l - R_{l'}} \sum_{l''} \int \phi_\mu(\mathbf{r} - \mathbf{R}_l) V_a(\mathbf{r} - \mathbf{R}_{l''}) \phi_{\mu'}(\mathbf{r} - \mathbf{R}_{l'}) d^3r. \qquad (106)$$

This approach has led to a further elaboration of the tight-binding picture in which neither potentials nor wave functions appear. Since one only needs the tight-binding matrix elements and their variation with interatomic separation the trend in recent years has been to obtain these quantities by fitting first-principles APW or KKR energy band calculations with a

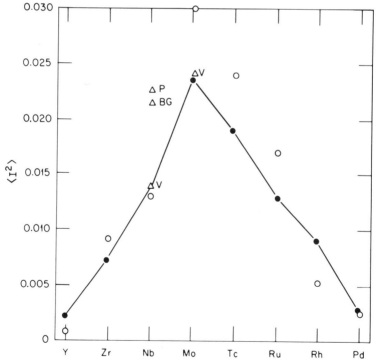

Fig. 11.   Empirical and calculated values of $\langle I^2 \rangle$ for the 4d transition metals. Open circles are derived from experimental data, triangles are results of modified tight-binding calculations (P: Peter *et al.*, 1977; BG: Birnboim and Gutfreund, 1975; V: Varma *et al.*, 1979) and connected closed circles are results of rigid muffin-tin calculations.

TABLE II

TIGHT-BINDING CALCULATIONS OF $\langle I^2 \rangle$

Material	Technique[a]	Reference	Comments[b]
18 Transition metals	MTB	Barišić et al. (1970)	a
14 Transition metals	TB	Bennemann and Garland (1972)	a
V, Nb	TB, MTB	Gutfreund and Birnboim (1973)	
V, Nb, Ta	MTB	Birnboim and Gutfreund (1974)	
V, Nb, Ta	FMTB	Birnboim and Gutfreund (1975)	
4d bcc alloys	FMTB	Birnboim (1976)	
V, Zr, Nb, Mo	MTB	Poon (1976)	a
Ta, W, Re, Ir			
Nb	FMTB	Peter et al. (1977)	b
4d bcc alloys	FMTB	Varma et al. (1979)	c

[a] MTB = modified tight binding, TB = (Bloch) tight binding, and FMTB = fitted modified tight binding.

[b] (a) Semiempirical calculation, (b) solved Eliashberg equation for T_c and (c) used nonorthogonal basis set.

parametrized tight-binding Hamiltonian. Different variations on the *fitted modified tight-binding* scheme have been used by Peter et al. (1974), Birnboim and Gutfreund (1975), Peter and Adam (1976), Peter et al. (1977), and Varma et al. (1979). Some of the results of the fitted modified tight-binding calculations for $\langle I^2 \rangle$ are shown in Fig. 11 together with results from the rigid muffin-tin approximation. In addition, a catalog of the tight-binding calculations known to me is given in Table II.

Ironically, it has recently been shown (Ashkenazi et al.; 1979, Varma et al., 1979) that the modified tight-binding scheme is entirely equivalent to the standard Bloch scheme of calculating the matrix element of the potential gradient. If carried to convergence with the same assumption about the change in crystal potential due to moving an atom, both schemes should give the same result. Unfortunately, the significance of this proof of equivalence is somewhat diluted by the fact that direct tight-binding calculations are seldom (if ever) carried to basis convergence and in the popular fitted tight-binding schemes one does not know what is being assumed about the crystal potential or the wave functions.

The recent calculations of Varma et al. (1979) deserve special comment since they used a fitting scheme which allowed for wave-function overlap and used an APW band calculation at reduced lattice spacing to help them deduce the gradients of the matrix elements with respect to interatomic spacing. Their calculations agree rather well with experiment and with the rigid muffin-tin calculations to be described below.

3. Rigid Muffin Tins

There is a second picture of electrons in transition metals which has led to a somewhat different view of the electron–phonon matrix elements. The electronic structure of transition metals is most readily calculated by using either APW or KKR techniques[†] which employ a "muffin-tin" approximation for the crystal potential. The muffin-tin approximation consists in spherically averaging the potential within nonoverlapping spheres surrounding each nucleus and volume averaging the potential in the remainder of the crystal. The constant interstitial potential can be taken to be zero[‡] without further approximation and the resultant potential can be written as

$$V_c(\mathbf{r}) = \sum_l V_{mt}(\mathbf{r} - \mathbf{R}_l), \tag{107}$$

where $V_{mt}(\mathbf{r})$ is a spherical potential which does not overlap with potentials on neighboring sites as atomic or screened ionic potentials would.

When calculating electron–phonon matrix elements in this picture, it is natural to assume that it is the muffin-tin potential which moves rigidly when an atom is displaced. Gaspari and Gyorffy (1972) popularized this scheme by deriving a very simple expression for $\langle I^2 \rangle$ in this "rigid muffin-tin" approximation,

$$\langle I^2 \rangle = \frac{2mE_F}{\pi^2 \hbar^2 N(0)^2} \sum_l 2(l + 1)\sin^2(\delta_l - \delta_{l+1})N_l^B N_{l+1}^B / N_l^1 N_{l+1}^1, \tag{108}$$

where δ_l is the phase shift of an electron at the Fermi energy for scattering off a muffin-tin potential. The quantities N(0) N_l^B and N_l^1 are all different kinds of Fermi energy densities of states. $N(0)$ is the usual Fermi energy density of states per atom spin, N_l^B is the component of $N(0)$ within the muffin-tin which can be attributed to components of the wave function with orbital angular momentum l, and N_l^1 is the same as N_l^B except that it is calculated for a single muffin-tin in a zero potential background rather than for the cyrstal. The quantities δ_l and N_l^1 can be obtained by solving the Schrödinger equation for a single potential and $N(0)$ and N_l^B can be generated without too much difficulty from the results of an APW or KKR energy band calculation.

The original derivation of Eq. (108) assumed a spherical electronic structure but it was shown by John (1973) that the first two terms are exact

---

[†] See Chapter 1 for a discussion of these techniques.

[‡] This is an arbitrary reference zero approximately equal to the maximum of the crystal potential. It should not be confused with the vacuum or absolute zero.

for the cubic elements. Later it was discovered that the largest contributions came from the third term (d $\leftrightarrow$ f transitions) but that the nonspherical corrections coming from the f states are small (Butler *et al.*, 1976; John *et al.*, 1978).

Table III lists some rigid muffin-tin calculations of $\langle I^2 \rangle$. Some of the earlier calculations were performed before it was realized that N_3^B/N_3^1 is a critical quantity which is substantially greater than unity for the transition

TABLE III

RIGID MUFFIN-TIN CALCULATIONS OF $N\langle I^2 \rangle$

Material	Reference	Comments[a]
Fe, W, Cu	Gaspari and Gyorffy (1972)	a
Cs	Stocks *et al.* (1972)	a
Nb, V	Evans *et al.* (1973a)	a, b
Zr, Nb, Mo, Ta	Evans *et al.* (1973b)	a
W, Re, Rh, Pd, Ir, Pt		
γCe, αCe	Mukhopadhyay and Gyorffy (1973)	a
V, Nb, Fe, Cu	Ratti *et al.* (1974)	a, b
V, Nb, NbC, TaC, HfC	Klein and Papaconstantopoulos (1974)	
W, PT, Pd, Ir, Pb, Al	Foulkes and Gomersall (1975)	a
TiFe	Papaconstantopoulos (1975)	
Pd, Pdh	Papaconstantopoulos and Klein (1975)	
Th_4H_{15}, ThH_2	Winter and Ries (1976)	
Metallic H	Switendick (1976)	
Nb	Butler *et al.* (1976)	c
V, Nb, Ta	Boyer *et al.* (1977a)	
V, Nb, Ta	Boyer *et al.* (1977b)	
V_3Si	Klein *et al.* (1977a)	
PdH_x, PdD_x	Klein *et al.* (1977b)	
Metallic H	Papaconstantopoulos and Klein (1977)	
32 metals with Z $\leq$ 49	Papaconstantopoulos *et al.* (1977)	
4d transition metals	Butler (1977)	c
4d transition metals	Pettifor (1977b)	d
PdH_x, PdD_x	Papaconstantopoulos *et al.* (1978a)	
PdH_x, PdD_x	Papaconstantopoulos *et al.* (1978b)	
Al, NbC	Rietschel (1978)	e
Mo_6Se_8, $PbMo_6Se_8$	Andersen *et al.* (1978)	d
Ti, V, Fe, Cu	John *et al.* (1978)	b, c
TiH_2	Kulikov (1978)	
TiH_2, NiH, NiH_2	Kulikov *et al.* (1978)	a
Zrc, YS, NbC, NbN	Winter *et al.* (1978)	
Nb, Ys, Th_4H_{15}	Ries and Winter (1979)	
10 A-15 compounds	Klein *et al.* (1979)	

[a](a) These calculations assumed $N_0^B/N_0^1 = N_1^B/N_1^1 = N_3^B/N_3^1 = 1$ and $N_2^B = N$. (b) Calculations were performed at reduced volume to investigate the effect of pressure on $\langle I^2 \rangle$. (c) These calculations contain "nonspherical corrections" to Eq. (122). (d) These calculations used a "rigid atomic sphere approximation." (e) These calculations contain "nonlocal corrections."

TABLE IV

CALCULATIONS OF λ USING THE RIGID MUFFIN-TIN
(AND RELATED) APPROXIMATIONS

Material	Ref.	λ_{calc}	λ_{emp}
V	[a]	1.2	$0.8^{[g]} \gtrsim 0.8^{[j]}$
Nb	[b]	1.2	$1.2^{[h]}, 1.0^{[j]}$
	[c]	1.1	
	[d]	1.6	
	[a]	1.3	
Ta	[a]	0.9	$0.7^{[i]}, (0.7 - 0.9)^{[j]}$
Mo	[e]	0.40	$0.44^{[k]}$
	[a]	0.4	
W	[a]	0.25	$0.24^{[k]}$
Pd	[e,f]	0.41	$0.38^{[j]}$
	[a]	0.5	
Pt	[a]	0.7	

[a] Glötzel et al. (1980).
[b] Butler et al. (1977).
[c] Butler et al. (1979).
[d] Harmon and Sinha (1977).
[e] Pinski et al. (1978a).
[f] Pinski et al. (1978b).
[g] Tunneling data of Zasadzinski et al., (1980).
[h] Tunneling data of Bostock et al., (1980).
[i] Tunneling data of Shen (1972).
[j] Estimate using resistivity, specific, heat, and critical field data.
[k] Estimates from T_c.

metals. A second note of caution is that although the spherical correction terms appear to be small for the cubic elements, they *may* be appreciable for materials such as the A-15's where the transition metal atom sits in a position of low symmetry. Although the rigid muffin tin is an ad hoc approximation, it *is* straightforward and it has the further advantage that (with some effort) one can use it to calculate a very large number of experimental properties without further approximation (outside of the usual approximations of band theory). Most importantly, the approximation seems empirically to yield the correct strength for the overall electron–phonon coupling for the transition metals, as measured by λ or the electrical resistivity ($\rho(T)$ (see Figs. 4 and 11 and Table IV).

The rigid muffin tin satisfies the "total force" sum rule, Eq. (90), but as was mentioned in Section II,A,1 this sum rule does not uniquely specify $\delta V/\delta R_l$. One limit in which the rigid muffin tin clearly fails is that of a weak ionic potential. In this limit, the proper form for $\delta V/\delta R$ is $-\nabla V_{sc}(r)$

where $V_{sc}(r)$ is the screened ionic potential.[†] This potential is substantially stronger and has substantially longer range than the corresponding muffin-tin potential even though both potentials may generate the same band structure. The two potentials correspond to different ways of breaking up the total crystal potential into contributions from each atom, a problem which has been discussed by Lee and Heine (1972). Thus, the rigid muffin-tin approximation gives λ values which are too small for the simple metals (Papaconstantopoulos et al., 1977). One can, however, fix it up to some extent by adjusting it so that it has the correct volume average (Nowak, 1972, Papaconstantopoulos et al., 1978c; Kahn et al., 1980). This may be done by extending the potential to the Wigner–Seitz radius and adding a negative constant to it which is chosen so that its phase shifts satisfy

$$\frac{4\pi\hbar^2}{V_a 2mk_F} \sum_l (2l + 1)\delta_l(E_F) = -Z/2N(0). \tag{109}$$

Such adjustments do not appear to be needed for the transition metals where the rigid muffin-tin λ's appear to already be slightly higher than experiment in most cases. In any event, Eq. (109) only applies to potentials which weakly perturb a free-electron gas.

Table IV shows calculated values of λ obtained in the rigid muffin-tin approximation and in slightly variant approximations. The calculations of Glötzel et al. are based on an approximate form of KKR theory in which slightly overlapping potentials are used. It is these potentials which they assume to move rigidly when an atom is displaced. The calculation of Harmon and Sinha assumes rigid muffin-tins but omits a contribution to the matrix element coming from the discontinuity at the muffin-tin sphere radius thus violating the sum rule, Eq. (90). Their results for λ seem substantially larger than other rigid muffin-tin values. Overall, the rigid muffin-tin approximation seems to work surprisingly well for the transition metals.

4. Physics of Electron–Phonon Matrix Elements for Transition Metals

According to Fig. 11, $\langle I^2 \rangle$ for the 4d transition metals is strongly peaked in the center of the series. A similar behavior is seen for the 3d's in the rigid muffin-tin calculations of Papaconstantopoulos et al. (1977) and in empirical estimates of $\langle I^2 \rangle$ for the 5d's, although the peak for the 5d's is probably shifted to higher e/a. This view of the e/a behavior of $\langle I^2 \rangle$ is also

[†]In practice $V_{sc}(r)$ must be a screened ionic *pseudopotential*. It is necessary to remove the core for linear screening to apply.

supported by the measurements of T_c for amorphous transition metals (Colver and Hammond, 1973) which show a single peak near the center of the series.

It is not surprising that the electron–phonon interaction should be strong near the center of the transition-metal series. In the modified tight-binding picture the electron–phonon matrix elements have the form $I_{\mu\mu'} = \nabla_R V_{\mu\mu'}'(R)$ where $V_{\mu\mu'}(R) = \langle \phi_\mu(r) | V_a(r) | \phi_{\mu'}(r - R) \rangle$. Assuming the atomic orbitals to decay exponentially as $\phi_\mu(r) \sim \exp(-q_0 r)$ for large r we have $I_{\mu\mu'} \sim -q_0 V_{\mu\mu'}$. Since the matrix elements $V_{\mu\mu'}$ determine the d-band width W_d we have

$$\langle I^2 \rangle \sim q_0^2 W_d^2. \tag{110}$$

The d-band width for the 4d transition metals *is* greatest for Mo (see Fig. 12) due to its small atomic volume and the extended nature of its wave functions (Pettifor, 1977a).

The variation of $\langle I^2 \rangle$ implied by Eq. (110) is very similar to that of simple formulae derived previously. Barišić *et al.* (1970) proposed that

$$\langle I^2 \rangle \sim q_0^2 E_c / N(0), \tag{111}$$

where E_c is the cohesive energy and $N(0)$ is the Fermi energy density of states. *If* there were no s band and only one d band then $N(0)$ would be proportional to W_d^{-1}, and since (for the 4d metals) $E_c = (0.74 \pm 0.07)W_d$, Eqs. (111) and (110) would be equivalent. In a real transition metal the d bands interact with each other and hybridize with the s band to form the highly structured densities of states typical of d-band metals (see, e.g., Chapter 1, this volume; see also Moruzzi *et al.* (1978)).

Varma *et al.* (1979) proposed as a simplified version of their theory that

$$\langle I^2 \rangle \sim \alpha \langle v^2 \rangle, \tag{112}$$

where $\langle v^2 \rangle$ is the mean square Fermi velocity and α is a "bond order parameter" which is less than unity for the bonding orbitals (bottom of the d band) and greater than unity for the antibonding orbitals. Neglecting the variation of α, Eq. (112) is approximately the same as Eq. (111) because $N(0)\langle v^2 \rangle \sim W_d$, the fine structure in $N(0)$ being quite effectively cancelled by the fine structure in $\langle v^2 \rangle$ (Allen, 1976; Klein *et al.*, 1980). Contrary to a notion which originated with McMillan (1968), I do not believe that there is either persuasive theoretical or empirical evidence that $\langle I^2 \rangle$ varies as $N(0)^{-1}$ except in so far as the strong d–d bonding which increases the electron–phonon matrix element also increases the d-band width.

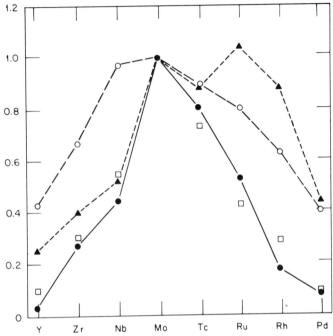

Fig. 12. Variation of some electron–phonon parameters with e/a for the 4d transition metals. Open circles are W_d^2, the square of the d-band width, triangles are $M\langle\omega^2\rangle$, the effective force constant, squares are f_3/V^2, closed circles are "experimental" values of $\langle I^2\rangle$. All quantities are normalized to be unity for Mo.

A reasonably accurate description of the e/a variation of $\langle I^2\rangle$ can be obtained from a simplified version of the Gaspari–Gyorffy formula (Butler, 1977)

$$\langle I^2\rangle \sim f_3/V_a^2, \tag{113}$$

where V_a is the atomic volume and f_3 is $N_3^B/N(0)$, the fraction of the Fermi energy density of states that comes from wave-function components which have $f(l = 3)$ symmetry when expanded about the center of the Wigner–Seitz cell. A similar formula has been obtained by Pettifor (1977a). Equation (113) assumes among other things that d ↔ f scattering predominates in the transition metals. This characteristic of the rigid muffin-tin results may seem strange until it is realized that the f states in Eq. (113) have nothing to do with atomic f orbitals but arise because the entire wave function is expanded about the center of the Wigner–Seitz cell. In the tight-binding approximation, the wave function at a given point will have contributions from atomic orbitals centered on many sites. The f-type states of KKR theory are concentrated near the boundary of the Wigner–Seitz cell and correspond to the tails of the d orbitals of tight-binding theory which "hang over" from adjacent cells. The physical mechanism for the increase in $\langle I^2 \rangle$ near the center of the transition-metal series is the same in both pictures— strong d–d intersite bonding.

As one fills the d-band complex by adding electrons the lattice parameter initially decreases rapidly due to the increased bonding. The smaller distance between neighbors causes an increased overlap of the d orbitals at the Fermi energy and thus a higher value of $\langle I^2 \rangle$. After the band is approximately one-half full, the lattice parameter no longer decreases so rapidly because the higher-energy orbitals are antibonding in character, i.e., they tend to vanish midway between neighbors and to be concentrated near the nuclei. These orbitals have weaker overlaps with neighboring sites and therefore generate a smaller $\langle I^2 \rangle$. An additional factor responsible for the increased localization of the Fermi energy d orbitals near the end of the series is the dropping of the d levels relative to the *s*.

The variation of $\langle I^2 \rangle$ with e/a as calculated from Eqs. (110) and (113) and as determined empirically as shown in Fig. 12. Also shown is $M\langle \omega^2 \rangle$. We can see that the extra variation of Eq. (113) as compared to that of Eq. (110) is quite important because $M\langle \omega^2 \rangle$ also peaks near the center of the series. The variation in $\langle I^2 \rangle$, however, overcomes the variation in $M\langle \omega^2 \rangle$ so that $\langle I^2 \rangle / M\langle \omega^2 \rangle$ is largest near the center of the transition-metal series. The e/a rule of Matthias (1957) which states that systems with e/a ratios of 4.5 to 5.0 and 6.5 to 7.0 have high T_c's is easily explained. These are the e/a ratios nearest the center of the series for which the Fermi energy density of states is high (see Hulm and Blaugher, 1972).

D. The Electron-Phonon Interaction in Transition-Metal Compounds

1. TRANSITION-METAL PHASES WITH HIGH T_c'S

Certain classes of transition-metal compounds exhibit higher transition temperatures than can be found among the elements or the solid solution alloys (see Table V). The two most important classes in terms of number of compounds with high T_c's are the B1 (NaCl) phase carbides and nitrides (Fig. 13a) and the A-15 phase compounds with nominal chemical composition A_3B where A is a transition metal and B is a simple metal (Fig. 13b). The most significant structural aspects of the two phases are the transition-metal octahedra which surround each of the nonmetal atoms in the B1 phase and the orthogonal, nonintersecting chains of transition-metal atoms in the A-15 phase.

Because of their high T_c's, some of these materials have been subjected to intense experimental and theoretical scrutiny. A number of anomalous properties have been discovered but it is not yet clear what the relationships are (if any) between the anomalies and the high T_c's. The high T_c B1 phase compounds (e.g., NbC or TaC) have strange wiggles in their phonon dispersion curves especially when compared with their low T_c counterparts (e.g., ZrC or HfC) (Smith and Gläser, 1970). It appears that these anomalies are indicative of strong electron–phonon coupling and that their magnitude and position in the Brillouin zone can be explained in terms of the electronic structure near the Fermi energy (Weber, 1980).

The high T_c A-15's have more anomalous properties than the B1's. Most of these anomalies have been reviewed by Weger and Goldberg (1973) and Izyumov and Kurmaev (1974). The magnetic susceptibilities of V_3Si (Clogston et al., 1964) and Nb_3Sn (Rehwald et al., 1972) have an anomalous temperature dependence decreasing by about 30% between 40 and 300 K. These same compounds also suffer lattice instabilities slightly above T_c (Batterman and Barrett, 1964; Mailfert et al., 1969). The lattice instability is characterized by a tetragonal distortion of the unit cell and a dimerization of the transition-metal chains (Shirane and Axe, 1971).

The resistivity of many A-15's has an anomalous temperature dependence (Lutz et al., 1976; Fisk and Webb, 1976; Williamson and Milewits, 1976; Bader and Fradin, 1976). Typically, the resistivity of a high T_c A-15 varies as $\rho_0 + AT^2$ at low temperatures (rather than as $\rho_0 + AT^5$ as predicted in Section I,B) and at high temperatures it tends to "saturate" at a more or less constant value of order 100 $\mu\Omega$ cm (rather than increase linearly). See Chapter 5 for a more extended discussion.)

It is possible to explain these anomalies by invoking models for the electronic density of states which have extremely sharp structure near the Fermi energy (e.g., Labbé and Friedel, 1966; Cohen et al., 1967; Labbé, 1967; Barišić, 1971; Gor'kov, 1973; Lee et al., 1977), but the details of most of the models are in conflict with recent band structure calculations (Mattheiss, 1975; Jarlborg and Arbman, 1977; Klein et al., 1978). It appears that a completely satisfactory explanation of the anomalous properties of the A-15's is not yet available.

TABLE V

THE HIGHEST SUPERCONDUCTING TRANSITION TEMPERATURES KNOWN
FOR VARIOUS CRYSTAL STRUCTURES

Crystal Structure	Compound	T_c (K)	Reference
A-15 (Cr_3Si)	Nb_3Ge	23	Gavaler (1973)
B1 (NaCl)	$Nb(N_{0.7}C_{0.3})$	17.8	Matthias (1953)
Pu_2C_3	$(Y_{0.7}Th_{0.3})_2C_{3.1}$	17	Krupka et al. (1969)
"Chevrel"	Mo_6PbS_8	14.7	Muhlratzer et al. (1976)
A3 (hcp)	$Mo_{0.3}Tc_{0.7}$	14.5	Alekseevskii et al. (1975)
A2 (bcc)	$Nb_{0.7}Zr_{0.3}$	11	Matthias (1953)

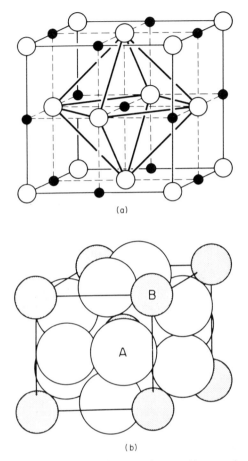

(a)

(b)

Fig. 13. (a) Bl crystal structure emphasizing the transition-metal octahedra. (b) A-15 crystal structure. The A atoms are transition metals which form orthogonal nonintersecting chains while the B atoms form a bcc lattice.

2. FIVE RULES FOR HIGH T_c IN TRANSITION METALS AND THEIR COMPOUNDS

In the remainder of this section, I shall try to rationalize the variation of T_c in the transition metals and their compounds in terms of their (imperfectly known) electronic structures. Although few of the ideas are entirely original (see Matthias, 1971) the reader should be warned that each point is likely to be considered controversial by at least some person working in this field.

TABLE VI

Transition Temperatures and e/a Values for
Transition-Metal B1 Carbo-Nitrides[a]

Compound	e/a	T_c	%NMV
TiC	4.0	< 0.05 K	~ 1
ZrC	4.0	< 0.05	~ 0
HfC	4.0	< 1.23 K	~ 0.5
VC	4.5	0.03	12
NbC	4.5	11.1	2
TaC	4.5	10.35	~ 0
MoC	5.0	14.3	~ 10
WC	5.0	10.0	~ 10
TiN	4.5	5.5	~ 0
ZrN	4.5	10.7	~ 0
HfN	4.5	8.8	~ 0
VN	5.0	8.5	≲ 1
NbN	5.0	17.3	< 5
TaN	5.0	8.7	~ 6

[a] Vacancies on the non-metal lattice site (NMV) have a
large negative effect on T_c. Data are from Roberts (1976,
1978) and references therein.

Table VI lists most of the B1 transition-metal carbo-nitrides and Table
VII lists most of the transition-metal–simple-metal A-15's.[†] There are
certain similarities in the variation of T_c which these classes of compounds
share with each other and with the pure transition metals. I believe that the
following generalizations can be drawn:

(1) *The rule of the half-filled d band.* High T_c's tend to be found in
those transition metal elements and compounds with approximately half-
filled d bands.

(2) *The rule of e/a.* Within a given crystal sturcture T_c correlates with
e/a, the average number of valence electrons per atom (Matthias, 1957).

(3) *The rule of 4d superiority.* For a given crystal structure and e/a the
compounds based on the 4d transition metal elements (e.g., Nb) tend to
have higher T_c's than those based on corresponding 3d (e.g., V) or 5d (e.g.,
Ta) elements.

(4) *The rule of togetherness.* For given crystal structure, e/a, and
period, electron–phonon coupling is enhanced if the transition metal atoms
are closer together.

[†] We omit discussion of those A-15's comprised of two transition metals because less is
known about their electronic structure.

TABLE VII

TRANSITION TEMPERATURES, e/a VALUES, AND LATTICE PARAMETERS FOR A_3B A-15
COMPOUNDS WHERE B IS A SIMPLE METAL

Compound	e/a	T_c (K)	a_0 (Å)	Reference
Ti_3Sb	4.25	5.7	5.22	Junod et al. (1970)
V_3Al	4.5	11.8	4.85	Schmidt et al. (1975)
V_3Ga	4.5	16	4.82	Junod et al. (1971); Blumberg et al. (1960)
V_3Si	4.75	17.1	4.72	Hardy and Hulm (1954)
V_3Ge	4.75	6.	4.78	Luo et al. (1970)
V_3Sn	4.75	3.8	4.96	Cody et al. (1964)
V_3Sb	5.00	0.8	4.94	Matthias et al. (1963)
Cr_3Ga	5.00	< 0.3	4.65	Flukiger et al. (1970)
Cr_3Si	5.50	< 0.015	4.56	Blaugher et al. (1969); Flukiger et al. (1970)
Cr_3Ge	5.50	< 1.2	4.63	Flukiger et al. (1970)
Zr_3Pb	4.00	0.8	5.66	Matthias et al. (1963)
Nb_3Al	4.50	18.8	5.19	Sahm and Pruss (1969)
Nb_3Ga	4.50	20.7	5.17	Flukiger and Jorda (1977); Webb et al. (1971)
Nb_3In	4.50	9.2	5.30	Banus et al. (1962)
Nb_3Si	4.75	19 [a]	5.03	Pan et al. (1975)[a]
Nb_3Ge	4.75	23	5.13	Gavaler (1973); Testardi et al. (1975)
Nb_3Sn	4.75	18	5.29	Matthias et al. (1963)
Nb_3Sb	5.00	0.2	5.26	Matthias et al. (1963); Knapp et al. (1976)
Mo_3Al	5.25	0.58	4.95	Matthias et al. (1963)
Mo_3Ga	5.25	0.76	4.94	Matthias et al. (1963)
Mo_3Si	5.50	1.3	4.89	Flukiger et al. (1974)
Mo_3Ge	5.50	1.45	4.93	Flukiger et al. (1974)
Mo_3Sn	5.50	< 1 K	5.09	Killpatrick (1964)
Ta_3Sn	4.75	6.4	5.28	Cody et al. (1964)
Ta_3Sb	5.00	0.7	5.26	Luo et al. (1970)

[a] Produced by megabar implosion; very little A-15 was present.

(5) *The rule of the perversity of nature.* For a given class of materials those with the highest T_c's tend to be less stable and more difficult to make.

3. THE REASONS FOR THE RULES

I believe that the same principles which were used to explain the variation of T_c in the transition metals (Section II,C) can be used to explain the above rules:

(1) For the transition metals we noted that both $\langle I^2 \rangle$, the Fermi surface averaged electron–phonon matrix element, and $M\langle \omega^2 \rangle$, the effec-

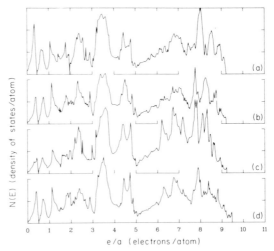

Fig. 14. Density of states of four A-15 compounds versus e/a[(a) Nb_3Sb, (b) Nb_3Al, (c) V_3Si, (d) Nb_3Sn]. These curves are based on the calculations of Klein *et al.* (1978) but the major features (e.g., low density of states at $e/a = 5$ and high density of states at slightly lower values of e/a) are in common with all recent calculations.

tive force constant or lattice stiffness paramenter, were largest when the d band is approximately one-half filled. The variation of $\langle I^2 \rangle$ is not completely cancelled by that of $M\langle\omega^2\rangle$, however, and the highest values of $\langle I^2 \rangle / M\omega^2$ are found near the center of the transition series. The variation of $\langle I^2 \rangle$ and of $M\langle\omega^2\rangle$ arises from variations in the strength of the d–d intersite bonds, and we expect these arguments which are based on calculations for the elements to be valid for transition-metal compounds as well if the states near the Fermi energy are primarily derived from d orbitals.[†] Thus, the compounds with the greatest potential for strong electron–phonon coupling are based on elements like Nb, Mo, and Tc.

(2)   Given that we confine our attention to compounds based on transition-metal elements near the center of the series in order to keep $\langle I^2 \rangle /M\langle\omega^2\rangle$ large, the remaining factor determining λ is $N(0)$ which we want to be as large as possible (Varma and Dynes, 1976). A high $N(0)$ raises λ directly since it enters as a multiplicative factor and indirectly since a high density of states in the vicinity of the Fermi energy tends to lower $M\langle\omega^2\rangle$ through renormalization of the phonon frequencies. The correlation of T_c with e/a is primarily a correlation of T_c with $N(0)$ which for a given crystal structure has a distinctive variation with e/a (see, e.g., Fig. 14).

(3)   The reason for the rule of 4d superiority is somewhat uncertain. $\langle I^2 \rangle$ and $M\langle\omega^2\rangle$ both increase as one descends a column (3d–4d–5d)

[†] This condition is very well satisfied by the A-15's but somewhat less so for the B1 carbo-nitrides.

because of increased d–d bonding. This increase in d–d bonding occurs because each time a node is added to the d-radial wave function its maximum is effectively pushed outward toward the interstitial region thereby increasing the d–d intersite overlap. This increase in d–d bonding also increases the d bandwidth which tends to decrease the Fermi energy density of states. An additional factor influencing T_c is the increase in atomic mass which decreases T_c by decreasing the factor $\omega_{\log}$ in Eq. (61). Somewhat more speculatively, it has been suggested (Butler, 1979; Rietschel and Winter, 1979) that Coulomb effects (e.g., spin fluctuations) may be more widespread and important in transition metals than generally believed. These effects should be more important in the narrow band V compounds and may be partly responsible for their lower T_c's. Thus, there are at least five factors affecting the period to period variation in T_c, and although we can predict the qualitative behavior of each one separately their net effect can only be determined empirically at present.

(4) The rule of togetherness arises from the need for strong d–d overlap to keep $\langle I^2 \rangle$ large. This rule is important in the A-15's where the simple metal atom influences T_c not only through its effect on e/a but also through its influence on the lattice parameter. That is why for a given e/a and A atom the smaller B atoms give the higher T_c's.

(5) Much has been written concerning the possible connection between lattice instabilities and strong electron–phonon coupling especially in the A15's (for reviews, see Testardi, 1975; Izyumov and Kurmaev, 1974, 1976). The instabilities in V_3Si and Nb_3Sn for example are caused by a soft low-frequency phonon mode whose sound velocity vanishes at the lattice

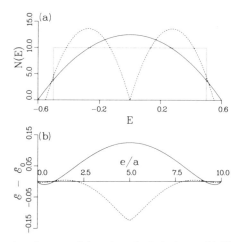

Fig. 15. (a) Density of states of three hypothetical phases. (b) The total energy of these phases as calculated from Eq. (114) versus e/a. The total energy of the phase with the rectangular density of states curve is denoted $\mathscr{E}_0$.

transition temperature. It is generally expected that soft phonons (i.e., anomalously low phonon frequencies) may help to increase the electron–phonon coupling by lowering $M\langle \omega^2 \rangle$ and thereby increasing λ.

I would like to emphasize here, however, a more mundane connection between T_c and phase stability. Rules (1) and (2) above indicate the importance for high T_c d-band superconductivity of having a phase with a half-filled d complex *and* a high Fermi energy density of states. Unfortunately, nature tends to avoid this combination. One reason for this tendency is illustrated in Fig. 15. Figure 15a shows three-model density of states curves chosen to represent three (highly idealized) phases in which a material may crystalize. The equilibrium phase will be the one with lowest energy.[†]

A simple procedure which appears to contain much of the physics of d-band cohesion (Friedel, 1969; Andersen *et al.*, 1977; Pettifor, 1979; Williams *et al.*, 1980) is to equate the total electronic energy $\mathcal{E}$ to the sum of the one-electron energies, i.e.,

$$\mathcal{E} = \int_{E_B}^{E_F} E N(E) \, dE, \tag{114}$$

where E_B is the bottom of the d band. In Fig. 15b $\mathcal{E}$ is plotted as a function of e/a, the number of d electrons per atom. There is clearly a negative correlation betwen stability and a high Fermi energy density of states especially near the center of the d band. These simple calculations are not of course a theory of phase stability for transition-metal compounds. They do, however, illustrate the close connection between the problems of phase stability and high T_c.

4. The Future

At least one experimentalist (Matthias, 1971) enjoys reminding the theorists that they have not been helpful in the search for new materials with higher superconducting transition temperatures. This situation may change during the 1980s because the theorists are at long last listening to the experimentalists and are beginning to understand the nature of the problems which the experimentalists confront and because the theorists are beginning to develop the computational tools necessary to treat the problems of electronic structure in complex lattices and phase stability. I am unable (and unwilling) to predict the maximum T_c in 1990, but I am sure

---

[†]We assume for simplicity that all three phases have the same chemical composition. Often a mixture of two phases of differing composition will minimize the energy.

that our understanding of the electron–phonon interaction in all of its many guises and with all of its interesting ramifications will be much advanced.

Acknowledgments

I am grateful to the editor of this volume, Frank Fradin, for his patience and helpful advice and especially to my colleagues P. B. Allen and F. J. Pinski from whom I have learned much about the electron–phonon interactions.

References

Abraham, J. M., and Deviot, B. (1972). *J. Less-Common Met.* **29**, 311–320.
Abramovitz, M., and Stegun, I. A. (1964). *In* "Handbook of Mathematical Functions," p. 258. Dover, New York.
Adams, G. B., Jr., Johnston, H. L., and Kerr, E. C. (1952). *J. Am. Chem Soc.* **74**, 4784–4787.
Alekseevskii, N. Y., Balakhorskiy, D. A., and Kirrillow, I. V. (1975). *Fiz. Met. Metalloved.* **40**, 50–54. [Eng. transl., *Phys. Met. Metallogr. (USSR)* **40**, 38 (1975).]
Allen, P. B. (1971a). *Phys. Rev.* **B 3**, 305–320.
Allen, P. B. (1971b). *Proc. Int. Conf. Low Temp. Phys., 12th Kyoto, 1970* p. 517.
Allen, P. B. (1972). *Phys. Rev.* **B 6**, 2577–2579.
Allen, P. B. (1976). *Phys. Rev. Lett.* **37**, 1638–1641.
Allen, P. B. (1980a). Unpublished lecture notes.
Allen, P. B. (1980b). *In* "Dynamical Properties of Solids," (G. M. Horton and A. A. Maradadin, eds.), Vol. 3. North Holland, New York and Amsterdam.
Allen, P. B., and Butler, W. H. (1978). *Phys. Today* **31**, 44.
Allen, P. B., and Cohen, M. L. (1969). *Phys. Rev.* **187**, 525–538.
Allen, P. B., and Dynes, R. C. (1975). *Phys. Rev.* **B 12**, 905–922.
Andersen, O. K., Madsen, J., Poulsen, U. K., Jepsen, O., and Kollar, J. (1977). *Physica (Utrecht)* **86b–88b**, 249–256.
Andersen, O. K., Klose, U., and Nohl, H. (1978). *Phys. Rev.* **17**, 1209–1237.
Ashcroft, N. W. (1968). *Phys. Rev. Lett.* **21**, 1748.
Ashkenazi, J., Dacorogna, M., and Peter, M. (1979). *Solid State Commun.* **29**, 181–184.
Bader, S. D., and Fradin, F. Y. (1976). *In* "Superconductivity in d- and f-Band Metals" (D. H. Douglass, ed.), pp. 567–582. Plenum, New York.
Banus, M. D., Reed, T. B., Gatos, H. C., Lavine, M. C., and Kaflas, J. A. (1962). *J. Phys. Chem Solids* **23**, 971–973.
Bardeen, J., Cooper, L., and Schrieffer, J. (1957). *Phys. Rev.* **108**, 1175–1204.
Barišić, S. (1971). *Phys. Lett.* **3**, 188.
Barišić, S. (1972a). *Phys. Rev.* **B 5**, 932–941.
Barišić, S. (1972b). *In* "Superconductivity in d- and f-band Metals" (D. H. Douglass, ed.), pp. 73–94. Am. Inst. Phys., New York.
Barišić, S., Labbe, J., and Friedel, J. (1970). *Phys. Rev. Lett.* **25**, 919–922.
Batterman, B. W., and Barrett, C. S. (1964). *Phys. Rev. Lett.* **13**, 390.
Bennemann, K. H., and Garland, J. W. (1972). *In* "Superconductivity in d- and f-Band Metals" (D. H. Douglass, ed.), pp. 103–137. Am. Inst. Phys., New York.
Birnboim, A. (1976). *Phy. Rev.* **B 1,4**, 2857–2864.

Birnboim, A., and Gutfreund, H. (1974). *Phys. Rev. B* **9**, 139– 146.
Birnboim, A., and Gutfreund, H. (1975). *Phys. Rev. B* **12**, 2682– 2689.
Blaugher, R. D., Hein, R. C., and Cox, J. E. (1969). *J. Low Temp. Phys.* **1**, 539– 561.
Bloch, F. (1928). *Z Phys.* **52**, 555.
Blumberg, W. E., Eisinger, J., Jaccarino, V., and Matthias, B. T. (1960). *Phys. Rev. Lett.* **5**, 149– 152.
Bostock, J., MacVicar, M. L. A., Arnold, G. B., Zasadzinski, J., and Wolf, E. L. (1980). *In* "Superconductivity in d- and f-Band Metals" (H. Sahl and M. B. Maple, eds.), pp. 152– 162. Academic Press, New York.
Boyer, L. L., Klein, B. M., and Papaconstantopoulos, D. A. (1977a). *Ferroelectrics* **16**, 291– 293.
Boyer, L. L., Papaconstantopoulos, D. A., and Klein, B. M. (1977b). *Phys. Rev. B* **8**, 3685– 3693.
Butler, W. H. (1977). *Phys. Rev. B* **15**, 5267– 5282.
Butler, W. H. (1979). *Bull. Am. Phys. Soc.* **24**, 234.
Butler, W. H., and Allen, P. B. (1976). *In* "Superconductivity in d- and f-Band Metals" (D. H. Douglass, ed.), pp. 73– 120. Plenum, New York.
Butler, W. H., and Williams, R. K. (1978). *Phys. Rev. B* **18**, 6483.
Butler, W. H., Olson, J. J., Faulkner, J. S., and Gyorffy, B. L. (1976). *Phys. Rev. B* **15**, 3823– 3836.
Butler, W. H., Smith, H. G., and Wakabayashi, N. (1977). *Phys. Rev. Lett.* **39**, 1004– 1007.
Butler, W. H., Pinski, F. J., and Allen, P. B. (1979). *Phys. Rev. B* **19**, 3708.
Caron, L. G. (1974). *Phys. Rev. B* **9**, 5025.
Clogston, A. M., Gossard, A. C., Jaccarino, V., and Yafet, Y. (1964). *Rev. Mod. Phys.* **36**, 170.
Cody, G. D., Hanak, J. J., McConville, G. T., and Rosi, F. D. (1964). *RCA Rev.* **25**, 338– 341.
Cohen, R. W., Cody, G. D., and Halloran, J. J. (1967). *Phys. Rev. Lett.* **19**, 840.
Colver, M. M., and Hammond, R. R. (1973). *Phys. Rev. Lett.* **30**, 92.
Cooke, J. F. (1979). "Progress Report of Oak Ridge National Laboratory, Solid State Division," ORNL-5486, pp. 13– 15. Oak Ridge Natl. Lab., Oak Ridge, Tennessee.
Cooper, L. N. (1956). *Phys. Rev.* **104**, 1189.
Craig, R. S., Krier, C. A., Coffer, L. W., Bates, E. A., and Wallace, W. E. (1953). *J. Am. Chem. Soc.* **76**, 238– 240.
Dagens, L., Rasolt, M., and Taylor, R. (1975). *Phys. Rev. B* **11**, 2726– 2734.
Evans, R., Gaspari, G. D., and Gyorffy, B. L. (1973a). *J. Phys. F* **3**, 39– 54.
Evans, R., Ratti, V. K., and Gyorffy, B. L. (1973b). *J. Phys. F* **3**, L199– L202.
Feynman, R. P. (1939). *Phys. Rev.* **56**, 340– 343.
Fisk, Z., and Webb, G. (1976). *In* "Superconductivity in d- and f-Band Metals" (D. H. Douglass, ed.), pp. 545– 550. Plenum, New York.
Flukiger, R., and Jorda, J. L. (1977). *Solid State Commun.* **22**, 109.
Flukiger, R., Heiniger, F., Junod, A., Miller, J., Spitzli, P., and Staudenmann, J. L.. (1970). *J. Phys. Chem. Solids* **32**, 459– 463.
Flukiger, R., Paoli, A., and Muller, J. (1974). *Solid State Commun.* **14**, 443– 447.
Foulkes, I. F., and Gomersall, I. R. (1975). *J. Phys. F* **5**, 153.
Friedel, J. (1969). *In* "The Physics of Metals" (J. M. Ziman, ed.), pp. 340– 408. Cambridge Univ. Press, London and New York.
Fröhlich, H. (1966). *In* "Perspectives in Modern Physics" (R. E. Marshak, ed.), pp. 539– 552. Wiley (Interscience), New York.
Gaspari, G. D., and Gyorffy, B. L. (1972). *Phys. Rev. Lett.* **28**, 801– 805.
Gavaler, J. R. (1973). *Appl. Phys. Lett.* **23**, 480– 482.
Gilat, G., and Nicklow, R. M. (1965). *Phys. Rev.* **143**, 487– 494.
Glötzel, D., Rainer, D., and Schober, H. (1980). *Z. Phys. B* **35**, 317– 326.

Glover, R. E., Moser, S., and Baumann, F. (1971). *J. Low Temp. Phys.* **5**, 519–536.

Gor'kov, L. P. (1973). *Zh. Eksp. Teor. Fiz.* **65**, 1658 [Engl. translation., *Sov. Phys.—JETP* **38**, 830 (1974).]

Grimvall, G. (1969). *Phys. Kondens. Mater.* **9**, 283–299.

Grimvall, G. (1976). *Phys. Scr.* **14**, 63–78.

Gupta, R. P., and Sinha, S. K. (1976). *In* "Superconductivity in d- and f-Band Metals" (D. H. Douglass, ed.), pp. 583–591. Plenum, New York.

Gutfreund, H., and Birnboim, A. (1973). *Pro. Conf. Low Temp. Phys. 13th, Boulder Colo.*, 1972 **4**, 247–252.

Gyorffy, B. L., and Stocks, G. M. (1979). "Electrons in Disordered Metals and at Metallic Surfaces." (P. Phariseau, B. L. Gyorffy, and L. Scheire, eds.), pp. 80–187. Plenum, New York.

Hardy, G. F., and Hulm, J. K. (1954). *Phys. Rev.* **93**, 1004.

Harmon, B. N., and Sinha, S. K. 1977). *Phys. Rev. B* **16**, 3919–3924.

Hedin, L., and Lundqvist, S. (1969). *Solid State Phys.* **23**, 1–181.

Heine, V., Nozieres, P., and Wilkins, J. W. (1966). *Philos. Mag.* **13**, 741–758.

Hellmann, H. (1937). "Einfuhrung in die Quantenchemie," p. 285f. Franz Deuticke, Leipzig.

Hohenberg, P., and Kohn, W. (1964). *Phys. Rev.* **136**, B864.

Hulm, J. K., and Blaugher, R. D. (1972). *In* "Superconductivity in d- and f-Band Metals" (D. H. Douglass, ed.), pp. 1–16. Am. Inst. Phys., New York.

Hultgren, R., Desai, P. D., Hawkins, D. T., Gleiser, M., Kelley, K. K., and Wagman, D. D. (1973). "Selected Values of the Thermodynamic Properties of the Elements." Am. Soc. Met., Metals Park, Ohio.

Izyumov, Y. A., and Kurmaev, É. Z. (1974). *Usp. Fiz. Nauk* **113**, 193–238. [Engl. transl., *Sov. Phys. Usp.* **17**, 356–380 (1974).]

Izyumov, Y. A., and Kurmaev, E. Z. (1976). *Usp. Fiz. Nauk* **118**, 53–100. [Eng. transl., *Sov. Phys. Usp.* **19**, 26–52 (1976).]

Jarlborg, T., and Arbman, G. (1977). *J. Phys. F* **7**, 1635.

John, W., (1973). *J. Phys. F* **3**, L231.

John, W., Hamann, D., and Urwank, D. (1978). *Phys. Status Solidi B* **86**, 569–577.

Junod, A., Heiniger, F., Muller, J., and Spitzli, P. (1970). *Helv. Phys. Acta* **43**, 59–66.

Junod, A., Staudenmann, J.-L., Muller, J., and Spitzli, P. (1971). *J. Low Temp. Phys.* **5** 25–43.

Khan, F., Pinski, F. J., and Butler, W. H. (1980). Unpublished.

Killpatrick, D. H. (1964). *J. Phys. Chem. Solids* **25**, 1499–1500.

Klein, B. M., and Papconstantopoulos, D. A. (1974). *Phys. Rev. Lett.* **32**, 1193–1195.

Klein, B. M., Papaconstantopoulos, D. A., and Boyer, L. L. (1977a). *Ferroelectrics* **16**, 299–302.

Klein, B. M., Economu, E. N., and Papaconstantopoulos, D. A. (1977b). *Phys. Rev. Lett.* **39**, 574–577.

Klein, B. M., Boyer, L. L., Papaconstantopoulos, D. A., and Mattheiss, L. F. (1978). *Phys. Rev. B* **18**, 6411.

Klein, B. M., Boyer, L. L., and Papaconstantopoulos, D. A. (1979). *Phys. Rev. Lett.* **42**, 530–533.

Klein, B. M., Boyer, L. L., and Papaconstantopoulos, D. A. (1980). *In* "Superconductivity in d- and f-Band Metals," (H. Suhl and M. B. Maple, eds.), pp. 455–464. Academic Press, New York.

Knapp, G. S., Bader, S. D., and Fisk, Z. (1976). *Phys. Rev. B* **13**, 3783.

Kohler, M. (1948). *Z. Phys. (Leipzig)* **124**, 772.

Kohn, W., and Sham, L. J. (1965). *Phys. Rev.* **140**, A1133.

Krupka, M. C., Girogi, A. L., Krikorian, N. H., and Szklarz, E. G. (1969). *J. Less-Common Met.* **19**, 113.

Kulikov, N. I. (1978). *Fiz. Tver. Tela* (*Lenningrad*) **20**, 2279–2282. [Engl. trans., *Sov. Phys.—Solid State* **20**, 1315–1317 (1978).]

Kulikov, N. I., Borzunov, V. N., and Zvonkov, A. D. (1978). *Phys. Status Solidi b* **86**, 83–91.

Labbé, J. (1967). *Phys. Rev.* **158**, 647.

Labbé, J., and Friedel, J. (1966). *J. Phys.* (*Paris*) **27**, 708.

Larose, A., and Vanderwall, J. (1974). "Scattering of Thermal Neutrons: A Bibliography (1932–1974)." Plenum, New York

Laubitz, M. J., and Matsumura, T. (1972). *Can. J. Phys.* **50**, 196.

Lee, M. J. G., and Heine, V. (1972). *Phys. Rev. B* **5**, 3839–3847.

Lee, T. K., Birman, J. L., and Williamson, S. J. (1977). *Phys. Rev. Lett.* **39**, 839.

Luo, H. L., Vielhaber, E., and Corenwitz, E. (1970). *Z. Phys.* (*Leipzig*) **320**, 443–448.

Lutz, H., Weismann, H., Gurvitch, M., Goland, A., Kammerer, O. F., and Strongin M. (1976). *In* "Superconductivity in d- and f-Band Metals" (D. H. Douglass, ed.), pp. 535–544. Plenum, New York.

Lyo, S. K. (1978). *Phys. Rev. B* **17**, 2545–2552.

McMillan, W. L. (1968). *Phys. Rev.* **167**, 331.

McMillan, W. L., and Rowell, J. M. (1969). *In* "Superconductivity" (R. D. Parks, ed.), Vol. 1, pp. 561–614. Dekker, New York.

Mailfert, R., Batterman, B. W., and Hanak, J. J. (1969). *Phys. Status Solidi* **32**, K67.

Martin, D. L. (1965). *Phys. Rev.* **139**, A150–A160.

Matsubara, T. (1955). *Prog. Theor. Phys.* (*Kyoto*) **14**, 351.

Mattheiss, L. F. (1975). *Phys. Rev. B* **12**, 2161.

Matthias, B. T. (1953). *Phys. Rev.* **92**, 874.

Matthias, B. T. (1957). *In* "Progress in Low Temperature Physics" (C. J. Gorten, ed.), pp. 138–150. North-Holland Publ., Amsterdam.

Matthias, B. T. (1971). *Phys. Today* **24**, 23.

Matthias, B. T., Geballe, T. H., and Compton, V. B. (1963). *Rev. Mod. Phys.* **35**, 1–22.

Migdal, A. B. (1958). *Zh. Eksp. Teor. Fiz.* **34**, 1438. [Engl. transl., *Sov. Phys.—JETP* **7**, 996 (1958).]

Mitra, T. K. (1969). *J. Phys. C* **2**, 52–60.

Morel, P., and Anderson, P. W. (1962). *Phys. Rev.* **125**, 1263.

Moruzzi, V. L., Janak, J. F., and Williams, A. R. (1978). "Calculated Electronic Properties of Metals." Pergamon, New York.

Muhlratzer, A., Nickl, J. J., Seeber, B., and Sprenger, H. (1976). *Solid State Commun.* **19**, 239.

Mukhopadhyay, G., and Gyorffy, B. L. (1973). *J. Phys. F* **3**, 1373–1379.

Nowak, D. (1972). *Phys. Rev. B* **6**, 3691–3700.

Pan, V. M., Alekseevskii, V. P., Popov, A. G., Beletskii, Y. I., and Yupko, L. M. (1975). *Zh. Eksp. Teor. Fiz. Pis'ma Red.* **21**, 494–496. [Engl. transl., *JETP Lett.* **21**, 228–230 (1975).]

Papaconstantopoulos, D. A. (1975). *Phys. Rev. B* **11**, 4801–4807.

Papaconstantopoulos, D. A., and Klein, B. M. (1975). *Phys. Rev. Lett.* **35**, 110–113.

Papaconstantopoulos, D. A., and Klein, B. M. (1977). *Ferroelectrics* **16**, 307–310.

Papaconstantopoulos, D. A., Boyer, L. L., Klein, B. M., Williams, A. R., Morruzzi, and Janak, J. F. (1977). *Phys. Rev. B* **15**, 4221–4226.

Papaconstantopoulos, D. A., Klein, B. M., Economu, E. N., and Boyer, L. L. (1978a). *Phys. Rev. B* **17**, 141–150.

Papaconstantopoulos, D. A., Klein, B. M., Faulkner, J. S., and Boyer, L. L. (1978b). *Phys. Rev. B* **18**, 2784–2791.

Papaconstantopoulos, D. A., Zdetsis, A. D., and Economou, E. N. (1978c). *Solid State Commun.* **27**, 1189–1192.

Peter, M., and Adam, G. H. (1976). *Rev. Roum. Phys.* **21**, 385–399.

Peter, M., Klose, W., Adam, G., Entel., and Kudla, E. (1974). *Helv. Phys. Acta* **47**, 807–832.

Peter, M., Ashkenazi, J., and Dacorogna, M. (1977). *Helv. Phys. Acta* **50**, 267–278.

Pettifor, D. G. (1977a). *J. Phys. F* **7**, 613–633.

Pettifor, D. G. (1977b). *J. Phys. F* **7**, 1009–1028.

Pettifor, D. G. (1979). *Phys. Rev. Lett.* **42**, 846–850.

Pickett, W. E., and Gyoffy, B. L. (1976). *In* "Superconductivity in d- and f-Band Metals" (D. H. Douglass, ed.), pp. 251–268. Plenum, New York.

Pinski, F. J., and Butler, W. H. (1979). *Phys. Rev. B* **19**, 6010.

Pinski, F. J., Allen, P. B., and Butler, W. H. (1978a). *J. Phys. (Paris)* **39**, Suppl. C6, 472–473.

Pinski, F. J., Allen, P. B., and Butler, W. H. (1978b). *Phys. Rev. Lett.* **41**, 431–434.

Pinski, F. J., Allen, P. B., and Butler, W. H. (1981). *Phys. Rev. B* (to be published).

Poon, S. J. (1976). *Solid State Commun.* **18**, 1489–1491.

Powell, B. M., Martel, P., and Woods, A. D. B. (1968). *Phys. Rev.* **171**, 727.

Prange, R. E., and Sachs, A. (1967). *Phys. Rev.* **158**, 672–682.

Rasolt, M., and Devlin, J. F. (1976). *Phys. Rev. B* **13**, 3290–3296.

Rasolt, M., and Taylor, R. (1975). *Phys. Rev.* **11**, 2717–2725.

Ratti, V. K., Evans, R., and Gyorffy, B. L. (1974). *J. Phys. F* **4**, 371–379.

Rehwald, W., Rayl, M., Cohen, R. W., and Cody, G. D. (1972). *Phys. Rev. B* **6**, 363–371.

Reissland, J. A. (1973). "The Physics of Phonons," p. 181. Wiley, New York.

Ries, G., and Winter, H. (1979). *J. Phys. F* **9**, 1589–1611.

Rietschel, H. (1978). *Z. Phys. B* **30**, 271–275.

Rietschel, H., and Winter, H. (1979). *Phys. Rev. Lett.* **43**, 1256–1260.

Roberts, B. W. (1976). *J. Phys. Chem Ref. Data* **5**, 581–821.

Roberts, B. W. (1978). *Natl. Bur. Stand. (U.S.), Tech. Note* No.983.

Sahm, P. R , and Pruss, T. V. (1969). *Phys. Lett. A* **28**, 707.

Scalapino, D. J. (1969). *In* "Superconductivity," (R. D. Parks, ed.), pp. 449–558. Dekker, New York.

Schiff, L. I. (1955). "Quantum Mechanics," 2nd ed., p. 399. McGraw-Hill, New York.

Schmidt, P. H., Bacon, D. D., Barz H., and Cooper, A. S. (1975). *J. Appl. Phys.* **46**, 2273–2343.

Schneider, T. (1969). *Helv. Phys. Acta* **42**, 957.

Shen, L. Y. L. (1972). *In* "Superconductivity in d- and f-Band Metals" (D. H. Douglass, ed.), pp. 31–44. Am. Inst. Phys., New York

Shirane, G., and Axe, J. D. (1971). *Phys. Rev. Lett.* **27**, 1803.

Shukla, R. C., and Taylor, R. (1976). *J. Phys. F* **6**, 531–544.

Sinha, S. K., and Harmon, B. N. (1976). *In* "Superconductivity in d- and f-Band Metals" (D. H. Douglass, ed.), pp. 269–296. Plenum, New York.

Smith, H. G., and Gläser, W. (1970). *Phys. Rev. Lett.* **25**, 1611.

Smith, H. G., Wakabayashi, N., and Mostoller, M. (1976). *In* "Superconductivity in d- and f-Band Metals" (D. H. Douglass, ed.), pp. 223–251. Plenum, New York.

Stocks, G. M., Gaspari, G. D., and Gyorffy, B. L. (1972). *J. Phys. F* **2**, L123–L128.

Switendick, A. C. (1976). *In* "Superconductivity in d- and f-Band Metals" (D. H. Douglass, ed.), pp. 593–604. Plenum, New York.

Taylor, R., Leavens, and Shula, R. C. (1976). *Solid State Commun.* **19**, 809–811.

Testardi, L. R. (1975). *Rev. Mod. Phys.* **47**, 637–648.

Testardi, L. R., Meek, R. L., Poate, J. M., Royer, W. A., Storm, A. R., and Wernick, J. H. (1975). *Phys. Rev. B* **11**, 4304–4317.

Tomlinson, P. G. (1979). *Phys. Rev. B* **19**, 1867.

Tomlinson, P. G., and Swihart, J. C. (1979). *Phys. Rev. B* **19**, 1893.

Varma, C. M., and Dynes, R. C. (1976). *In* "Superconductivity in d- and f-Band Metals" (D. H. Douglass, ed.), pp. 507–529. Plenum, New York.

Varma, C. M., and Weber, W. (1979). *Phys. Rev.* **19**, 6142–6154.

Varma, C. M., Blount, E. I., Vashishta, P., and Weber, W. (1979). *Phys. Rev. B* **19**, 6130–6141.

Vilenkin, A., and Taylor, P. L. (1979). *Phys. Rev. Lett.* **42**, 597–599.

Webb, G. W., Vieland, L. J., Miller, R. E., and Wicklund, A. (1971). *Solid State Commun.* **9**, 1769–1773.

Weber, W. (1980). *In* "Superconductivity in d-f-Band Metals" (H. Suhl and M. B. Maple, eds.), pp. 131–142. Academic Press, New York.

Weger, M., and Goldberg, I. B. (1973). *Solid State Phys.* **28**, 2–177.

Williams, A. R., Gelat, C. D., Jr., and Moruzzi, V. L. (1980). *Phys. Rev. Lett.* **44**, 429–433.

Williamson, S. J., and Milewits, M. (1976). *In* "Superconductivity in d-and f-Band Metals" (D. H. Douglass, ed.), pp. 551–66. Plenum, New York.

Winter, H., and Ries, G. (1976). *Z. Phys. B* **24**, 279–284.

Winter, H., Rietshel, G., Ries, G., and Reichardt, W. (1978). *J. Phys. (Paris)* **39**, Suppl. C6, 474–476.

Young, J. A., and Koppel, J. U. (1964). *Phys. Rev.* **134**, A1476–A1479.

Youngblood, R., Noda, Y., and Shirane, G. (1979). *Phys. Rev. B* **19**, 6016–6019.

Zasadzinski, J., Schubert, W. K., Wolf, E. L., and Arnold, G. B. (1980). *In* "Superconductivity in d-f-Band Metals" (H. Suhl and M. B. Maple, eds.), pp. 159–164. Academic Press, New York.

Ziman, J. M. (1960). "Electrons and Phonons," Chap. 7. Oxford University Press, London and New York.

4

Elastic Properties of Transition Metals*

S. G. STEINEMANN

University of Lausanne, Lausanne, Switzerland

and

E. S. FISHER

Argonne National Laboratory, Argonne, Illinois

I. Introduction

In recent years there have been several reports of the unusual response of elastic properties of transition metals to alloying with substitutional solute atoms. Most of the work pertains to the bcc solid solutions of the Nb-, V-,

*Work supported by the U.S. Dept. of Energy.

and Ta-based alloys, but includes some studies of bcc Fe alloys and fcc Pd- and Ni-based alloys. The common thread in all the pertinent results is the very marked influence of relatively small changes in chemical composition on one or more of the single-crystal shear moduli and/or their temperature and volume derivatives. It was recognized early that these important changes were often accompanied by parallel effects on magnetic properties in both paramagnets and ferromagnets and were, therefore, probably associated with the electronic structure and its variation with alloying and temperature. The close relationship to Fermi surface topology was, however, not evident until the detailed measurements on the binary alloys of the Zr–Nb–Mo systems were completed. It is clear from these data that the elastic shear moduli not only respond to changes in densities of states and electron–phonon coupling parameters that are indicated by electronic specific heat, superconducting T_c, and phonon dispersion studies, but respond, also, to Fermi surface changes that may be deduced from band structure calculations and rigid-band behavior. The purpose of this chapter is to present a coherent review of the available data, together with a framework for a quantitative theory. The objective is to convince the metallurgist and the physicist that the mechanical properties of these important alloys can in part be interpreted and predicted from the electron band structure. The relevance also includes the understanding of other thermal properties of the alloys, such as thermal expansion and lattice heat capacity, and a basis for understanding phase stability in terms of the Fermi surface topology and elastic moduli.

The chapter is not intended as a complete tutorial on the mathematics of elasticity theory; the definitions of the elastic parameters are, however, given in Section II with an outline of how they relate to Hooke's law and to the dynamic methods for experimental evaluation from acoustic velocity measurements. The capability to obtain accurate and precise data is an important aspect of the use of elastic properties as a probe in Fermi surface changes with alloying.

As a background to the central theme, a brief review of other experimental evidence for the response of electrons to strains is presented in Section III followed by the concept of a deformation potential and its evaluation from band structure parameters for a bcc lattice. This concept is related to the strain dependence of the chemical potential μ, and thus to the free energy so as to derive the electronic contribution to shear moduli. This part of Section III is also important in the interpretation of the temperature and volume derivatives of the elastic moduli in terms of the lattice and electronic free-energy terms.

Because of the preponderance of data regarding the effects of alloy composition on the temperature and volume derivatives in the 4d, bcc alloys, these data, given in Section IV, are the primary evidence for a convincing argument relating the compositional dependence of the elastic shear moduli to Fermi surface changes and the anomalies in temperature dependence to relatively small degeneracy temperatures. Since there are also considerable data for the V- and Ta-based 3d and 5d bcc alloys, these results follow, in brief, in Section IV, as further evidence for the application of the rigid-band model and for the theme of relating the compositional dependence to Fermi surface changes. The data are given in sufficient detail so as to emphasize the similarities as well as the differences between the 4d alloys and their isoelectronic and isostructural alloys in the 3d and 5d series.

There is also relatively clear evidence for strong electronic effects on the shear moduli of Pd and its alloys with Ag and Rh. These data, as well as the corresponding evidence in the hcp transition metals and the A-15 structures, are briefly presented as evidence that the strong electronic components exist for shear moduli in general and are not restricted to bcc structures. At the present time, however, the data for the fcc and hcp solid solutions are nowhere near as extensive as in the case of the bcc alloys. The A-15 anomalies and their relation to the high densities of d states and to the martensitic phase transitions in the high T_c superconductors are indeed a well-known example of the electron contributions to the temperature dependence of shear moduli.

II. Definition and Measurement of Elastic Moduli

A. Definition of Stiffness Moduli

Under the action of forces, a solid will exhibit deformation, i.e., change its volume and its shape. These changes are described by the displacements **u** of any point in the body. The gradients of **u**, i.e., $u_{mn} = \partial u_m / \partial x_n$ (x refers to position and m,n to coordinates), specify the configuration of the strained solid and they compose the strains

$$\eta_{mn} = \tfrac{1}{2}\left(u_{mn} + u_{nm} + \sum_o u_{om} u_{on} \right). \tag{1}$$

The strains are related to the stresses through the linear Hooke's law

$$\sigma_{kl} = \sum_{mn} c_{klmn} \eta_{mn} \quad \text{or} \quad \sigma_i = \sum_j c_{ij} \eta_j, \tag{2}$$

where Voigt's notation is introduced instead of the tensorial scheme. By convention, $k = l = i = 1,2,3$ for the diagonal elements of the strain (or stress) tensor and $k \neq l$, $9 - k - l = i = 4,5,6$ for the shear components.

c_{klmn} or c_{ij}, respectively, are the stiffness moduli or elastic constants. c_{11} relates the compressional stress σ_1 (along x) to the compressional strain η_1; c_{12} relates σ_1 to η_2, c_{44} relates the shear stress $\sigma_{23} = \sigma_4$ to the shear strain η_4, etc.

The rotational invariance of η and symmetry reduce the number of independent c_{ij} in the matrix (see Nye, 1957; Landau and Lifshitz, 1959). Furthermore, there exist combinations of the c_{ij} that are evidently suitable for treating thermodynamic properties of materials. Such combinations are related to the lattice stability conditions and are simply the eigenvalues of the elastic constant matrix with an associated deformation that has the symmetry determined by the eigenvector of that eigenvalue (Born and Huang, 1954; Cowley, 1976). For example, in the case of a cubic crystal, there are three eigenvalues: $c_{11} + 2c_{12} = 3K$, which corresponds to a change in volume; $c_{11} - c_{12} = 2c'$, which is connected to an expansion along one cube axis and equivalent contractions along a second cube axis; and $c_{44} = c$, which is also a change of shape of the cell, with the expansion occurring along the direction [111]. The two latter constants are associated with pure shear strains. Shear moduli only determine the elastic anisotropy; in the cubic case the anisotropy ratio is $A = c/c'$. Table I collects these data for the isotropic solid and the cubic and hexagonal classes in which transition metals crystallize.

B.   Relation of Elastic Constants to Acoustic Wave Propagation

One way to study the elastic properties of crystals is to measure the speed of sound. The displacements in an acoustic wave with wave vector $\mathbf{q}$ and angular frequency ω are given by (Thurston and Brugger, 1964).

$$\mathbf{u}(\mathbf{r}) = \mathbf{u}_o \exp\{i(\mathbf{q} \cdot \mathbf{r} - \omega t)\} \tag{3}$$

and a component of strain generated by this wave is

$$u_{mn} = \frac{\partial u_m}{\partial x_n} = iq_n u_{om} \exp\{i(\mathbf{q} \cdot \mathbf{r} - \omega t)\}. \tag{4}$$

In the absence of external stresses, the equation of motion is then

$$\rho \ddot{u}_m - \sum_n \sum_{kl} c_{mknl} \frac{\partial^2 u_n}{\partial x_k \partial x_l} = -\sum_n \left(\rho \omega^2 \delta_{mn} + \sum_{kl} c_{mknl} q_k q_l \right) u_n = 0. \tag{5}$$

ρ is the density of the crystal and $u_m = \delta_{mn} u_n$. The relation applies to isothermal and adiabatic conditions. Since ω is a homogeneous function of degree one in the components q_k, the propagation is dispersion-free and the

TABLE I

HOMOGENEOUS DEFORMATIONS, ELASTIC CONSTANTS, AND ASSOCIATED STRAINS

Crystal class	Combinations of c_{ij}'s (eigenvalues)	Strains (eigenvectors)	Symmetry of strain (Schoenflies symbol)
Isotropic	$3c_{12} + 2c_{44} = 3K$	$\eta_1 = \eta_2 = \eta_3$	K_h
	$c_{44} = G$	$\left.\begin{array}{l}\eta_1 = -\eta_2 \\ \frac{1}{2}\eta_3 = -\eta_1 = -\eta_2 \\ \eta_4, \eta_5, \eta_6\end{array}\right\}$	$D_{\infty h}$
Cubic	$c_{11} + 2c_{12} = 3K$	$\eta_1 = \eta_2 = \eta_3$	0_h
	$c_{11} - c_{12} = 2c'$	$\left.\begin{array}{l}\eta_1 = -\eta_2 \\ \frac{1}{2}\eta_3 = -\eta_1 = -\eta_2\end{array}\right\}$	D_{4h}(tetragonal)
	$c_{44} = c$	η_4, η_5, η_5	D_{3d}(trigonal)
Hexagonal	$(c_{11} + c_{12})c_{33} - 2c_{13}^2$	$\eta_3, \eta_1 = \eta_2$	D_{6h}
	$c_{11} - c_{12} = 2c_{66}$	$\eta_1 = -\eta_2, \eta_6$	D_{2h}(rhombohedral)
	c_{44}	η_4, η_5	C_{2h}(monoclinic)
	$\left[\frac{1}{9}(2c_{11} + 2c_{12} + c_{33} + 4c_{13}) = K\right]$	(change of c/a at constant volume)	

velocity

$$v_s = \omega/|q| \quad \text{or} \quad v_s = (c/\rho)^{1/2} \tag{6}$$

is a function of the direction of propagation, but not of the frequency of the wave.

The set of the three homogeneous equations for the unknowns u_n, i.e., Eq. (5), is known as Christoffel's equations. For each wave vector, $\sum c_{mknl} q_k q_l$ (occasionally called propagation matrix) has three eigenvalues and eigenvectors corresponding to the three acoustic modes for that wave vector. In general, these eigenvectors do not coincide with the combinations of the strains that are the eigenvectors of the elastic constant matrix nor are the eigenvalues of the two matrices the same. In particular, no acoustic modes may be excited for the eigenvectors of Table I, which conserve the symmetry of the unstrained lattice. However, the orientation of the crystal might be chosen in a way which permits the measurement of all eigenvalues of the c_{ij} matrix with a minimum number of samples. The propagation modes for these preferred orientations are given in Table II for the isotropic, cubic, and hexagonal classes.

Tables I and II include the isotropic solid, which applies to polycrystalline metals strictly when the crystallites themselves are isotropic. Otherwise,

TABLE II

PREFERRED PROPAGATION MODES AND ASSOCIATED ELASTIC CONSTANTS

Crystal class	Acoustic wave		Eigenvalue of propagation matrix	Ratio of shear to total strain
Isotropic	Unilateral	$q\|u$	$K + \frac{4}{3}G$	$\dfrac{4G}{3K + 4G}$
		$q \perp u$	G	1
	Uniaxial	$q\|u$	$\dfrac{9KG}{3K + G} = E$	$\dfrac{3K}{3K + G}$
		$q \perp u$	G	1
Cubic	$q\|[110]$	$u\|[110]$	$K + \frac{1}{3}c' + c$	$\dfrac{c' + 3c}{3K + c' + 3c}$
		$u\|[1\bar{1}0]$	c'	1
		$u\|[001]$	c	1
Hexagonal	$q \perp [001]$	$u\|q$	c_{11}	—
		$u \perp [001]$ and q	$\frac{1}{2}(c_{11} - c_{12}) = c_{66}$	1
		$u\|[001]$	c_{44}	1
	q at angle $\theta \neq 0$ from [001]	$u \neq \perp [001]$	$f(c_{11}, c_{33}, c_{44}, c_{13}, \theta)$	
	$q\|[001]$	$u\|q$	c_{33}	

"quasiisotropic" moduli are obtained from procedures such as Voigt's (1889) or Reuss's (1929), and Hill's (1952) average. Under certain conditions, the single-crystal c_{ij} may be extracted from polycrystalline measurements. This has, for example, been done by Lenkkeri and Lahteenkorva (1978) for vanadium alloys.

Table II mentions also the uniaxial case where guided wave conditions apply. These moduli are also obtained in static and low-frequency resonance experiments. It is important to note that in these cases the longitudinal mode reflects strongly the shear properties of the material, while in high-frequency experiments the volume strains are more important.

C. Experimental Techniques

Various techniques for measuring elastic constants are capable of high precision: (i) the stress techniques, with strain gauges as sensor elements; (ii) the resonance techniques where bars of circular or rectangular shape are excited for longitudinal, bending, or torsional vibrations (Huntington, 1958); and (iii) the ultrasonic techniques (McSkimin, 1964, Truell et al., 1969). The stress techniques and sonic resonances generally require large specimens and a careful machining of the samples. The most flexible are the sound velocity measurements for which two types of sample are used; at lower frequencies, slender specimens with lateral dimensions much smaller than the wavelength (uniaxial stress condition) are convenient to determine the compliance moduli s_{ij}, whereas, at frequencies of 1 MHz and above the specimen dimensions are chosen to be much larger than the wavelength (unilateral stress condition) and are thus suitable for direct determination of the c_{ij}.

Different transducer-to-sample coupling arrangements are used for measuring the elastic constants under pressure, for low and high temperatures, in magnetic fields, etc. (Fig. 1):

(i) In the simplest form, one or two transducers are bonded to the oriented crystal to obtain the "time of flight" in transmission or reflection; this setup is convenient for the measurements under pressure and in the range of low temperatures to about 800 K.

(ii) With a "stepped crystal," the difference in time of flight gives the elastic properties at one end of the crystal that may be brought to high temperature while the other end of the crystal is cooled (Lowrie and Gonas, 1967).

(iii) Equally suitable for measurements at high temperatures is the "buffer rod technique." The sample is bonded to fused silica glass or to a ceramic or metallic wave guide by a cement (Fisher and Renken, 1964), or a molten and solidified salt by brazing or diffusion (Ashkenazi et al., 1978).

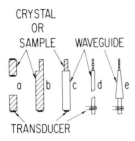

CRYSTAL
OR
SAMPLE   WAVEGUIDE

TRANSDUCER

Fig. 1. Setup for sound- velocity measurements: (a) crystal with two or one piezoelectric transducer; (b) "stepped crystal" for high temperatures; (c)–(e) "buffer rod" or waveguide technique for large and slender specimens [including impedance matching in (e)].

This technique permits one to measure very thin samples and a variant of it is used for sound-velocity determinations in liquid metals (Keita and Steinemann, 1978).

The transducers for wave excitation in longitudinal and transverse modes are either a piezoelectric type or a magnetostrictive device where an axial magnetic bias gives extensional waves and a circumferential magnetic bias gives a torsional excitation.

The frequency of the excited wave is determined by the carrier frequency of a rf pulse train and resonance of the transducer. Various techniques for measuring the wave velocity are known: superposition of echoes in delayed sweep, pulse superposition, phase comparison, etc. (Truell *et al.*, 1969). All are capable of a resolution of less than 1% and better and may, in extremes, be brought to a relative precision of 10^{-7} ["sing-around" technique by Forgacs (1960)]. The ability to obtain high precision is a definite advantage for the study of thermodynamic properties because of the possibility to obtain not only the elastic constants but especially their derivatives with respect to parameter temperature, pressure, and stress.

III.   Theory

A.   *Historical Development of Theory*

1.   EARLY THEORIES

Elastic constants and phonon spectra of metals are frequently analyzed in terms of a Born–van Karman lattice model, where the parameters are pair-interaction force constants (see, for example, Born and Huang, 1954; Huntington, 1958; Ludwig, 1967; Wallace, 1972). It is commonly found that a fit of phonon spectra of metals involves interactions with far distant neighbors, indicating the existence of terms in energy that depend on volume and shear strain. A prior knowledge of such dependencies is required. These questions have first been treated by Fuchs (1936a, b) and Fuchs and Peng (1942), with calculations of the elastic constants of the alkali metals and of copper. In the case of these simpler metals, a clear subdivision between bonding or valence electrons and ion-core electrons can be made. The cores are of small size and their displacements make weak perturbations of the movement of the conduction electrons. Under these premises, the important interactions can be treated as direct and indirect, via conduction electrons, between ions represented as pseudopotentials including screening and exchange. The success of the pseudopotential concept in treating the elastic constants is remarkable (Heine and Weaire, 1970).

2. Variations in Cohesive Energy and Polycrystalline Elasticity in Transition Series

The cohesion energy E_c of transition metals (Fig. 2) shows a regular variation peaking to a maximum for the refractory metals in the middle of the series. The width and magnitude of this peak clearly shows that it must be related to d-band formation in the solid, and the regularity of the peak is a sign that the property is neither sensitive to crystal structure nor to the details of the band structure (Friedel, 1969; Ducastelle, 1970). This is especially clear for the second and third series; in the first transition-metal series, deviations are due to what Friedel calls "magnetic complications." The bulk modulus follows a rather similar trend (Fig. 2b). The peaking is

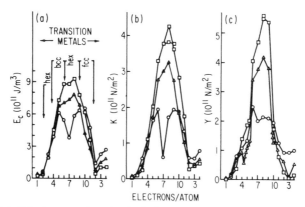

Fig. 2. (a) Cohesion energies E_c, (b) bulk moduli K, and (c) Young's moduli E for elements and binary alloys (of neighboring elements) in period 4 (O, K through Ge), period 5 ($\triangle$, Rb through Sn), and period 6 ($\square$, Cs through Pb). Scales are drawn in the ratios $E_c \sim (\frac{1}{3})K \sim (\frac{3}{11})E$, according to the model of Ducastelle (1970). Data from Gschneidner (1964), Fischer et. al. (1969), and other sources.

still present for Young's modulus E (Fig. 2) and is obviously of electronic origin; however, its regularity is lost. The moduli of V, Nb, Ta, and their alloys, and those of Ti, Zr, Pd, and Pt are rather low. These anomalies are clearly related to a "softening" of the shear modulus G, which, as explained in Section II, B, is the major contributor to E.

3. RELATIONS BETWEEN MAGNETIC AND ELASTIC PROPERTIES

Strong evidence of band structure contributions to elastic properties is seen when the variation of Young's modulus is related to the variation of the magnetic susceptibility. A similar trend goes through each series with peaks of the susceptibility around electron-per-atom ratios of 5 and 10 and a valley at a ratio of 4. Classical reviews on the subject (Kriessman and Callen, 1954; Gladstone et al., 1969) all relate these variations with the behavior of the density of states. Fischer et al. (1969) have established the coincidence of high magnetic susceptibilities and low shear moduli. In either of these properties, details of the band structure become important. A similar effect is absent in the bulk modulus, which depends essentially upon the filling of the electron band.

The temperature dependencies of χ and E provide further evidence for band structure effects over wide temperature spans. Strong negative and positive $d\chi/dT$ occur whenever the Fermi level lies in a peak region of the density of states or falls in a valley, respectively (Kriessman and Callen, 1954). The best examples are Nb with a negative $d\chi/dT$, and Zr with a positive temperature derivative in the hexagonal phase. For Mo, $d\chi/dT$ is small because the density of states is small and only weakly dependent on electron concentration. The same metals have rather different temperature dependencies of E: $dE/dT > 0$ in Nb, strongly negative in Zr, and slightly negative ("normal") in Mo.

B. Strains and Electrons

1. DEFORMATION POTENTIAL

The most useful concept to describe strain effects is the deformation potential. Bardeen and Shockley (1950) introduced it to treat the electron mobility in semiconductors. Herring and Vogt (1956) generalized the approach to include interaction of conduction electrons with transverse as well as longitudinal lattice-vibration modes. The fundamental idea of the deformation potential approach is that effective potentials can be found that correspond to the position of electron band states as these are affected by a gradually varying deformation of the lattice. A first theorem (referred to as the method of effective mass) states that strain $\eta(\mathbf{r})$ produces a perturbing potential $\delta V = D\eta(\mathbf{r})$, and a second theorem shows that the scattering matrix element $\langle \mathbf{k} | \delta V | \mathbf{k}' \rangle$, comprising this locally varying potential, describes the electron–lattice interaction.

It is in line with the (classical) deformation potential theorem to write down a perturbation expansion for the energy levels $\varepsilon(\mathbf{k})$ of a band in terms of the strain η as

$$\varepsilon[\mathbf{k}(\eta),\eta] = \varepsilon(\mathbf{k},0) + \frac{d\varepsilon(\mathbf{k},\eta)}{d\eta}\bigg|_{\eta=0} \eta + \frac{d^2\varepsilon(\mathbf{k},\eta)}{d\eta^2}\bigg|_{\eta=0} \frac{1}{2}\eta^2 + \cdots$$

$$= \varepsilon(\mathbf{k},0) + D_1(\mathbf{k},\varepsilon)\eta + D_2(\mathbf{k},\varepsilon)\frac{1}{2}\eta^2 + \cdots. \qquad (7)$$

The expression defines the local (or wave-number–dependent) first-order, second-order, etc., deformation potentials that are the coefficients of the respective terms. It is noted that these quantities have the reduced symmetries of the strained crystal.

The shifts of a given energy level per unit strain, i.e., the local deformation potentials, may be readily obtained using quantum mechanical perturbation methods known as the Hellmann–Feynman theorem and the quantum curvature theorem (Hellmann, 1933, Feynman, 1939; Löwdin, 1963, 1964; Landsberg, 1969). To find a direct interpretation of these potentials, reference to classical analogies may help.

The fields that act on electrons may be internal or external. The former is the result of the periodic potential in the crystal; a velocity $\nabla_{\mathbf{k}}\varepsilon = \hbar v(\mathbf{k})$ and inverse mass tensor $\partial^2\varepsilon(\mathbf{k})/(\partial k_j \partial k_j) = \hbar^2 m_{ij}(\mathbf{k})$ describe particles in the internal field (Ziman, 1960, Landsberg, 1969). The external field is, in the present context, the mechanical stress. It produces a deformation of the direct space $\mathbf{r}(\eta)$, which is related to a deformation of wave-number space $\mathbf{k}(\eta)$ through an invariance relation

$$\mathbf{k}(\eta)\cdot\mathbf{r}(\eta) = \mathbf{k}(0)\cdot\mathbf{r}(0) \qquad (8)$$

for any strain (Peter et al., 1974; Posternak, 1976). This relation determines the strain derivatives of $\mathbf{k}$ for a given type of strain, so that the effect of stress is that it changes the momentum $\hbar\mathbf{k}$ of a state by the "strain momentum" $\hbar d\mathbf{k}/d\eta$ to first order, by $(\hbar d^2\mathbf{k})/d\eta^2$ to second order, etc.

The derivatives of Eq. (7) can now be written down. From the expression for $D_1(\mathbf{k},\varepsilon)$, after rearrangement of terms, it is

$$\hbar\frac{dk_\perp}{d\eta} = \frac{D_1(\mathbf{k},\varepsilon)}{|v(\mathbf{k})|} - \frac{\partial\varepsilon(\mathbf{k})/\partial\eta}{|v(\mathbf{k})|}, \qquad (9)$$

where $\perp$ means the component along the normal to ε = const. The second term comes from the intrinsic dependency of ε upon η. This result may be stated in different ways: (1) a variation in strain must be accompanied by a variation in wave number whose component normal to an equienergetic surface is proportional to the ratio of the local deformation potential and the electron velocity; or (ii) the projection of the strain momentum onto the velocity of the state equals the deformation potential. A scattering picture emerges, in that the strain changes the kinematics of electrons; the first-order deformation potential reflects the strength of the scattering.

Continuing to second order and using as above the descriptive quantities, it is

$$\hbar \frac{d^2 k_\perp}{d\eta^2} = \left[D_2(\mathbf{k}, \varepsilon) - \sum_{ij} m_{ij}^{-1}(\mathbf{k}) \left(\hbar \frac{dk_i}{d\eta} \right) \left(\hbar \frac{dk_j}{d\eta} \right) \right] |\mathbf{v}(\mathbf{k})|^{-1}$$

$$- \left[\frac{2 \, \partial \mathbf{v}(\mathbf{k})}{\partial \eta} \cdot \left(\hbar \frac{d\mathbf{k}}{d\eta} \right) + \frac{\partial^2 \varepsilon(\mathbf{k})}{\partial \eta^2} \right] |\mathbf{v}(\mathbf{k})|^{-1}. \tag{10}$$

It is noted that in general $\mathbf{k}$ and $\mathbf{v}$ will be by no means parallel nor will $\mathbf{v}$ and $d\mathbf{k}/d\eta$ or $d^2\mathbf{k}/d\eta^2$ be to any other vector. Considering the two extrinsic terms in the first bracket, it is seen that $D_2/|\mathbf{v}|$ may be looked at, like above, as a scattering, however, to second order. The second term including the effective mass and the first-order strain momentum represents the expense in energy to change the inertial properties of the electrons. In similar wording as above, the result may be stated as follows: The projection of the second-order strain momentum of a state onto its velocity, plus an inertial term involving the first-order strain momentum, equals the local second-order deformation potential. Whereas $D_1(\mathbf{k}, \varepsilon)$ is related to the kinematics of the electron, all second-order changes in the energy due to strain clearly are connected with the dynamics of the electrons.

2. EXPERIMENTS TO DETECT STRAIN RESPONSE

Strain modulation produces structure in reflectance whenever a singularity in the electron dispersion coincides with either a filled initial state or an empty final state in an optical transition. A large range of energies in a band may be probed with such piezoreflectance measurements as were made on Cu by Gerhardt (1968), and on Ag and its alloys by Nilsson and Sandell (1970).

The strain response of the Fermi surface may be explored by measuring the frequency shifts of de Haas–van Alphen oscillations or by relating the amplitudes of quantum oscillatory effects such as magnetostriction, torque, sound velocity, or magnetization (see Griessen, 1978). Molybdenum metal

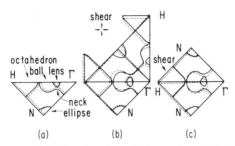

Fig. 3. Strain response of the Fermi surface of Mo: (a) in a (100) plane for a dilation of 10%; (b) in the (100) plane containing the shear axis and in the plane perpendicular to the shear axis for a tetragonal shear of 4%; (c) in the (001) plane perpendicular to shear axis for an angular shear of 3°.

has been extensively studied by these methods (Posternak *et al.*, 1975; Posternak, 1976; Griessen *et al.*, 1977) and results are shown in Fig. 3; the experimentally determined cross-sectional areas A and strain derivatives $d\ln A/d\eta$ are fitted by aid of a KKR calculation. The deformation of the Fermi surface is rather isotropic for volume strain but anisotropic for shear, even for small strains. In the first case, the derivatives are seen to be all negative. For shear strains, they change sign for the different orbits, and from the relation $d\ln A/d\eta = m^* \cdot D$ (m^* is the cyclotron effective mass), it is found that the local deformation potentials are at most 0.4 Ry/unit strain.

3. CALCULATED DEFORMATION POTENTIALS FOR ISOTROPIC STRAINS

Under hydrostatic strain the energy shifts are the sum of two terms, one due explicitly to the change in the lattice parameter and the other, more difficult to formulate, due to the change in the crystal potential within a unit cell. Gray *et al.* (1975) and Segall *et al.* (1978) treated the case of Cu.

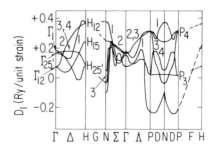

Fig. 4. Local deformation potential for volume strain of Nb (Silva and Ferreira, 1978). Numbers label the bands and the strain is an expansion.

Silva and Ferreira (1978) have calculated the local deformation potentials along symmetry directions in the Brillouin zone for Nb. The authors use the Green's function method and a perturbation scheme similar to Segall et al. (1978). The dispersion curves coincide with those of Mattheiss (1970), given in Section IV, B, 2. The deformation potentials, shown in Fig. 4, are far from being constant in wave-number space; extreme values for d states differ by about 0.4 Ry/unit strain but an average value lies at about 0.15 Ry/unit strain. It is particularly interesting to follow band 3, which crosses through the Fermi energy along $\Gamma\Sigma N$; the deformation potential increases first and then sharply drops. This result is apparently related to the singularity in this branch and the change of character resulting from hybridization with sp states.

4. CALCULATED DEFORMATION POTENTIALS FOR SHEAR STRAINS

In the pure-shear case it is assumed that the perturbation leaves unchanged the crystal potential within a unit cell (Juras and Segall, 1972, 1973; Griessen et al., 1977). Band structure shifts and splittings of degeneracies are due solely to the changes in symmetry and interatomic distances. These effects can be very large (Posternak and Steinemann, 1970).

Calculations for Cu have been performed by Juras and Segall (1972, 1973) with the Green's function method and by Gray et al. (1975, 1976) by the transformed plane-wave method. Rather similar results have been obtained and this is expected because the effective potential problem does not appear. It is interesting to extract from their work some values for the d-band states. For tetragonal deformation (relating to c', see Table I), D_1 is up to 0.3 Ry/unit strain at Γ and L and 0.07 Ry/unit strain at the X high-symmetry location. For trigonal deformation (relating to c_{44}), D_1 is 0.8 and 0.7 Ry/unit strain at Γ and L, respectively. It is seen that deformation potentials come out to be highly anisotropic. For trigonal deformation also, $|D_1| \approx 2.3$ Ry/unit strain for the hybrid state at $L_{2'}$ just below the Fermi level; this state has essentially plane-wave character and reflects with its large value the general behavior that $D \sim$ bandwidth as suggested by Weger and Goldberg (1973).

Pfersich et al. (1979) calculated the deformation potential for bcc lattices, starting from Pettifor's model Hamiltonian with only the d block retained. The best available parameters for bcc Mo were used, i.e., those of Posternak et al. (1975), which were found to fit successfully the stress dependence of the Fermi surface.

The objective of Pfersich et al. (1979) was to evaluate the thermodynamically relevant quantities, such as the Brillouin zone sums of the local deformation potentials and the strain dependence of the chemical potential. It is a model calculation that can justify the neglect of hybridization.

To first order, the condition for particle conservation is

$$\frac{dn_{el}}{d\gamma} = \sum_{k} \frac{dg_k}{d\gamma} = \sum_{k} \frac{dg_k}{d\varepsilon} \frac{d\varepsilon(k)}{d\gamma} = 0, \tag{11}$$

where n_{el} is the number of electrons, γ is now explicitly the shear strain and g_k the ground-state evaluation of the (statistical average) occupation number $f_k = [\exp\{\varepsilon(k) - \mu\}/kT\} + 1]^{-1}$. Then

$$D_1(\varepsilon) = \langle D_1(k,\varepsilon) \rangle = \sum_{k} \frac{\partial g_k}{\partial \varepsilon} \frac{d\varepsilon(k)}{d\gamma} \Big/ \sum_{k} \frac{\partial g_k}{\partial \varepsilon} = 0 \tag{12}$$

is the Brillouin zone average for the first-order deformation potential.

Sums, such as in relation (12) can be calculated by aid of the Gilat–Raubenheimer method. The deformation potentials are always tied to the symmetry of the strained lattice, whereas the derivative of the occupation number $dg_k/d\varepsilon$ has the symmetry of the unstrained lattice. Considering this fact, Pfersich et al. (1979) calculate separately the sum for any cubic irreducible zones ($\frac{1}{48}$ of the volume of the Brillouin zone), which compose the larger irreducible zones of the strained lattice. In the case of tetragonal strain, there are three zones, and for trigonal strain four cubic zones make out the irreducible zone of the strained structure. It is seen in Fig. 5 that the partial sums have different sign and energy dependence; their total sum, however, is always zero.

The interaction of electrons with phonons or a strained lattice is sometimes described as "redistribution of states" or "transfer," referring to strain induced shifts in local electronic occupations. The picture is misleading: the sum rule as exposed here shows nothing different than the anisotropy of

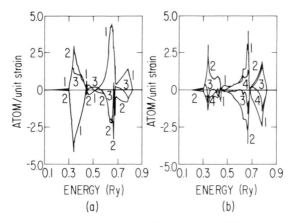

Fig. 5. The sums $\Sigma(\partial g_k/\partial \varepsilon)(d\varepsilon(k)/d\gamma)$ for the cubic irreducible zones: (a) 3 in the case of tetragonal shear and (b) 4 for trigonal shear; the total sum is always zero. The band structure is that of a d band with parameters of Mo.

scattering of Bloch electrons in the sheared lattice. As explained in Section III, B, 1, strain may modify the kinematic variable (velocity of the electron) in one or the other way, and this change equals a local deformation potential of variable size and sign.

Pfersich *et al.* (1979) have also calculated the Brillouin zone averaged square of the first-order deformation potentials, i.e.,

$$D_1^2(\varepsilon) = \langle D_1^2(\mathbf{k},\varepsilon) \rangle = \sum_{\mathbf{k}} \frac{\partial g_{\mathbf{k}}}{\partial \varepsilon} \left(\frac{d\varepsilon(\mathbf{k})}{d\gamma} \right)^2 \bigg/ \sum_{\mathbf{k}} \frac{\partial g_{\mathbf{k}}}{\partial \varepsilon}. \tag{13}$$

This quantity is related to the dynamics of the electrons and appears in one of the Fermi terms of the electronic contribution to elasticity. Its root, i.e., $[D_1^2(\varepsilon)]^{1/2}$, is shown in Fig. 6. A relation of this root with the direct average evidently does not exist.

The Brillouin zone averaged second-order deformation potentials, i.e.,

$$D_2(\varepsilon) = \langle D_2(\mathbf{k},\varepsilon) \rangle = \sum_{\mathbf{k}} \frac{\partial g_{\mathbf{k}}}{\partial \varepsilon} \frac{d^2\varepsilon(\mathbf{k})}{d\gamma^2} \bigg/ \sum_{\mathbf{k}} \frac{\partial g_{\mathbf{k}}}{\partial \varepsilon} \tag{14}$$

are shown in Fig. 7. This potential appears in the second Fermi term of the electronic contribution to elasticity. As seen above, this potential is connected to dynamical changes. It changes signs for bonding and antibonding states in the two halves of the band. In addition, it has opposite variations with energy for the two independent shears.

For the two shears, the deformation potentials do not have the same magnitude nor do they follow the same energy dependence. This is an expected anisotropy. It is further evident that the energy dependence is not similar to that of the density of states. In consequence, these potentials will play the dominant role for thermodynamic properties such as the variation

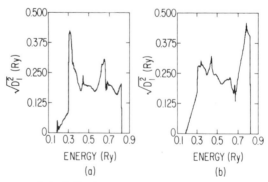

Fig. 6. Square root of the Brillouin zone averaged squared first-order deformation potential for the (a) tetragonal case and (b) trigonal strain.

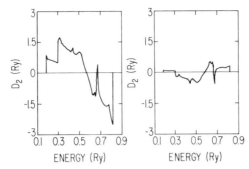

Fig. 7. Second-order deformation potential for (a) tetragonal shear and (b) trigonal shear.

of the elastic constants with the electron-per-atom ratio (alloying) and their temperature and pressure dependence.

Some remarkable features of the deformation potentials are the narrow peaks and shallow valleys, which can all be traced back to singularities in the energy dispersion. This sensitivity to the details of the band structure is expected since the velocities and effective masses of the electrons take extremal values at singularities. Thus, a shoulder in the density of states for 0.45 Ry and pronounced peaks in all deformation potentials are clearly linked with the singularity at $\Gamma_{25'}$ (two saddle points and one minimum). The peak heights (above background) are about 0.05 and 0.08 Ry/unit strain for $\sqrt{D_1^2}$ and about 0.16 and 0.2 Ry/(unit strain)^2 for D_2 in the case of tetragonal and trigonal deformations, respectively. The energy of $\Gamma_{25'}$ nearly corresponds to the band occupation of Nb, which has anomalous properties of the shear elastic constants and deformation potential for c_{44} of the above size. This conclusion seems reasonable even if the model calculation neglects hybridization, because locally the states have d character and the disposition curves are rather similar to those of the full calculation of Mattheiss (1970).

5. CHEMICAL POTENTIAL IN SHEARED CRYSTALS

As a function of shear strain, the chemical potential of a system of independent electrons (fermions) is

$$\mu(\gamma) = \mu(0) + \frac{d^2\mu}{d\gamma^2}\bigg|_{\gamma=0} \frac{1}{2}\gamma^2 + \cdots. \tag{15}$$

The result is valid for high-symmetry shears (i.e., those shears that leave unchanged the symmetry of the structure, regardless of the sign of the stress that produces them). The first-order derivative is zero by virtue of the sum

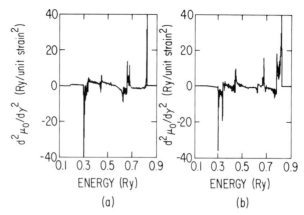

Fig. 8. Second-order strain derivative for the chemical potential for (a) tetragonal shear and (b) trigonal shear. The derivative is the sum of $D_2(\varepsilon)$ of Fig. 7 and the energy derivative of ND_1^2, which was obtained by numerical derivation. The band structure is that of a d band with parameters of Mo.

rule [Eq. (11)]. From the particle conservation condition, the second derivative is

$$\frac{d^2\mu}{d\gamma^2} = D(\mu) - \frac{1}{N(\mu)} \left\{ \frac{\partial}{\partial \varepsilon} \left[N(\varepsilon) D_1^2(\varepsilon) \right] \right\}_{\varepsilon = \mu} \tag{16}$$

The density of state formalism instead of the energy-level formalism is used for convenience; to do so, the definitions of Eqs. (13) and (14) are used, together with

$$\sum \partial f_{\mathbf{k}} / \partial \varepsilon \equiv 2 \int d\varepsilon (\partial f / \partial \varepsilon) N(\varepsilon) = -2N(\mu).$$

The second-order strain derivative of the Fermi level is clearly the most sensitive quantity to use in finding the energy regions where elastic anomalies will exist and to weigh the importance of the deformation potentials (Pfersich 1980). For the model calculation of a d band, the result is shown in Fig. 8. It is seen that the dominant term is the second term of relation (16), including D_1^2. Any singularity of the energy dispersion is marked by a positive or negative peak, depending on the type of the singularity, and the type of strain determines which singularities are affected because it determines what degeneracies are lifted. The height of the peaks and their number differ between the two shears. In the region between about 0.4 and 0.6 Ry, corresponding to the occupation of a stable bcc structure, one anomaly related to crossing through $\Gamma_{25'}$ exists which is stronger for the shear corresponding to c_{44}.

6. ELECTRONIC CONTRIBUTIONS TO THE SHEAR ELASTIC CONSTANTS

As for the free energy $F = n_{el}\mu$, electronic contributions to the shear elastic constants c_{el} can be derived by taking $d^2\mu/d\gamma^2$ over occupied states so that

$$c_{el} = \frac{d^2F}{d\gamma^2} = \sum_k g_k \frac{d^2\varepsilon(k)}{d\gamma^2} + \sum_k \frac{\partial g_k}{\partial \varepsilon}\left(\frac{d\varepsilon(k)}{d\gamma}\right)^2$$

$$= 2\int d\varepsilon\, g(\varepsilon)N(\varepsilon)D_2(\varepsilon) + 2N(\varepsilon)D_1^2(\varepsilon). \tag{17}$$

The result was obtained by Fischer *et al.* (1969), Peter *et al.* (1974), Fletcher (1978), and Steinemann (1979) and is of identical content as the equations that apply to special cases of the band structures of semiconductors (Keyes, 1967) and of A-15 superconductors (Barisic and Labbe, 1967).

Pfersich *et al.* (1979) and Pfersich (1980) have evaluated the two terms of Eq. (17) shown in Fig. 9. The first term of c_{el} varies roughly parabolically with the filling of the band by bonding and antibonding states. It is of a repulsive type for the c' elastic constant and attractive for c_{44}; the difference may be associated with the fact that bonds stretching between second nearest neighbors are essential for the tetragonal deformation but only nearest-neighbor overlaps are important for c_{44}. The second term is always negative and goes in the direction of a "softening." Its strong variation with energy is clearly related to the details of the band structure, especially

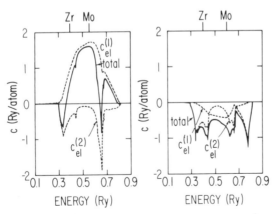

Fig. 9. Calculated electronic contributions to the elastic constants (a) $\frac{1}{2}(c_{11} - c_{12})$ and (b) c_{44}. The model is that of a d band with parameters of Mo.

around 0.45 Ry where the change of Fermi surface topology produces an anomaly.

The bcc phase is stable between approximately 0.41 and 0.59 Ry (an equivalent occupation for the single d band is $n_d \approx 2.5-4.9$). When comparing Fig. 9 with the experimental elastic constants shown later in Figs. 16 and 17, it is seen that the calculation essentially reproduces the variations; c_{44} has a pronounced minimum and c' increases with filling of the band and peaks near its middle. The absolute values clearly will not match because a lattice part must be added.

Peter et al. (1974) and Ashkenazi et al. (1978) calculated the elastic constants in a tight-binding scheme. In the former work, transformations are developed that simplify an expansion in energy. The calculations yield electronic contributions to the elastic constant of fcc and bcc metals that are of the expected magnitude. In the second work, a hybridized band is fitted to the augmented plane-waves calculations of Mattheiss (1970) for Nb. These results reproduce correctly the temperature dependences and alloying variations for the elastic constants of bcc Nb–Zr in a rigid-band picture. An equally satisfactory agreement between theory and experiment is found in a recent study of Nb–Mo alloys (Bujard et al., 1980).

C. Separation of Electronic and Lattice Contributions to Elasticity

Experimental data on elastic properties and their temperature and pressure dependence may give a good deal of information about the lattice-dynamical and electronic properties of a solid. A principal question then arises: Are the thermodynamic quantities strictly separable as sums of the electronic and lattice contributions? One can show that this is indeed so for the elastic moduli. The approach requires a rigorous thermodynamic treatment of the total energy, the lattice energy, and the electronic excitation energy.

1. Total Energy of the Crystal

The total energy of a crystal is the sum of the kinetic energies of positive ions and negative electrons and of the interaction energies between ions, electrons, and ion–electron pairs. Following standard theory (see Wilson, 1965; Wallace, 1972), the large difference in ion–electron masses enables the use of the adiabatic approximation that considers electrons and phonons independently moving. As a result, the adiabatic potential may be written as

$$\phi_{tot} = V_{io} + U_{el} = V_{io} + \sum_k g_k \varepsilon(k_1 \eta) - V_H - V_{xc}, \qquad (18)$$

where V_{io} is the bare ion–ion Coulomb interaction and U_{el} the total electronic energy, which is a parametric function of the ionic coordinates and, therefore, strain. In the second expression, this latter energy is replaced by the sum of self-consistent one-electron eigenvalues and the electron–electron interaction energy double counted in the sum is explicitly subtracted off. V_H is the Hartree energy and V_{xc} the exchange and correlation energy.

Each of the terms in Eq. (18) has a clear identity and may be grouped for convenience. At zero temperature, the above potential equals the total energy and the second derivative with respect to strain is

$$\frac{d^2 U_{tot}}{d\eta^2} = c = \frac{d^2}{d\eta^2}(V_{io} - V_H - V_{xc}) + \frac{d^2}{d\eta^2}\sum_k g_k \varepsilon(\mathbf{k}_1\eta). \qquad (19)$$

For the shear strains the use of relation (17) and definition (7) gives

$$c = c_{io} + \sum_k g_k D_2(\mathbf{k},\varepsilon) + \sum_k \frac{\partial g_k}{\partial \varepsilon}D_1^2(\mathbf{k},\varepsilon)\Big|_{\varepsilon=\mu}$$

$$= c_{io} + c_{el}^{(1)} + c_{el}^{(2)} = c_{la} + c_{el}. \qquad (20)$$

In grouping $c_{io} + c_{el}^{(1)}$ for a lattice part c_{la}, the sum of all changes in the electronic energy that are second order in the displacements is added to the ion–ion interactions. $c_{el}^{(1)}$ varies smoothly with the filling of the band as was seen in the preceding section, and so does c_{io} due to the repulsive interaction of ions embedded in a Hartree field. c_{el} is of a different kind; it reflects changes at the Fermi level that are first order in the strains. Hence, an energy $c_{el} \cdot \frac{1}{2}\gamma^2$ might be looked at as a "mechanical" polarization (Weinmann, 1977). It is further noted that this term is associated with the inertial effects in the strain response [Eq. (10), second term].

Varma and Weber (1977) do a similar grouping of terms for the dynamical matrix. They show that the ion interaction, double-counting correction, and second-order response together produce a short-range force field that can be parametrized by a few near-neighbors coupling constants. The remaining term, however, is long range and essentially produces the anomalous features in phonon dispersion whenever singularities near the Fermi level are present. Calculations agree remarkably with neutron scattering data for Nb, Mo, and their alloys and also predict the softening of the transverse mode for long wavelengths in Nb and Nb-rich alloys.

Varma and Weber use a fit with experimental dispersion curves for the short-range forces. On the other hand, no satisfactory theoretical approach to the lattice part of the elastic constants of transition metals were known until recently when Ashkenazi et al. (1980) showed that a variational procedure, together with the density functional, can be used to find the electronic ground-state energy and the effective charges for the Madelung energy in a strained crystal. The elastic constants can be extracted; the calculated c' for the nonmagnetic fcc transition metals are in good agreement with measured ones.

2. DUCASTELLE'S MODEL FOR THE ELASTIC CONSTANTS

The essential contribution to cohesion in transition metals is the electronic energy (Friedel, 1964). A repulsive part that assures equilibrium of the crystal is, in absolute value, only a small fraction of U_{el}. Ducastelle (1970) assumes that this repulsive energy U_{re} is given by sum of pair potentials of Born–Mayer type. On the other hand, the subtle association of a Gaussian with the second moment, m_2, of a tight-binding d band gives an equally simple form for the electronic energy in which the strain derivatives can be calculated as if they were dependent on pair interactions. The cohesive energy is

$$U_{co} = -U_{el} - U_{re} = B\beta_0 \exp\{-qR\}\exp\{-X_F^2/2\} - A\exp\{-pR\},$$

$$(21)$$

where B, β_0, A are constant, R the distance between first neighbors, $X_F = E_F/\sqrt{m_2}$ measures the filling of the band and q and p are coefficients. The equilibrium interatomic distance and the quasiisotropic elastic constants come out to be

$$R_0 = \frac{1}{p-q}\left(\frac{X_F^2}{2} + \log\frac{p^A}{qB\beta_0}\right),$$

$$(22)$$

$$K = \left(\tfrac{1}{9}\right)pqR_0^2 U_{co} = \left(\tfrac{1}{9}\right)\left[1 - (q/p)\right]p^2 R_0^2 U_{re},$$

$$(23)$$

and

$$G = \tfrac{1}{15}\left[1 - 2(q/p)\right]p^2 R^2 U_{re}$$

$$(24)$$

The ratio K/U_{co} is 2 to 2.5 at the left of the transition metal series and increases to 4 to 4.5 at the right with the antibonding states; an average value for the whole series is 3.4. On the other hand, the product of Slater's overlap parameter and the equilibrium interatomic distance, i.e., qR_0, is ≈ 3. Thus $pR_0 \approx 10$ and $q/p = -U_{re}/U_{ba} \approx 0.3$. With these rough figures, it is found that $U_{co}/U_{el} \approx \tfrac{2}{3}$ and, because q/p measures the importance of the band part, the latter has a much stronger influence in G than in K. The parabolic variations of the lattice parameter and the elastic constant throughout the series are reproduced by this simple analysis.

3. ELASTIC CONSTANTS FROM THE LATTICE ENERGY

The adiabatic potential of Eq. (18) may be cast in a form suitable for the discussion of the elastic constants and their temperature and configuration dependence. Excitations of the lattice and of the partially filled electron band have to be considered. To split off the latter, it may be written as

$$\phi_{tot} = V_{io} + U_{el}(T = 0) + \left[U_{el}(T) - U_{el}(T = 0)\right]$$

$$= V_{io} + U_{el}(T = 0) + \sum_{\mathbf{k}} (f_{\mathbf{k}} - g_{\mathbf{k}})\varepsilon(\mathbf{k},\eta)$$

$$= \phi_{1a} + \sum_{\mathbf{k}} (f_{\mathbf{k}} - g_{\mathbf{k}})\varepsilon(\mathbf{k},\eta). \tag{25}$$

Now, ϕ_{1a} expresses the total crystal energy when the electrons are in their ground state.

Lattice thermodynamics is based on an expansion of ϕ_{1a} in terms of the ion displacements $\mathbf{U}$

$$\phi_{1a}(\mathbf{R} + \mathbf{U}) = \phi_0(\mathbf{R}) + \phi_1(\mathbf{U}) + \phi_2(\mathbf{U}_l\mathbf{U}) + \phi_3 \cdots, \tag{26}$$

where the ϕs are the products of coupling parameters and displacements in increasing order. In particular, ϕ_0 is the total crystal energy when all the ions are in their equilibrium position and in the second-order term, $-\sum_j [\partial^2\phi(M,N)/\partial U_i(M)\partial U_j(N)]U_j(N)$ is the force acting on ion M in i direction if the ion N is displaced in j direction. This restoring force determines the lattice excitations of lowest order, i.e., the harmonic phonons, and a harmonic elastic constant c_{ha}. The free energy is the harmonic free energy $F_{ha}(\phi_2)$.

The ion positions that make the potential energy a minimum may not be the equilibrium positions, e.g., if external forces are applied or if there is a finite temperature. Force and thermal expansion change the configuration and, to find the new equilibrium positions, one has to take derivatives of the ϕs with respect to strains that involve the anharmonic coefficients. Furthermore, the phonon occupation number $n_q(T)$ implies an explicit strain dependence of the phonon frequency, ω_q, through $\partial n_q/\partial\eta = (\partial n_q/\partial\omega_q)$ $(\partial\omega_q/\partial\eta)$. The higher-order coupling parameters are related to the higher-order elastic constants and with $\partial\omega_q/\partial\eta$, define the generalized Gruneisen parameters $\gamma_q = -\partial\ln\omega_q/\partial\eta$. On the other hand, the anharmonic frequency shifts, the finite lifetime of the phonon and the free energy $F_{an}(\phi_3^2,\phi_4,\text{etc.};$ $\phi_1^2,\phi_1,\phi_3,\text{etc.})$ all have an explicit temperature dependence as T, T^2, T^3, etc.

These intricate theories have been worked out in detail by Leibfried and Ludwig (1961), Ludwig (1967), and Wallace (1972), but other references should be named such as Born and Huang (1957) and Cowley (1967). The results obtained for the elastic constants may be cast in some simple formula:

(i)   At low and intermediate temperatures, the quasiharmonic and full anharmonic theories lead to

$$c(T) = c_{ha} - a\sum_{q} f(\gamma_q, c)\varepsilon(\omega_q, T) \tag{27}$$

for the adiabatic elastic constants, where a is a constant, c is related to high-order elastic constants, γ_q are first- and second-order generalized Gruneisen parameters, and $\varepsilon(\omega_q, T) = \hbar\omega_q[n_q(T) + \frac{1}{2}]$ is the average thermal energy of an oscillator.

(ii)   If mean values for the Gruneisen parameters and the nondispersive Debye model are introduced, then

$$c(T) = c_{ha}[1 - af(T/\theta_D) + bp], \tag{28}$$

where $f(T/\theta_D)$ is the thermal function going as T^4 at low temperatures and as T for $T > \theta_D$ with an intercept at $T = 0$ equal to $-0.37(T/\theta_D)$. The last term represents configuration variations under pressure p and may, for example, compensate the thermal expansion.

(iii)   For high temperatures,

$$c(T) = c_{ha}(1 - aT - bT^2 - cT^3 \cdots + dp), \tag{29}$$

where $b_1 c$ comprise generalized Gruneisen parameters and elastic constants of various higher orders.

It is of further interest to know the behavior, with respect to temperature, of the Gruneisen parameter

$$\gamma(T) = \gamma(1 + bT + \cdots), \tag{30}$$

which depends on temperature only in ranges where the elastic constants have a sensible T^2 contribution.

The approximation of Eq. (28) gives quite good results in nearly all metals (Leibfried and Ludwig, 1961; Lakkad, 1971; Wallace, 1972). An exception is noted in Ti, where c_{66} has a maximum at 18 K and a $-T^4$ dependence (or T^2, due to electrons) is not found (Alers and Karbon, 1968). Ludwig (1967) explains that the approximation (28) may indeed be inapplicable. At low temperature, the γs need not be positive for all modes so that the sum in the full expression (27) may become negative.

Relation (29) may be tested with known and recent measurements carried to very high temperatures. In Table III, experimental data are fitted to the expansion Eq. (29) as a function of the reduced temperature $T' = T/T_m$. Molybdenum and W have low electronic density of states so that the elastic

TABLE III

TEMPERATURE DEPENDENCE OF ELASTIC CONSTANTS[a]

Metal	Measured up to T'	T_m [K]	Elastic constant	c_{ha} [10^{11} N/m^2]	Coefficients			References
					a'	b'	c'	
V	0.97	2170	c'	0.584	0.461	0.089	0.018	Walker (1978)
Nb	0.92	2741	c'	0.597	0.353	0.084	0.060	Talmor et al. (1977)
Ta	0.96	3270	c'	0.532	0.450	0.078	0.009	Walker and Bujard (1980)
Mo	0.81	2883	c'	1.656	0.683	small		Bujard et al. (1980)
			c	1.113	0.270	0.101	0.045	
W	0.56	3683	c'	1.650	0.442	0.358	—	Lowrie and Gonas (1967)
			c	1.635	0.207	0.170	—	
			K	3.156	0.207	-0.015	—	

[a] $c(T') = c_{ha}(1 - a'T' - b'T'^2 - c'T'^3)$ with $T' = T/T_m$ for various metals.

constants essentially reflect the lattice thermodynamical properties. In V, Nb, and Ta, the electronic excitation contribution to c' is negligible. The anharmonic effects are strong in all these metals. It is seen that the low-order anharmonic variation of the cs is valid up to $T/T_m \approx 0.5$ and that near $T/T_m \approx 1$, the coefficients decrease roughly as $a'/b'/c' \sim 1/\frac{1}{3}/\frac{1}{10}$ and stronger. This is not necessarily an acceptable sequence for a perturbation expansion.

4. Electronic Excitation Contribution to the Shear Constants

The part of the elastic constants associated with the electronic excitation free energy $\Sigma(f_{\mathbf{k}} - g_{\mathbf{k}})\varepsilon(\mathbf{k}, \gamma)$ in Eq. (25) reads

$$c_{\mathrm{el}}(T) = \sum_{\mathbf{k}} (f_{\mathbf{k}} - g_{\mathbf{k}}) \frac{d^2\varepsilon(\mathbf{k})}{d\gamma^2} + \sum_{\mathbf{k}} \left(\frac{\partial f_{\mathbf{k}}}{\partial \varepsilon} - \frac{\partial g_{\mathbf{k}}}{\partial \varepsilon} \right) \left(\frac{d\varepsilon(\mathbf{k})}{d\gamma} \right)^2. \quad (31)$$

In principle it can be computed directly for any temperature. But, an analytical treatment is preferable because it helps to explain the phenomenologically difficult question for the relative importance of the deformation potential and can indicate the correct grouping of terms in Eq. (20).

Equation (31) may be translated into density of states formalism, using Eqs. (13) and (14), and then be executed by the Sommerfeld expansion to give (Steinemann, 1979)

$$c_{\mathrm{el}}(T) = 2 \int d\varepsilon (f - g) N(\varepsilon) D_2(\varepsilon) + \int d\varepsilon \left(\frac{\partial f}{\partial \varepsilon} - \frac{\partial g}{\partial \varepsilon} \right) N(\varepsilon) D_1^2(\varepsilon)$$

$$= -2 N D_1^2 \frac{a_{\mathrm{el}}}{|a_{\mathrm{el}}|} \frac{T^2}{T_F^2} + O(T^4), \quad (32)$$

where

$$\frac{\pi^2 k^2}{6} \left[-\frac{D_2'}{D_1^2} + \frac{N''}{N} - \left(\frac{N'}{N} \right)^2 + 2 \left(\frac{N'}{N} \right) \left(\frac{D_1'}{D_1} \right) + 2 \left(\frac{D_1'}{D_1} \right)^2 + \frac{D_1''}{D_1} \right] = a_{\mathrm{el}}.$$

The abbreviations $N = N(\mu_0)$, $N' = [\partial N(\varepsilon)/\partial \varepsilon]_{\varepsilon=\mu_0}$, $D_1^2 = D_1^2(\mu_0)$, $D_1' = |[\partial D_1(\varepsilon)/\partial \varepsilon]_{\varepsilon=\mu_0}|$, etc., are used and refer to the chemical potential at zero temperature. The derivatives characterize the band structure and its strain response and, together with $T_F = |a_{\mathrm{el}}|^{-1/2}$, define a local degeneracy or effective Fermi temperature. a_{el} itself may take positive or negative values according to the sign of N'', D_1', etc., so that the incoherent thermal excitation produces a decreasing or increasing elastic constant. This is exactly what happens in the case of Zr and Nb, respectively.

From Eqs. (31) or (32) it is obvious that an excitation in the Fermi term $c_{el}^{(1)}$ will be observable only insofar as D_2 varies strongly with energy; a glance at the results of the model calculation, in Fig. 7, shows that this is not the case. Thus, the inclusion of $c_{el}^{(1)}$ with the lattice term is justified [Eq. (20)].

If N_1', N_1'', D_1', etc., are large, i.e., if the band has structure, then T_F is small. Lipton and Jacobs (1970) have calculated the effective Fermi temperature for Pd whose Fermi level lies near a peak. Taking into account only $N''/N - (N'/N)^2$, they find $T_F \sim 1000$ K. A measure for T_F might also be obtained from the width of peaks in $\sqrt{D_1^2}$ of Fig. 6 or $c_{el}^{(2)}$ in Fig. 9; at 0.45 Ry the peaks have a width $\Delta W \approx 0.02$ Ry and thus $\Delta W/k \sim 3000$ K $= T_F$. A thermodynamic property will then include variable electronic effects, but, for temperatures exceeding T_F, the excitation fades out because the essential features of the band are no longer "sensed" by the fermions.

The peaks and valleys in the band and its strain response are associated with flat energy versus wave-vector regions and the presence of singularities in the dispersion. But otherwise, a smooth background has typically a width $W \approx 0.4$ Ry and $T_F = W/k \sim 20T_m$. The structureless part of the band will thus not lead to temperature-dependent properties.

Following this discussion, Eq. (20) may be formulated for any temperature as the electron renormalized elastic constant

$$c(T) = c_{io} + c_{el}^{(1)} + c_{el}^{(2')}(T_F \gg T_m) + c_{el}^{(2'')}(T_F \lesssim T_m)$$

$$= c_{la}(T) + c_{el}(0)\left(1 + \frac{a_{el}}{|a_{el}|}\frac{T^2}{T_F^2}\right). \tag{33}$$

The variation of the first three terms together is described by the phonon thermodynamical relations of the preceding section. For the last term $c_{el}^{(2'')} = c_{el}(T) = 0$ for $T > T_F$, and can otherwise be combined with Eq. (32). It might also be recalled that the lowest-order Sommerfeld expansion is satisfactory for $T \lesssim \frac{1}{2}T_F$ (Wilson, 1965).

The effect of lattice vibrations upon the band and its strain response might be taken into account by adding to the free energy the nonadiabatic, or electron–phonon interaction terms. Wallace (1972) finds the free energy augmented by the electron–phonon interaction that has an explicit temperature dependence as $T^2(1 + k_1/T - k_2T)$ for $T > \theta_D$. k_1 and k_2 are the configuration dependencies of the interaction. This variation is retained for the elastic constants. Whereas a T^2 dependence of the excitation contribution may be verified for c_{44} in Nb and Pd, other variations do not appear; the renormalization effects on the elastic constants are apparently small.

$c_{el}^{(1)}$ and D_2 cannot be extracted from experiments, but c_{el} and $\sqrt{D_1^2}$ may be under certain conditions. One could use the correlation of shear elasticity with magnetic susceptibility. The well-known result for the spin susceptibility is (Wilson, 1965)

$$
\chi_{sp}(T) = -\mu_B^2 \sum_k \frac{\partial f_k}{\partial \varepsilon} = -2\mu_B^2 \int d\varepsilon \, \frac{\partial f}{\partial \varepsilon} N(\varepsilon)
$$

$$
= 2\mu_B^2 N(\mu_0)\left(1 + \frac{a_{sp}}{|a_{sp}|} \frac{T^2}{T_F^2}\right) + O(T^4), \tag{34}
$$

with $a_{sp} = (\pi^2/6)k^2[(N''/N) - (N'/N)^2]$ and $|a_{sp}|^{-1/2} = T_F$; μ_B is the Bohr magneton. It is noted that χ_{sp} has the same form as c_{el}, except for the opposite sign, and that phenomenologically both quantities are connected to a polarization. Setting $a_{sp} \approx a_{el}$, it is

$$
c_{el} = -(D_1^2/\mu_B^2)\chi_{sp}. \tag{35}
$$

Normally χ_{meas} has sizable diamagnetic and orbital parts. However, these have negligible temperature dependence and

$$
\frac{dc_{meas}}{dT} - \frac{dc_{la}}{dT} = -\frac{D_1^2}{\mu_B^2} \frac{d\chi_{meas}}{dT} \tag{36}
$$

now refers to the excitation contribution c_{el}. By this procedure it is found that $\sqrt{D_1^2}$ is 0.08 Ry/unit strain for c_{44} in Nb, 0.12 Ry/unit strain for c_{66} in Ti and Zr (Posternak et al., 1968), and about 0.04 Ry/unit strain (when corrected for exchange enhancement) for c_{44} in Pd (Weinmann, 1977).

Some early studies of the T dependence of the c_{ij} in several metals were carried out by Alers and Waldorf (1961) with high-resolution acoustic velocity measurements. The experiments were designed to measure the very small changes in c_{ij} due to the normal-to-superconducting transition. In the normal states, they found a variation of the elastic constants as $-\alpha T^2$. In Nb and V, the measured coefficients give $T_F = (\alpha)^{-1/2} \sim 1000$ K, but, curiously, the sign cannot explain the temperature dependency at medium temperatures. It was also found that the decrease of c_{44} in the superconducting state was 3 to 4 times the effect on c' for both Nb and V. This work was instrumental in prompting the calculations by Bernstein (1963, 1965), who extended the "electron transfer" model of Leigh (1951) to include temperature dependency and changes with the superconducting transition.

IV.   Experimental Data on Elastic Properties of Transition-Metal Alloys

A.   Introduction

As noted in Section III, C, the electronic contributions to elastic moduli c_{el} are indicated primarily in the measured temperature derivatives, as deviations from the normal quasiharmonic lattice behavior. In order to establish clearly the existence of significant c_{el} terms, it is necessary, however, to measure the variations in dc_{ij}/dT with alloy composition in simple systems where the chemical changes can be related to electrons-per-atom (e/at.) ratio. This type of correlation becomes even more convincing when e/at. changes can be reasonably associated with Fermi surface changes, as, for example, in systems where rigid-band behavior over a wide range of composition is indicated in experimental data. The interpretation is, however, still somewhat clouded if changes with chemical composition cannot be separated into effects of pure volume changes and electronic changes at constant volume, as discussed in Section III, C. Consequently, the presence of experimental data on dc_{ij}/dV is an essential factor in deriving the final quantitative separation of c_{la} and c_{el}.

Such ideal data are now available for the bcc solid-solution system Zr–Nb–Mo–Re, where the composition dependence of the c_{ij} and their temperature and pressure derivatives have been measured for the binary alloys Zr–Nb, Nb–Mo, and Mo–Re, extending from 80 at.% Zr to 35 at.% Re. This system has the further advantage of having measured values of the Gruneisen constant, as deduced from low-temperature thermal expansion data. These precise low-temperature data allow a separation of the electronic γ, γ_e, from the lattice γ, γ_l. In addition, the phonon dispersion curves are known for six alloys of Nb–Mo, for pure Zr in the α(hcp) and β(bcc) phases and in Zr–Nb alloys with 8 and 20 at.% Nb. On the theoretical side, the electronic band structure is well known in considerable detail for Nb and Mo, and a study of departures in the alloys from rigid-band approximation has been carried out with the CPA procedure. Thus, correlations of singularities in the composition dependence of the c's with Van Hove singularities can be reasonably well specified.

B.   BCC Transition-Metal Systems

For the three transition-metal series, the bcc phase is stable for the groups 5 and 6 elements and their alloys in the range between about 4.3 and 6.3

e/at. This structure has minimal energy for a similar electronic dispersion throughout the series (Pettifor, 1970).

1. Experimental Results for Zr–Nb–Mo–Re Binary Solid Solutions

According to the phase diagram of Nb–Zr given in Hansen and Anderko (1958), a single-phase solid solution extends completely across the entire range of composition at high temperature. It decomposes below 1135 K into a hcp + bcc structure at the Zr end and into two bcc solid solutions below 1243 K for alloys with a Zr content between 83 and 13%. However, this β decomposition is quite sluggish or may even not occur in alloys with interstitial contents below a solubility limit of some 100 ppm (Berghout, 1962; Hillmann and Pfeiffer, 1967). Nb and Mo are mutually soluble in the solid state, but Van Torne and Thomas (1966) suspect a segregation tendency because the alloys easily cleave. The solubility of Re in Mo extends to about 35%.

The major contributions to the study of elastic properties of this system are as follows: (i) temperature dependencies for Zr–Nb from $4 \leq T \leq 300$ K by Goasdoue et al. (1972), Walker and Peter (1977) and from $300 \leq T \approx T_{\text{melt}}$ by Ashkenahi et al. (1978); for Nb–Mo from $83 \leq T \leq 375$ K by Hubbell and Brotzen (1972); for Mo–Re from $83 \leq T \leq 473$ K by Davidson and Brotzen (1968); (ii) hydrostatic pressure dependencies at 300 K for Zr–Nb alloys by Weiss (1979), and for Nb–Mo by Katahara et al. (1977). This wealth of data will be the base for an extended discussion of lattice and electronic contributions to elasticity.

a. Temperature Variation of Elastic Constants. Because of the wide range of dc_{ij}/dT in the Zr–Nb and Nb–Mo alloys, the terminology used in the following description needs a brief explanation. If the modulus decreases steadily with temperature, as is the case for harmonic and anharmonic excitations of the lattice (see Section III,C,2), the behavior will be called "normal." On the other hand, the "anomalous" variation is connected to the electronic excitations as explained in Section III,C,3.

The measurements of c_{44} over the large temperature range of Fig. 10 show a marked anomaly of this elastic constant in all the Nb–Zr alloys, except for the alloy with 30% Zr. This is to be compared with the $c' = \frac{1}{2}(c_{11} - c_{12})$ versus T plots given in Fig. 11. In this case, the measurements show very little change in slope, which is always negative as expected for the "normal" behavior expected from quasiharmonic lattice theory. These remarkable measurements, so near to the melting temperatures of the alloys, are extremely useful in qualitatively defining the electronic contributions and

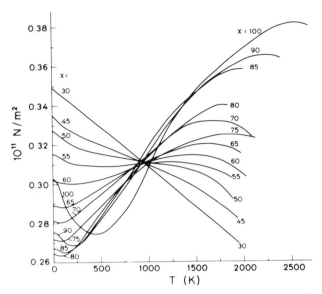

Fig. 10. Temperature and composition dependence of c_{44} in Nb–Zr alloys following Walker and Peter (1977) and Ashkenazi *et al.* (1978). x is the Nb concentrations in at.%.

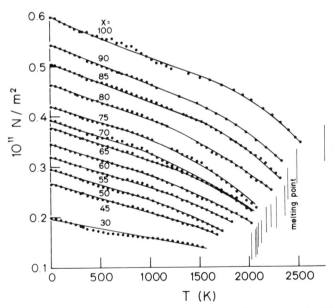

Fig. 11. Temperature and composition dependence of $c' = \frac{1}{2}(c_{11} - c_{12})$ for Nb–Zr alloys following Walker and Peter (1977) and Ashkenazi *et al.* (1978). x is the Nb concentration.

the lattice anharmonicities by reverse extrapolation from the temperature derivatives at high T. This reverse extrapolation uses Eq. (33) and Eq. (29); for Nb, $c_{1a} \approx 0.53 \times 10^{11}$ N/m^2 at 0 K and the electronic softening c_{el} amounts to about 0.22×10^{11} N/m^2.

The bulk modulus of the Nb–Zr alloys (Fig. 12) has a small negative variation. This is an expected behavior because in the absence of symmetry-breaking strains no temperature-dependent polarization effects can occur. Only interband electron transfers can have an effect, but these are more likely to happen in compounds with some ionic character.

No high-temperature data exist for Nb–Mo and Mo–Re alloys. Nevertheless, electronic effects upon the elastic constants may be seen in c_{44} (Fig. 13) and c' (Fig. 14). dc_{44}/dT has a marked positive trend for four alloys with a maximum for Nb34Mo. These alloys also show a clear deviation from a normal variation in dc'/dT, the maximum "anomalous" variation occurring for Nb23Mo. Between 52% Mo in Nb and 27% Re in Mo or

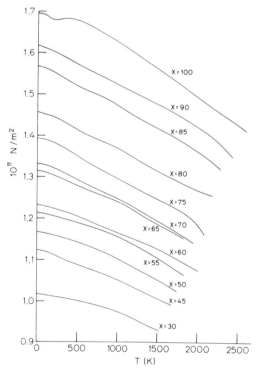

Fig. 12. Temperature and composition dependence of $K = \frac{1}{2}(c_{11} + 2c_{12})$ for Nb–Zr alloys following Walker and Peter (1977) and Ashkenazi et al. (1978). x is the Nb concentration.

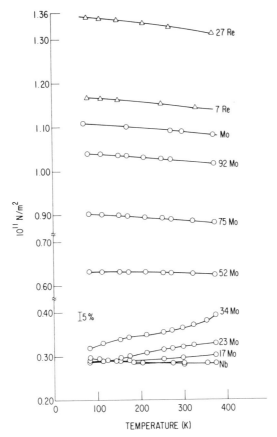

Fig. 13. Temperature and composition dependence of c_{44} for Nb–Mo ($\bigcirc$) (Hubbell and Brotzen, 1972) and Mo–Re alloys ($\triangle$) (Davidson and Brotzen, 1968).

5.52 < e/at.< 6.27, the anomaly is absent and there are essentially no electronic excitation contributions.

The bulk modulus of Nb–Mo alloys has an anomalous temperature dependence for 34 and 52% Mo, but the small effects may be the result of the indirect determination of K from the c_{12} calculation (Fig. 15). The larger slope of the Mo27Re alloy is probably due to anharmonicity connected to the heavy Re atom.

b. Composition Dependence of the Elastic Constants. The $c = c_{44}$ elastic constant (Fig. 16) shows a very large variation with composition at $T < 300$ K and, in particular, falls to low values at electron-per-atom ratios where anomalous temperature variations occur, i.e., between electron-per-atom ratios of 4.5 to 5.5. At low temperature, the negative electronic excitation contribution, i.e.,$-2ND_1^2$, is then fully effective or has its degenerate value.

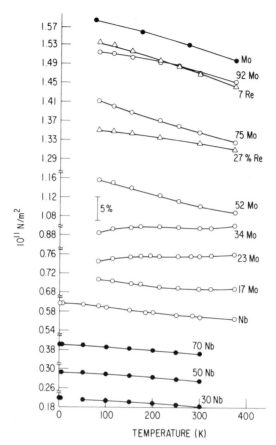

Fig. 14. Temperature and composition dependence of $c' = \frac{1}{2}(c_{11} - c_{12})$ for Nb–Mo ($\bigcirc$) (Hubbell and Brotzen, 1972), Mo–Re alloys ($\triangle$) (Davidson and Brotzen, 1968), and Zr–Nb ($\bullet$) alloys.

As the elastic constants of Nb–Zr alloys are known up to the melting point (Fig. 10), the constant without the excitation electronic part, i.e., c_{la}, may be constructed by the reverse extrapolation explained above, Section III, B, 1, a. The results for several Nb–Zr alloys are indicated in Fig. 16. We also include the value of c_{44} for bcc Zr at 300 K, as extrapolated from phonon dispersion data by Stassis *et al.* (1978). It appears that c_{la} has a smooth variation with composition and that the excitation part makes a large difference to the measured elastic constants. An estimate for the effective Fermi temperature can now also be made with a simple procedure derived from Eq. (33), i.e., the position of the minimum in an anomalous $c(T)$. T_F is of the order of 2000 K for Nb and Nb-rich alloys

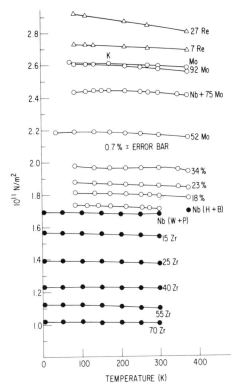

Fig. 15. Temperature and composition dependence of the bulk modulus K for Nb–Mo ($\bigcirc$) (Hubbell and Brotzen, 1972), Mo–Re ($\triangle$) (Davidson and Brotzen, 1968), and Zr–Nb ($\bullet$) alloys. (H + B), Hubbell and Brotzen (1972); (W + P), Walker and Peter (1977).

and falls to about 1000 K for the equiatomic alloys. In fact, around these temperatures the slopes dc/dT begin to decrease, which indicates that degeneracies are lifted. A similar analysis cannot be done for Nb–Mo alloys but a guess using the correlation between susceptibility and shear elastic constants suggests the tentative curve given. The electronic excitation contribution to the elastic constants is apparently even larger than for Nb–Zr alloys. At higher e/at., the rapid and nearly linear increase of c_{44} continues through pure Mo and the 27% Re alloy; it is associated with the filling of the band and the volume contraction as explained by Ducastelle's model (see also Section IV, B, 6).

The measurements of the $c' = \frac{1}{2}(c_{11} - c_{12})$ shear constants show a very strong increase from Zr to Mo (Fig. 17). This strong effect of e/at. on c' is evidently causally related to the phase stability of the bcc crystal structure as is discussed in connection with experimental data in bcc Ti (see Section

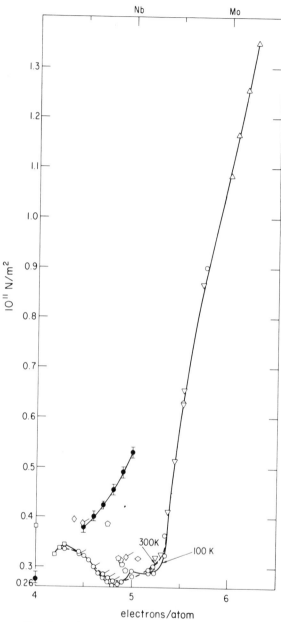

Fig. 16. Composition dependence of c_{44} for Zr–Nb–Mo–Re and Nb–Ti, Nb–Hf, Nb–W alloys. The black line represents the Nb–Zr series lattice c_{44}, at 300 K, in absence of the electronic excitation contribution. △, Mo–Re (Davidson and Brotzen, 1978); ○, Nb–Mo (Hubbell and Brotzen, 1972); ▽, Nb–Mo (Katahara *et al.*, 1979a); ⊘, Zr–Nb (Walker and Peter, 1977); □, Zr–Nb (Goasdoue *et al.*, 1976); ◇, Ti–Nb (Reid *et al.*, 1973); ⟨⟩, Ti–Nb (calculated from polycrystals); ◇, Hf–Nb (300 K); ◇, Nb–W (300 K); ⱡ, extrapolated from $T > 900$ K.

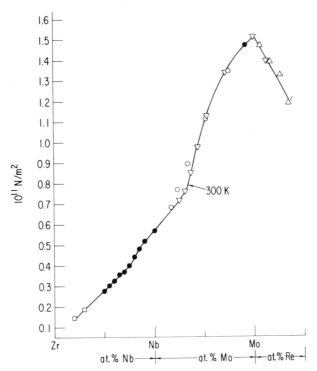

Fig. 17. Composition dependence of c′ at 300 K for Zr–Nb–Mo–Re alloys. △ (Smith, 1960); △ (Davidson and Brotzen, 1968); ▽ (Katahara *et al.*, 1979b); ○ (Hubbell and Brotzen, 1972); ● (Walker and Peter, 1977); □ (Goasdoue *et al.*, 1976).

IV, B, 6). As observed above, this modulus also has an electronic excitation contribution although considerably smaller than in c_{44}. From the evidence of the temperature derivatives, Fig. 11, it occurs for the 17, 23, and 34% Mo in Nb alloys. In Fig. 17, the c' values of these alloys also fall below a smooth curve that connects the Nb–Zr data with the Mo-rich region, a difference being about 0.1×10^{11} N/m². This would be the electronic excitation part. For e/at.$\gtrsim 5.5$, c' further increases up to the sharp maximum for Mo. The abrupt change of composition dependence at e/at.> 6.0, where c' falls with further alloying, is an important effect that is not understood; it might be related to the admixture of 5d electrons of Re or simply be due to the antibonding states becoming relevant above e/at.$= 6$.

The first-order deformation potential is the relevant quantity for temperature anomalies in the shear elastic constant and $2ND_1^2$ can make up a large part of c_{44} at $T = 0$. But, D_1 does not essentially determine the variation of c' with composition or with the filling of the band, as was discussed in Section III, B, 2. Here the second-order deformation potential is the relevant quantity. The summation of the occupied state D_2 contributions, i.e.,

$\int^{E_F} d\varepsilon N(\varepsilon) D_2(\varepsilon)$, plus a smaller negative polarization contribution coming from degenerate states with $T_F \gg T_{melt}$, determines, in fact, the increase of the c's with the filling of the band. But, the variation should not be the same for the two independent shear constants because the lattice deforms differently. Distances of the first nearest-neighbors change for trigonal shears and t_{2g} orbitals are then important while, e_g orbitals are more important for the tetragonal distortion where the second nearest-neighbors atoms take non-equivalent positions.

The variations of the bulk modulus with composition are shown in Fig. 18. The Zr–Nb–Mo alloy data at 100 K show a broad "peak" between

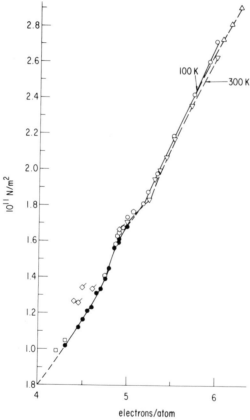

Fig. 18. Composition dependence of K for Zr–Nb–Mo–Re and Nb–Ti, Nb–Hf, Nb–W alloys. $K = \frac{1}{3} (c_{11} + 2c_{12})$. $\triangle$, Mo–Re (Davidson and Brotzen, 1968); $\bigcirc$, Nb–Mo (Hubbell and Brotzen, 1972); $\triangledown$, Nb–Mo (Katahara *et al.*, 1979b); $\bullet$, Zr–Nb (Walker and Peter, 1977); $\square$, Zr–Nb (Goasdoue, *et al.*, 1976); $\diamondsuit$, Nb–W (Frey *et al.*, 1978); $\diamondsuit$, Nb–Hf (Frey *et al.*, 1978); $\diamondsuit$, Ti40Nb (Reid, *et al.*, 1973); $\diamondsuit$, Ti–Nb (polycrystals); (Nedyukha and Chernyy, 1966).

4.75 < e/at.< 5.2. This is confirmed by the Nb–Hf and Nb–W data as well as the results of Katahara *et al.* (1979a), where a sharp change of slope occurs near e/at.= 5.2. The "peak" must be due to band structure effects and the relatively high-volume deformation potentials for bands 2 and 3 might be an explanation (see Section III, B, 3).

   c. Effect of 3d and 5d Solutes in Nb.   Some very important data exist due to Frey *et al.* (1978) and to Reid *et al.* (1973), which show the very pronounced effects of mixing 5d and 3d elements with 4d Nb. These data given in Fig. 16 show that either 5d or 3d elements decrease the magnitude of the electronic excitation effect on c_{44} as well as dc_{44}/dT. The Ti–Nb data are obtained from two sources: (i) single crystal of Ti40Nb, and (ii) calculation of c_{44} from polycrystalline G, assuming that c' can be evaluated from the e/at. ratio as discussed below. The values of c_{44} are severely increased over the Zr–Nb–Mo plot when either W is added to increase e/at.> 5 or Hf or Ti are added to decrease e/at. The data show, however, that dc_{44}/dT is affected by Hf in Nb in the same way as Zr. This similarity is shown in the temperature T_{min} at which $dc_{44}/dT = 0$. For pure Nb, $T_{min} \sim 420$ K, whereas with 10% Zr or Hf, $T_{min} = 185$ and 200 K, respectively. A similar parallel is noted for 15% Zr and 13% Hf, where T_{min} is shifted to 110 and 170 K, respectively. The reduction of the magnitude of the electronic contribution with addition of Ti, Hf, or W appears to be related with a distortion of the electron band, and the overall temperature variation will in consequence be smaller. The shift in T_{min} indicates further that the effective Fermi temperatures are increased when alloying with 3d or 5d solutes, which is consistant with a "blurring" of the electronic structure at the Fermi level.

   d. Variation of Young's Moduli in Nb–Zr and Nb–Ti Alloys.   An interesting aspect of Fig. 18 is the significant difference between K in the Ti- and Zr-rich alloys, respectively. Furthermore, the addition of Ti to Nb does not produce the softening in c_{44} as is the case at e/at.= 4.8 for the Nb–Zr system. This difference is also observed in Young's moduli as measured in polycrystalline Nb–Ti and Nb–Zr alloys and also as calculated from the single-crystal c_{ij} of Walker and Peter (1977) in Nb–Zr. These data, shown in Fig. 19, show the interesting effect of the increasing c_{44} upon initially adding Ti, and the very marked decrease from the maximum at $\sim 85\%$ Nb caused by the rapidly decreasing c' and K with increasing Ti.

   e. Pressure Derivatives of the Elastic Constants.   Interpretation of the temperature and composition dependence of the elastic moduli requires as a primary step the separation of the volume dependence from the intrinsic effects, i.e., $(\partial c_{ij}/\partial T)_v$ and $(\partial c_{ij}/\partial X)_v$, where X is an alloy component. In

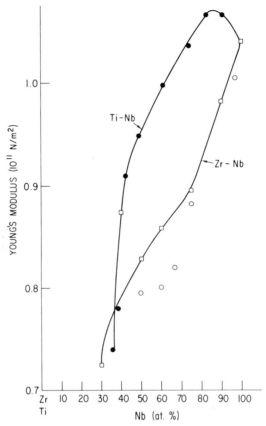

Fig. 19. Comparison of Young's modulus for polycrystalline Nb–Zr and Nb–Ti alloys. E_{avg}: □, calculated from c_{ij}; ○, measured, Zr–nb; ●, measured, Ti–Nb (Nedyukha and Chernyy, 1966).

this respect the studies of the variation in c_{ij} under pure hydrostatic pressure are indispensible when the band structure effects are an important component. One finds, however, in the bcc transition metal that the measured dc_{ij}/dP are in themselves significantly dependent on the band structure and are, therefore, perhaps even more useful than the c_{ij} in detecting critical compositions. This shows quite clearly in Fig. 20 for the 4d alloys when the volume derivative of the shear moduli

$$\pi_{ij} = d\ln c_{ij}/d\ln V = -\left(dc_{ij}/dP\right)\left(K/c_{ij}\right) \qquad (37)$$

are plotted as a function of e/at. The large variations of π_{44} from a maximum at e/at.$= 4.7$ to a sharp minimum at e/at.$= 5.4$ correlate with two, possibly even three changes of the Fermi surface topology. $\pi_{c'}$ has a

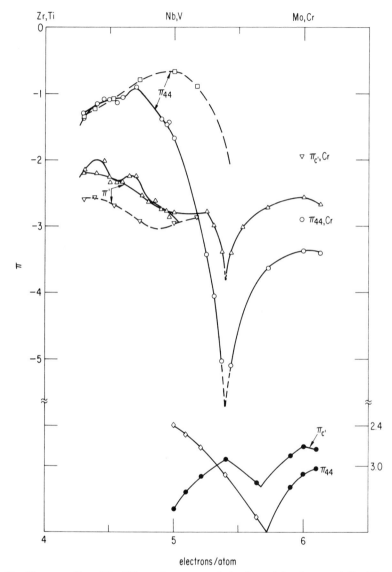

Fig. 20. $\pi_c = d \ln c / d \ln V$ from the pressure derivatives of the shear moduli of Zr–Nb–Mo–Re and Ti–V–Cr bcc alloys (upper curves) and Ta–W–Re bcc alloys (lower curves). Upper curves: ——, Zr–Nb–Mo–Re; ---, Ti–V–Cr.

minimum also at precisely e/at. = 5.4 and two smaller maxima on the Nb–Zr side.

In contrast, there is only slight evidence of a direct interaction between $dK/dP = -d\ln K/d\ln V$ and the Fermi surface changes noted below.

2. ELECTRONIC STRUCTURE OF ZR–NB–MO ALLOYS AND ITS RELATION TO THE ELASTIC PROPERTIES

a. Interrelation of Temperature and Volume Derivatives, Thermal Expansion, and Deformation Potentials. The elastic properties of the bcc Zr–Nb–Mo–Re alloys are highly illustrative of the statement of Lifschitz (1960) that changes in the topology of the Fermi surface produce anomalies in the thermodynamic properties. In the preceding section, such anomalies have been shown to exist. The shear elastic constants have some remarkably sharp variations with alloying, and their temperature and pressure derivatives have rather distinct extremes. These derivatives probe the explicit temperature dependence and the configuration dependence of the thermodynamic functions of the system.

The pressure derivatives are given in Fig. 20 and the temperature coefficients at 300 K in Fig. 21. It should be remarked that the temperature derivatives of the electronic excitation and the lattice part are not directly separable in this coefficient, which is

$$\frac{1}{c}\frac{dc}{dT} = \frac{d\ln c}{dT} = \frac{1}{c_{la} - c_{el}}\left[\frac{dc_{la}}{dT} - c_{el}(0)\frac{a_{el}}{|a_{el}|}\frac{2T}{T_F^2}\right] \tag{38}$$

from Eqs. (32) and (33). But if there is no electronic excitation, then $(d\ln c)/dT$ gives the lattice part. It is further seen that the electronic part in the coefficient depends on both $c_{el}(0) = 2ND_1^2$ and T_F^2.

Thermal expansion data also give evidence for electronic effects. The elastic moduli and the volume thermal expansion coefficient α are related in two ways through the Grüneisen parameter (Slater, 1939):

$$\gamma = \frac{\alpha V K_v}{c_v} = -\frac{d\ln\omega}{d\ln v}, \tag{39}$$

where V, K, and c_v are the volume, bulk modulus, and heat capacity, respectively, and ω is some average of the acoustical phonon mode frequencies. The mode γ's $\gamma_i = \partial\ln\omega_i/d\ln V$ describe the anharmonic character of the individual mode that has frequency ω_i, and

$$\gamma_{ij} = -\frac{1}{6} - \frac{1}{2}\frac{\partial\ln c_{ij}}{\partial\ln V} \tag{40}$$

is a parameter that describes the low-frequency mode i on a given branch j

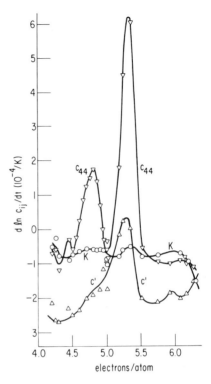

Fig. 21. Temperature coefficients of K, c_{44}, and c' at 300 K for Zr–Nb–Mo–Re alloys.

(Slater, 1939; Smith *et al.*, 1965). In the thermal-expansion coefficients, the γ's are weighed by the phonon occupation numbers and summed. At low temperature, γ is essentially a weighted average of the γ_{ij}, where the weights are inversely related to ω. This γ_{la} is thus heavily weighted by the smaller shear moduli and should reflect the electronic contributions to c_{44} in Nb alloys. As shown in Fig. 22, there is indeed a minimum and a maximum in γ_{la}, although not so pronounced as in π_{44} and $\pi_{c'}$. This γ_{la} is a coefficient of one contribution to the thermal expansion at low temperature, i.e., $\alpha = \alpha_{el} + \alpha_{la} = a\gamma_{el}T + b\gamma_{la}T^3$, where a and b are numerical constants. Of greater importance to the present context is the electronic gamma $\gamma_{el} = \partial \ln N / \partial \ln V$. It has extremes at e/at.$= 4.7$ and 5.4, just like the πs. But there are differences in the variations of γ_{el} and of π_{44} that may be understood when splitting the terms by writing

$$\frac{d \ln c}{d \ln v} = \frac{1}{c}\left[\frac{\partial c_{la}}{\partial \ln V} - c_{el}\left(\frac{\partial \ln N}{\partial \ln V} + \frac{\partial \ln D^2}{\partial \ln V} \right) \right]. \qquad (41)$$

The electronic contribution to π includes the volume derivative of D_1^2,

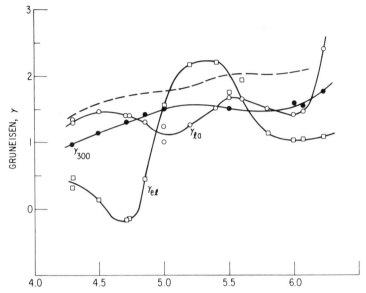

Fig. 22. Electronic and lattice Grüneisen parameters for Zr–Nb–Mo–Re alloys (Smith and Finlayson, 1976; White *et al.*, 1978). ---, from $-(\partial \ln K)/(\partial \ln V) - \frac{1}{6}$. From thermal expansion: ●, γ_{300}; ○, γ_{1a}; □, γ_{el}.

whereas γ_{el} represents only the volume derivative of the density of electronic states at E_F.

The existence of a negative value of γ_{el} at e/at. ~ 4.7 should be emphasized here. It is a direct result of a careful analysis by Smith and Finlayson (1976) of the low-temperature thermal expansion coefficients in a Nb28.7Zr alloy in both the as-cast state and in the annealed state. The data for the α/T versus T^2 plot consist of 35 data points between 11 and 18 K. Thus the existence of the strong electronic contribution to the dc_{44}/dP data in Nb–Zr alloys is confirmed by an entirely different measurement.

b. Elastic Moduli and Fermi Surface Singularities. There exist good arguments to show that the Zr–Nb–Mo(Re) series can be described by a rigid-band model. In fact, the renormalized d-band energies of the component elements are very similar so that the scattering strength is small [according to Hodges *et al.* (1972), renormalized energies are Zr—0.071 Ry, Nb—0.057 Ry, Mo—0.075 Ry]. Also, changes in the phonon dispersion of the alloy series are readily explained with changes of a common electronic structure (Powell *et al.*, 1968). However, the most direct affirmation is given by the calculations, with the coherent potential approximation by Giuliano *et al.* (1978) and Colavita *et al.* (1979).

The Nb-band structure as calculated by Mattheiss (1970) is taken as reference in the following discussion. Considering the energy dispersion and electronic density of states shown in Fig. 23, the relativistic calculations by Elyashar and Koelling (1977) indicate some corrections, e.g., that splittings instead of accidental degeneracies exist for the crossing branches G_4–G_3, Δ_2–Δ_5, Δ_2–$\Delta_{2'}$, etc. If the stable range of the bcc extends over electron per atom ratios of 4.3–6.3, this corresponds to positions of the Fermi level of 0.67–0.84 Ry in a rigid-band picture. Within this energy range there are several singularities: a minimum on line $\Sigma 1$ (band 3, saddle point), 3 singularities at and around Γ_{25} (bands 2 and 3, saddle points; band 4, minimum), a crossover Δ_2 and Δ_5 (in fact, split), a crossover $\Delta_{2'}$ with Δ_2 (in fact, split), a minimum on line G_1. Important changes of the Fermi surface topology result when the Fermi energy crosses through these singularities because of alloying. These changes are summarized as follows, with reference to the cross sections given in Fig. 24:

Jungle-gym and ellipsoid merge	for e/at.$\lesssim$ 4.7
Nb-type	4.7 $\lesssim$ e/at.$\lesssim$ 5.4
Jungle-gym broken up; Fermi surface is of Mo-type without lens in central octahedron	5.4 $\lesssim$ e/at.$\lesssim$ 5.8
Mo-type	5.8 $\lesssim$ e/at.$\lesssim$ 6.3
Electron bubble on H–N	for e/at.$\lesssim$ 6.3

For Nb, the important sheets are the second-zone octahedron with a partial density of states $\langle N \rangle = 1.37$ states of one spin/Rydberg atom, the third-zone jungle-gym with an individual $\langle N \rangle = 4.42$, and the third-zone ellipsoid with $\langle N \rangle = 4.10$. These sheets are shown in the middle of Fig. 24. At a lower electron-per-atom ratio, the jungle-gym connects with the ellipsoids and at higher e/at., the jungle-gym is disrupted to form a central octahedron and the octahedron centered on H; the ellipsoids centered on N become smaller. These changes affect the two (hole) sheets with large individual densities of states, and so the scattering is drastically changed. Furthermore, the strain response proper, i.e., the local deformation potentials, are certainly large in the regions of singularities as was discussed and demonstrated by numerical calculation for a d band in Section III, B, 4. Both effects together are at the origin of anomalous variations of elastic constants and especially in their temperature and pressure derivatives. Katahara et al. (1977) provide a more detailed interpretation for anomalies of Nb–Mo alloys as well as the V–Ti and Ta–W alloys. For the 4d alloy, anomalies of the temperature coefficients exist for both c_{44} and c' at e/at.$= 5.3$ and 4.8. Those of π occur at e/at.$= 5.4$ and 4.7 for both c_{44} and c'. The slightly different locations of extremes for $d\ln c/dT$ and π are not unexpected; a_{el} of c_{el} comprises the functions $N''/N - (N'/N)^2$ and

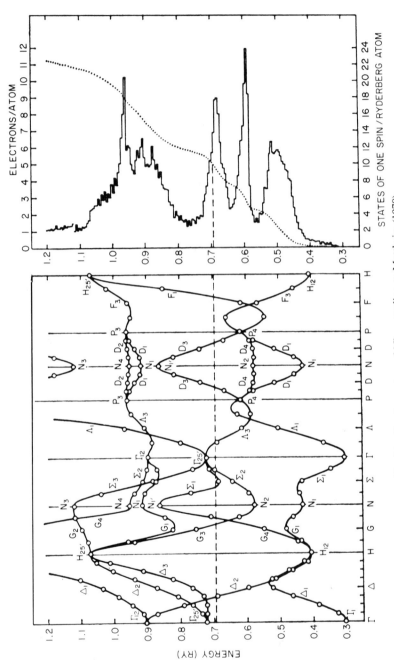

Fig. 23. Band structure of Nb according to Mattheiss (1970).

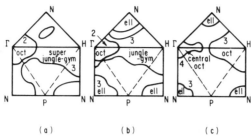

(a) (b) (c)

Fig. 24. Central (100) and (110) cross sections of the Fermi surfaces (a) for electron-per-atom ratios $e/at. \approx 4.6$; (b) for Nb with $e/at. = 5$; and (c) for $e/at. \approx 5.6$. The labels on the surfaces indicate the corresponding energy band. Because of alloying and thermal vibrations, the Fermi surfaces clearly are "blurred."

$D_1''/D_1 + (N'/N)^2$, which take extremal values in between the locations of a peak in N or D_1^2. On the contrary, no such shift affects the pressure derivatives or π.

The bonding part of the Nb–Mo band structure is rigid band–like as shown by Giuliano *et al.* (1978) and Colavita *et al.* (1979). It can reasonably be expected that this also applies to Nb–Zr alloys. Alloys, however, have a pseudoband structure where the energy levels are spread out in a spectral function. For Nb–Mo, the full width at half maximum of the spectral function is 0.015–0.025 Ry, which corresponds to a width in the electron-per-atom ratio of about 0.4 units for $e/at. \lesssim 5.4$. The widths of the maxima and minima in the temperature coefficients (Fig. 21) and pressure derivatives (Fig. 20) are of this magnitude. Thermal excitations of the lattice are another source of the smearing out of energy levels. Chen *et al.* (1972) have considered the effect in the coherent potential approximation; for a d band the level broadening is given as $\langle \Delta \varepsilon^2 \rangle \approx 0.075 \, (T/T_{melt})$ Ry. At $T \sim \Theta_D$, the thermal broadening will thus be of the order of the above alloy broadening of levels, but exceeds considerably the latter near melting point.

In summary, it appears clear that the anomalies in the elastic constants can be explained in some detail by changes in the electronic structure with alloying. In the case of Zr–Nb–Mo–Re alloys, the agreement between experimentally observed effects and theory is extremely good.

3. Phonon Dispersion and Zero-Sound Elastic Constants for the bcc Zr–Nb–Mo System

a. General Relationship. Inelastic neutron-scattering experiments have revealed anomalies in the phonon spectra of transition metals that are generally attributed to strong electron–phonon interaction. The incentive for this work has been generated by (1) apparent correlations between high-temperature superconductivity and the presence of these anomalies, (2) by relations between lattice dynamics and phase transitions that appear to be electronically driven, and other topics. It is now well established that the dynamics of a lattice essentially determine its stability in opposition to the classical criterion that the independent elastic constants (eigenvalues of c matrix) be positive definite (Voigt, 1889). Elastic constants refer essentially to "statics" of a lattice and so do the anomalies of electronic origin; consideration of the counterpart, especially of small wave-vector phonons, cannot be left out. Furthermore, zero-sound elastic constants have been measured for various interesting metals and phases.

  b. Phonon Dispersion of Nb and Nb–Mo Alloys. The anomalies in the temperature, pressure, and composition dependence of c_{44} in the bcc alloys have their counterpart in the anomalous $[00\zeta]T$ and $[\zeta\zeta0]T_2$ branch of the phonon dispersion curves for Nb and Nb-rich Nb–Mo alloys. For $[\zeta\zeta0]T_2$, the branch has a small slope near the origin and increases markedly above 0.1 of the reduced wave vector and, in $[00\zeta]T$, the slope decreases first and then increases with a marked upward curvature. For Nb, measurements were done for different temperatures (Fig. 25). For better visualization and easy comparison with the sound velocities, the "group velocities" $v_s(\nu) = \Delta\nu/\Delta\zeta$ are useful [after Miller and Brockhouse (1968) for Pd]. The zero-sound (i.e., phonon) velocities correspond reasonably well with the first-sound (ultrasonic) data and their temperature variation. But phonon frequencies at $0.1 < \zeta\zeta0 < 0.5$ have a positive temperature dependence that is even more

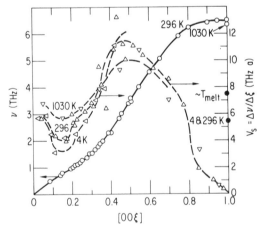

Fig. 25. $[00\zeta]T$ phonon branch of Nb at room temperature and the "group velocity" $v_s(\nu) = \Delta\nu/\Delta\zeta$ (dashed lines) for 4, 296, and 1030 K (data for 296 K from Nakagawa and Woods, 1965; for 4 K from Shapiro *et al.*, 1975; for 1030 K from Powell *et al.*, 1972). The low temperature and very high temperature first-sound velocities (●) are given on the ordinate at right.

anomalous than the first-sound velocity. Powell *et al.* (1972) show that variations of this part of the phonon dispersion are related to drastic changes of the higher-than-order-2 coupling parameters and involve long-range interactions via the conduction electrons. The anomaly at the small wave vectors is not correlated with any feature of the Fermi surface which might, for example, lead to nesting interactions. The long-range interactions via the conduction electrons may well be significantly temperature-dependent and the explanation of the phonon anomaly is thus analogous to the picture of the electronic contribution to the elastic constants.

The $[00\zeta]T$ and $[\zeta\zeta0]T_2$ branches of Nb–Mo alloys with 15.35 and 42% Mo show, in decreasing strength, the same anomaly as pure Nb. Above this concentration, the dispersion curves are smooth and the intersection of various branches that occurs in the pure metal is absent. As for Nb, when the temperature is varied, the higher-than-order-2 coupling parameters show variations and even change signs for Nb42Mo. That is the electron-per-atom ratio where the Fermi level crosses through a major singularity in the band structure and where the long-range interactions via the conduction electrons are drastically changed.

Anharmonic interactions influence the sound waves differently in the collision-free zero-sound and the collision-dominated first-sound regions, and give rise to different velocities of propagation and different attenuations (Cowley, 1967). Theoretical predictions give small discrepancies between the two sound velocities that are not beyond the experimental errors on the zero-sound measurements. One case, the Fe–Ni invars (Endoh, 1979), is known where the zero-sound constants differ widely from the ordinary elastic constants. Another example seems to be the Nb–Mo alloys measured by Powell *et al.* (1968). The two sound velocities coincide for the elements Nb and Mo, but not for the alloys. The c_{44} constant taken from the slopes of the $[00\zeta]T$ and $[\zeta\zeta0]T_2$ branches and $c' = \frac{1}{2}(c_{11} - c_{12})$ from the $[\zeta\zeta0]T_1$ branch are up to 60% higher than the ordinary elastic constants. Since the atomic masses are nearly identical, the effect cannot be attributed to mass disorder. A force constant disorder, affecting the low-order coupling parameters, must be the cause.

c. Phonon Dispersion and Transverse Elastic Constant of bcc Zr.

Stassis *et al.* (1979) studied the lattice dynamics of bcc Zr (at 1400 K) and found a massive softening of the $\frac{2}{3}[\zeta\zeta\zeta]L$ mode indicating the inclination of this metal to undergo a structural transformation to the trigonal ω phase. It should be noted, however, that the bcc → hcp transformation of this metal is of martensitic type; Burger's orientation relation involves, from the point of view of lattice dynamics, a long-wavelength instability of the $\langle 112 \rangle$ shear, followed by a short-wavelength (zone boundary) [110] $\langle 1\bar{1}0 \rangle$ shear.

An anomaly of the $[\zeta\zeta 0]L$ branch in the vicinity and at the zone boundary may just be related to this latter shear.

Stassis et al. (1979) further find that the two transverse $[\zeta\zeta 0]$ branches are degenerate, i.e., the bcc Zr would be isotropic with regard to the propagation of elastic waves. This result is unexpected because a large c' cannot give the long-wavelength instability required for the martensitic transformation and would be inconsistent with the empirical law of Fisher and Dever (1970b). But the transverse constants can also be evaluated from other branches. The zero-sound c' elastic constant extracted from $[\zeta\zeta\zeta]T$ together with $[\zeta\zeta 0]T$ is 0.075×10^{11} N/m² and from $[\zeta\zeta\zeta]T$, together with $[00\zeta]T$, it is 0.042×10^{11} N/m². A value of $c' \approx 0.07 \times 10^{11}$ N/m² was also measured by Fisher and Dever (1970a) in bcc Ti and is found by extrapolation of the measurements at 1400 K by Ashkenazi et al. (1978). A similar extrapolation gives $c_{44} = 0.25 \times 10^{11}$ N/m² for bcc Zr, which compares well with the phonon data.

The most probable explanation for the conclusion, by Stassis et al., that $c_{44} = c'$ is that the c' branch was not observed because of heavily damped phonons and that several domains with parallel (110) planes, but rotated $\langle\zeta\zeta 0\rangle$ directions, are produced by the hcp $\rightarrow$ bcc transformation.

4. Brief Summary of Experimental Studies in Ti–V–Cr, Ta–W, and Isoelectronic Nb–V and Nb–Ta Solid Solutions

A considerable quantity of data now exists that shows the striking similarities between the bcc solid solutions of the 3d and 5d elements and the 4d Zr–Nb–Mo characteristics described. The similarities that are rather clearly derived from the similar band structures of V, Nb, and Ta (Boyer et al., 1977; Laurent et al., 1978) and of Cr, Mo, and W are as follows:

    (i) The c_{44} versus T data for V and Ta at high T show anomalous positive curvatures similar to Nb (Fig. 10), but considerably smaller in degree (Walker 1978; Walker and Bujard, 1980).

    (ii) Unpublished data by Carpenter and Shannette (1979) show anomalous dc_{44}/dT values for Ta–W solid solution with $5.2 < $ e/at.$ < 5.7$, as in Nb–Mo (Fig. 13).

    (iii) Data for c_{44} at 300 K versus e/at., show relatively abrupt changes at $5.3 < $ e/at.$ < 5.8$ [Fig. 26, data for Ti–V–Cr and Ta–W from Katahara et al. (1979a), Lenkkerri and Lahteenkorva (1978); Fisher and Dever (1970a)].

    (iv) Almost identical values of c' versus e/at. are measured for the 3d and 4d solutions for $4.0 < $ e/at.$ < 6.0$, with very similar c' values for the Ta–W alloys, as shown in Fig. 27 and discussed in Section IV,B,6, below.

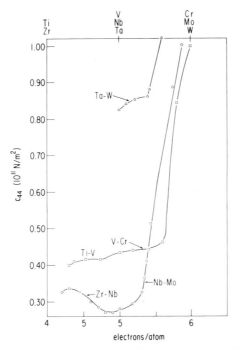

Fig. 26. Comparison of the c_{44} versus e/at. dependence for the 3d, 4d, and 5d alloys, at 300 K. References to data given in Section IV,4 of text.

(v)   Data obtained by Katahara *et al.* (1979a) show similar anomalous values of π_{44} for Ti–V and Zr–Mo alloys at $4.3 < $ e/at.$ < 5$, and sharp singularities in both π_{44} and $\pi_{c'}$ for Ta–W at e/at. ~ 5.7, as shown in Fig. 20.

The elastic moduli of the isoelectronic solid solutions of Nb–V and Nb–Ta single crystals (Fisher *et al.*, 1980) indicate that the uniquely small value of c_{44} in Nb is indeed related to a unique band structure feature in Nb that is gradually erased by adding V or Ta (Fig. 28). The changes in c' in Nb–V alloys suggest that atomic ordering may occur in this system, with dissimilar next-nearest neighbors, because of the large atomic size differences.

5.   Summary and Interpretation of Volume-Dependent Effects in bcc Transition-Metal Alloys

In this section an attempt is made to summarize the observations of temperature and alloying effects on the c_{ij} by separating the volume-change contribution from the intrinsic electronic contributions. The equations used

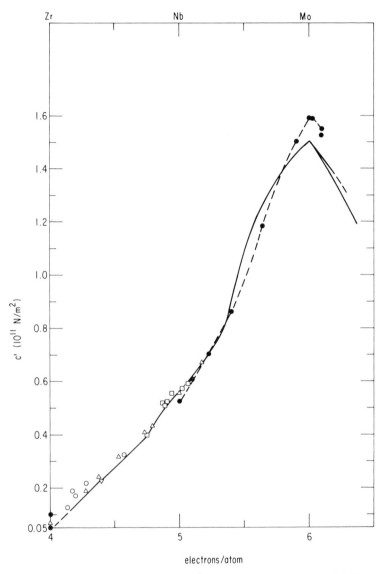

Fig. 27. Composition dependence of $c' = \frac{1}{2}(c_{11} - c_{12})$ in 3d, 4d, and 5d bcc solid solutions at 300 K. Base——: Zr–Nb, Nb–Mo, Mo–Re, 4d. △: Ti, Ti–V, V–Cr, Cr, 3d. ○: Ti–Cr, 3d; ●---: Ta, Ta–W, W–Re, 5d. ▽: Ti40Nb. □, □': Nb–Hf, Nb–W.

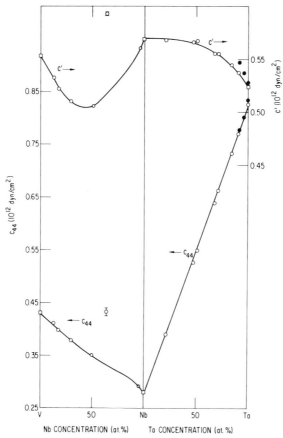

Fig. 28. Composition dependence of the shear elastic shear moduli in V–Nb–Ta alloys. (Fisher *et al.*, 1980.) □, Fisher *et al.* (1975); ○, Fisher *et al.* (1980). ●, Armstrong and Mordike (1970).

for this separation are

$$\frac{d\ln c_{ij}}{dT} = \left(\frac{\partial \ln c_{ij}}{\partial T}\right)_v + \alpha \pi_{ij}, \tag{42}$$

$$\frac{d\ln c_{ij}}{dX} = \left(\frac{\partial \ln c_{ij}}{\partial X}\right)_v + \frac{d\ln V}{dX} \pi_{ij}, \tag{43}$$

where α is the volume thermal-expansion coefficient and X is a unit denoting composition of an alloy component. The constant volume derivatives are the intrinsic components that are associated primarily with the temperature-dependent but volume-independent electron energies. It should be noted, however, that because the π factors for these transition-metal alloys are also very significantly affected by the Fermi surface topology, the total electron contribution to the temperature derivatives will also contain a volume-dependent part that can only be estimated by assuming some model for the lattice contribution to π, i.e., $\pi \sim$ form of interatomic repulsive potential.

   a.   *Separation of Temperature and Volume Effects.*   Because the measurements of π_{ij} are so far restricted to 300 K, this discussion is restricted to values of $d\ln c_{ij}/dT$ at 300 K. From Eq. (42) it can be quickly established that the positive values of dc_{ij}/dT, which are of primary interest to this chapter, are not derived from the volume term $\alpha \pi_{ij}$, since all of the measured π values are negative and α is always positive for the bcc metals and alloys of groups IV–VI. The relative components to $d\ln c_{ij}/dT$, for c_{44}, c', and K are quantitatively compared in Table IV, using values of α that are

TABLE IV

COMPONENT OF TEMPERATURE DERIVATIVES FOR c_{ij} OF 3d, 4d, AND 5d ALLOYS

	c_{44}			c'			$\langle K \rangle$		
	1 $\dfrac{d\ln c_{44}}{dT}$	2 $\alpha_v \pi_{44}$	1–2 Intr.	3 $\dfrac{d\ln c'}{dT}$	4 $\alpha\pi'$	3–4 Intr.	5 $\dfrac{d\ln K}{dT}$	6 $\alpha_v \pi_K$	5–6 Intr.
Zr30Nb	−1.33	−0.23	−1.1	−2.7	−0.40	−2.3	−0.7	−0.7	0
58Nb	0	−0.19	+.19						
80Nb	+1.9	−0.24	+2.14	−1.8	−0.50	−1.3	−0.6	−0.7	~0
Nb	−0.3	−0.34	~0	−1.7	−0.50	−1.2	−1.0	−0.8	−0.2
Nb17Mo	+1.8	−0.51	+2.3	−0.2	−0.50	+0.30	−0.8	−0.8	0
22Mo	+4.5	−0.64	+5.14	+0.3	−0.50	+0.80	−0.7	−0.8	+0.1
34Mo	+6.0	−0.81	+6.8	0	−0.54	+0.54	−0.5	−0.7	+0.2
51Mo	−0.6	−0.8	+0.2	−0.8	−0.51	−0.3	−0.8	−0.8	0
75Mo	−0.9	−0.6	−0.3	−0.5	−0.43	~0	−0.8	−0.7	−0.1
MO	−0.9	−0.5	−0.4	−0.5	−0.40	−0.1	−0.8	−0.7	−0.1
Ti29V	−0.46	−0.30	−0.1	−2.0	−0.61	−1.4	−0.93	−0.91	~0 +.1
39V	−1.2	−0.28	−0.90	−2.7	−0.61	−2.1	−0.85	−0.91	+ ~.1
53V	−0.7	−0.26	−0.44	−2.9	−0.63	−2.3	−0.84	−0.94	+ ~.1
73V	−1.2	−0.19	−1.0	−2.6	−0.68	−1.9	−0.85	−0.98	+ ~.1
V	−1.9	−0.16	−1.7						
Ta	−2.0	−0.48	−1.5	−1.1	−0.73	−0.4	−0.8	−0.77	
10W	−3.4	−0.51	−2.9	−2.9	−0.68	−0.4	−1.6		
22W	−2.5	−0.55	−2.0	−4.2	−0.63	−3.6	−0.4		
40W	+0.7	−0.63	+0.70						
43W	+0.3	−0.64	+0.67						
73	−0.3	−0.80	+0.50						
83	−0.8	−0.71	−0.10	−2.4	−0.60	−1.8	−0.8		
90	−0.8	−0.67	−0.1	−3.3	−0.57	−2.7			

probably correct to within 15%. It should be noted from columns 5 and 6 that the experimental $d\ln K/dT$ can be completely accounted for by the volume thermal expansion, i.e., the intrinsic $(d\ln K/dT)$ values are not significantly different from zero. The calculated $(\partial\ln c_{44}/dT)_v$ values, plotted in Fig. 21 for the 4d alloys contain the wide deviations from zero in the range of $4.5 < $ e/at.< 5.5, but even for c_{44} the temperature derivatives at e/at. approaching 6 can be accounted for by the volume change. The large positive deviations for the 4d series are evidently derived from the Fermi function, as is also the case for the smaller positive values for the 5d series at e/at.> 5.4. The relatively large negative values for the 3d and 5d near e/at.$= 5$ are, however, not understood. The most direct interpretation would be a volume-independent anharmonic term that also affects the c' shear in the 3d and 4d alloys at e/at.< 4.8.

The consistently near-zero values for $(d\ln K/dT)_v$ confirm the conclusions derived from the absence of anomalies over the wide temperature range of K measurements, namely, that Fermi surface topology has no significant effect on K. A similar conclusion can reasonably be applied to the c_{44} and c' values for the 4d alloys at e/at.> 5.55.

b. Separation of Composition and Volume Effects. One of the primary tasks in understanding the effects of compositional changes on the elastic moduli of the bcc alloys is to separate the total effects into the basic contributions of the electrostatic attraction, the repulsive ion-core and the Fermi surface effects. With the electrostatic and core-repulsion contributions grouped together as c_{la}, the lattice term, and c_{el}, as the Fermi surface contribution, Eq. (43) can be expanded to

$$\frac{d\ln c_{ij}}{dX} = \frac{1}{c_{ij}} - \left[\left(\frac{\partial c_{la}}{\partial X} \right)_v + \left(\frac{\partial c_{el}}{\partial X} \right)_v \right] + \frac{\pi_l c_{la} + \pi_e c_{el}}{c_{ij}} \frac{d\ln V}{dX}, \tag{44}$$

$$\frac{d\ln c_{ij}}{dX} - \pi \frac{d\ln V}{dX} = \frac{1}{c_{ij}} \left[\left(\frac{\partial c_{la}}{\partial X} \right)_v + \left(\frac{\partial c_{el}}{\partial X} \right)_v \right]. \tag{45}$$

Therefore, the sum of the known values on the left consists of the derivatives with respect to X of the intrinsic or volume-independent lattice and electronic terms. If $(\partial c_{la}/\partial X)_v$ is relatively small compared to the Fermi surface effects at constant volume, then Eq. (45) gives a realistic measure of $(\partial c_{el}/\partial X)_v$.

Figure 29 gives the composition dependence of the total volume and the intrinsic contributions to c_{44} in the Zr–Nb–Mo bcc binary alloys based on Eq. (43). Since we are plotting the change in c_{44}, the reference composition is pure bcc Zr. Although this is an arbitrary reference choice, the results suggest that the constant volume Δc_{44}, $(\Delta c_{44})_v$ plot does in fact represent the constant volume Δc_{el}, $(\Delta c_{el})_v$.

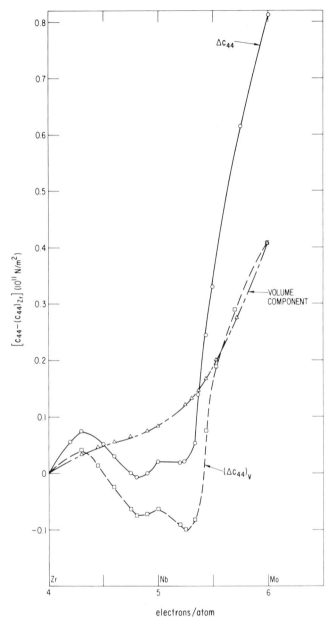

Fig. 29. Volume and intrinsic contributions to the c_{44} constant in the Zr–Nb–Mo system. ——, total Δc_{44}; —·—, volume component; — —, $(\Delta c_{44})_v$, intrinsic component.

c. Intrinsic Contributions to c_{44} with e/at.   There are four features of interest that appear semiquantitatively related to the total c_{44} and dc_{44}/dT variations in the 4d alloys:

(1)   The initial increase in c_{44} has equal volume and intrinsic components.

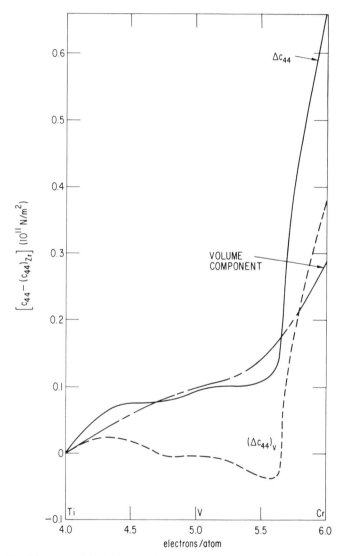

Fig. 30.   Volume and intrinsic contributions to the c_{44} constant in the Ti–V–Cr system. ——, total; —·—, volume component, ---, $(\Delta c_{44})_v$, intrinsic component.

(2)   The intrinsic component decreases at Zr30Nb and becomes a negative contribution between approximately Zr40Nb and Nb40Mo. Thus, the negative intrinsic c_{44} coincides with the negative $(c_{44})_{el}$ component that is assumed to explain the anomalous c_{44} versus T data of Fig. 10.

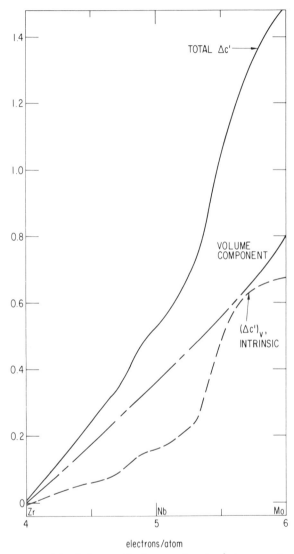

Fig. 31. Volume and intrinsic contributions to the $c' = \frac{1}{2}(c_{11} - c_{12})$ constant in the Zr–Nb–Mo system. ——, total; —·—, volume component; ---, $(\Delta c')_v$, intrinsic component.

(3)   Since the volume component is a smoothly increasing function of e/at., the structure in the total c_{44} versus e/at. plot is indicative of the band structure contributions due to Fermi surface changes.

(4)   The shift to a positive $(\Delta c_{44})_v$ component at Nb40Mo is equally striking in that it rejoins the volume component as an equal part to the total c_{44} at compositions greater than 60 at.% Mo. It appears that both components would be nearly equal at all compositions between bcc Zr and Mo35Re in the absence of the Fermi surface singularities.

For the 3d alloys, beginning with V additions to bcc Ti, the intrinsic component to Δc_{44} also starts as a positive component, as noted in Fig. 30. There is no negative intrinsic component at V compositions up to 50 at.% and relatively small negative component to 100 at.% V. This relative absence of a negative component is consistent with the absence of anomalies in dc_{44}/dT of Ti–V alloys. The negative component is more significant in V–Cr alloys to 40 at.% Cr and thus anomalies in dc_{44}/dT may be expected.

The abrupt increase in the intrinsic component to c_{44} at > 40 at.% Cr indicates the same type of Fermi surface changes with e/at. that occurs in the 4d alloys. As in the Mo-rich Nb–Mo alloys, the volume and intrinsic c_{44} contributions are nearly equal in the Cr-rich alloys.

d.   Intrinsic Contributions to c′.   As further evidence for the assumption that the intrinsic $(\Delta c_{ij})_v$ component is dominated by the electronic contributions, the separation of the $\Delta c′$ components from Zr → Mo is plotted in Fig. 31. The volume-change component is very near a linear relation to e/at., whereas the intrinsic components, although positive, show the perturbations due to Fermi surface changes at e/at. < 5.4. In the high-Mo alloys, however, $(\Delta c′)_v$ is about the same as the volume component, indicating that in the absence of the Fermi surface changes, the electronic components would also be linear with e/at.

6.   The *c′* Modulus and Implications Regarding the Stability of the bcc Structure

In all three series, the *c′* elastic constant tends to zero with e/at. ~ 4. For this constant, the contribution of the second-order deformation potential D_2 is essential and $\int d\varepsilon N(\varepsilon)D_2(\varepsilon)$ is positive and increases with the filling of the band, (see Fig. 7); on the other hand, the repulsive contribution to this constant is small. The similarity of the band structures in the 3d, 4d, and 5d bcc metals suggests further that *c′* will not be very different in the periods. This was, in fact, observed by Fisher and Dever (1970a), who showed that, to within 15%, from below e/at.= 4.3 to e/at.= 6, the c' constant at 300 K follows the relation

$$c' = 1.391 \times 10^{10} \, (e/at. - 2)^{3.34} \quad dyn/cm^2. \tag{46}$$

The implications of $c' \to 0$ at e/at.≈ 4 are clearly associated with the mechanical stability of the bcc structure in pure Ti, Zr, and Hf and in alloys based on these elements with e/at.< 4.30. The similarity of the values of the c' moduli suggests similar thermochemical data; in fact, the martensitic transformation temperatures and the heat of transformation are very close in Ti and Zr, and phase diagrams, such as those of Ti–V, Ti–Nb, Ti–Mo, Zr–Nb (> 600 C), Hf–Ta, etc., have similar limits of bcc phase stability as functions of e/at.

C. FCC Transition-Metal Series

During the past 20 years there have been several investigations of the anomalies in the temperature dependence of the shear moduli in Pd, Pt, and solid-solution alloys of Pd with its two neighboring elements in the periodic table, Rh and Ag. The original investigation of Rayne (1960) showed the anomalous variations in dc_{44}/dT and dc'/dT at $T > 50$ K. The c_{44} anomalies were, however, considerably more significant, with positive values at dc_{44}/dT at $T > 100$ K. Rayne suggested that this behavior is due to the change with temperature of the contribution of holes in the almost filled d band and applied the transfer model of Leigh (1951) to find an effective degeneracy temperature $T_F \approx 1500$ K. Subsequent measurements (MacFarlane *et al.*, 1965) have shown a similar c_{44} anomaly in Pt, the 5d element in the same group as Pd. In both Pd and Pt, the values of the c_{44} modulus differ by only $\sim 1\%$, whereas the c' modulus for Pt is about twice that of Pd and the K value of Pt is about 1.5 that of Pd. Figure 32 shows the more recent data for c_{44}, c', and K of Pd up to $T = 825$ K; the c_{44} and c' values for Pt to $T = 300$ K, are plotted in Fig. 33. The anomaly is smaller than that in the 4d bcc transition metals, and the temperatures at which the electronic effects are smeared out are also lower than in Nb. This is expected because the peak in the density of states of these metals is narrow. From an analysis similar to Section III,C,3 for Pd, $c_{el} \approx 0.03 \times 10^{11}$ N/m², or about 40% of $c_{44}(0)$, and $T_F \approx 500$ K. From the relation between the magnetic susceptibility and elastic constants of Pd [Eqs. (35) and (36)], Weinmann and Steinemann (1974) found for Pd, $D_1(c_{44}) \approx 0.015$ Ry/unit strain and $D_1(c') \approx 0$.

The effects on c_{44}, shown in Fig. 32, by adding Rh or Ag to solid solutions of Pd are indicative of electronic effects that are similar in some respects to the bcc 4d alloys but are considerably more complex. In fact, the

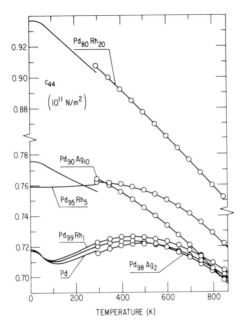

Fig. 32. The c_{44} elastic constant of Pd and Rh–Pd–Ag alloys according to Weinmann and Steinemann (1977) and Walker et al. (1970). —— (Walker, 1970); -o-o-o, present work.

rigid-band hypothesis is not expected to apply in these alloys. Weinmann (1977) found that electronic effects are maximum at Pd and disappear on alloying with either 20% Rh and somewhere above 10% Ag, the elements to left and right of Pd in the periodic table.

A relevant question in regard to Pd-alloy systems is the eventual exchange enhancement of the "mechanical polarization." Fletcher (1978) found that these effects are small or absent, on the basis that $d\mu/d\gamma = 0$ and that lattice rules make $dN/d\gamma = 0$ (Steinemann, 1979).

An anomalous phonon branch has also been reported for both Pd and Pt but, unlike the Nb results, this anomalous phonon dispersion is associated with the c' elastic mode that shows, essentially, no anomalous T dependence in Pt. As in Nb, the kink in the dispersion curve becomes less pronounced with increasing T and it may be argued that its disappearance is a result of normal anharmonicity causing the decrease in c' with increasing T. Several other interpretations have, however, been proposed and an explanation in terms of a combination of Kohn anomalies related to the Fermi surfaces appears to be most consistent with observed shift in the kink to lower ζ with higher T. These Kohn anomalies may play a role in the lattice part of c', which shows a downward curvature and flattens out above 300 K. The $[\zeta\zeta0]T_2$ branch of the phonon dispersion of Pd, on the other hand, shows no significant anomaly (Miller and Brockhouse, 1968; Dutton et al., 1972).

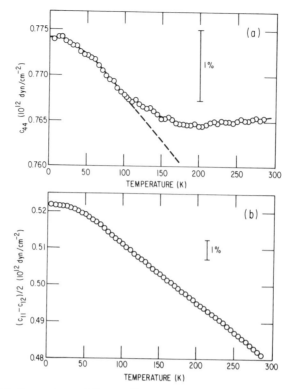

Fig. 33. Elastic constants of Pt according to MacFarlane *et al.* (1965) plotted for c_{44} (a) and c' (b).

D. Hexagonal Transition Metals

The hcp structure is quite prominent among the elements of the *d*- and f-band transition series. Elastic modulus studies are, however, limited to the pure metals; therefore, the absence of the chemical composition variable that is so important in detecting and understanding electron Fermi surface contributions limits the extent of the review. There are, however, certain observations that indicate the importance of electronic effects that may be uncovered in future research. These effects are briefly noted in this section.

1. TITANIUM, ZIRCONIUM, AND HAFNIUM

The group IV transition elements are at the center of interest among the hcp metals because of the possible relation of the c_{ij} to the temperature-induced structural transformations to the bcc phases at high temperatures and the composition- and pressure-induced transitions to the omega ω phases. There are two elastic moduli that are of importance in relation to either plastic deformation or structural phase changes; the c_{44} shear shifts basal (0001) planes relative to each other and the c_{66} shear shifts prism (hk.0) planes, causing a distortion of the sixfold symmetry. The c_{44} is relatively weak when the structural $c/at.$ ratio is greater than that for ideal close packing of atoms and is, therefore, of great importance in deforming materials such as graphite, Zn, and Cd, whereas the c_{66} modulus is generally of greater importance where $c/at. < 1.66$. Transformation mechanisms from hcp $\rightleftarrows$ bcc generally involve the prism planes and the T dependence of c_{66} in the many metals showing this transition was of interest as a possible indicator of a "soft-mode" phase transition, at high temperature [see Axe et al. (1975) for application to bcc $\rightleftarrows$ ω].

The results of the measurements of ultrasonic velocities in Ti, Zr, and Hf to $\sim$ 1200 K are shown in Fig. 34. There are three features of these results that suggest a strong Fermi surface influence on c_{66} of Ti and Zr: (i) the values of c_{66} in Ti and Zr at all temperatures are almost identical, whereas there are wide differences for the c_{44} values; (ii) c_{66} decreases by almost a factor of 5 from 4 $\rightarrow$ 1156 K, whereas thermal expansion and π_{66} (volume derivate) values (Fisher et al., 1970) account for only 1% decrease; (iii) there is a strong positive curvature in c_{66} versus T that suggests a gradual decrease in the ratio of electronic-to-lattice components with increasing T, i.e., a smearing of the Fermi surface in k space. There is a fourth feature here that indicates a close relation between the decrease in c_{66} and the hcp $\rightarrow$ bcc transition; the transformation temperatures for Ti and Zr differ by only 23 K. The phonon measurements (Moss et al., 1973; Stassis et al., 1978) confirm the very large decreases in c_{66} with T, but do not show any anomalous softening to zero frequencies that would establish a causal relation of phonons to phase changes. No Kohn anomalies are observed that would serve to relate the c_{66} behavior to critical Fermi surface wave vectors. It is for this reason that studies of c_{ij} in Ti–Sc and Zr–Y very dilute solid solutions would be of special importance as a starting point for understanding the apparently strong electronic components to c_{66}. The system Sc–Zr is probably the most promising with respect to obtaining stable single crystals of concentrated solid solutions.

It should be noted that c_{ij} measurements over wide temperature ranges have been carried out for a relatively large number of hcp metals, including most of the heavy rare-earth elements (Fisher and Dever, 1967). No indications of strong electronic effects, such as found in Ti and Zr, have been found in other metals, with exception for the anomalies in Sc metal discussed below and those related to the magnetic ordering in the rare-earth metals, which is outside the scope of our review.

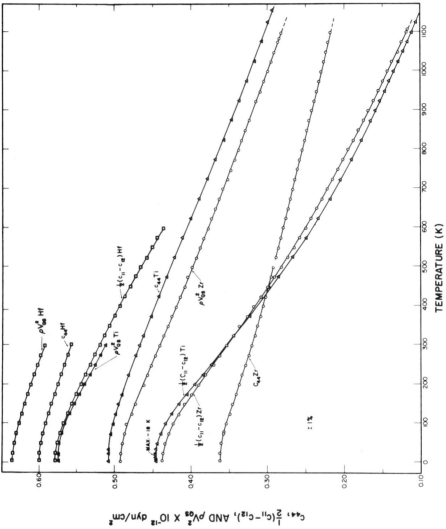

Fig. 34. c_{44} and c_{66} elastic constants in Ti, Zr, and Hf according to Fisher and Renken (1964).

2. SCANDIUM AND YTTRIUM

Scandium metal is extremely difficult to obtain in a form that is relatively free of impurities. It is only very recently that highly purified Sc metal has been obtained for heat-capacity studies (Tsang *et al.*, 1976). Several years ago, however, an attempt was made (Fisher and Dever, 1968) to derive the best c_{ij} values for "pure" Sc from measurements of single crystals from several sources with total impurity levels ranging from 1500 to 950 ppm by weight. The results showed a 1–2% variation with impurity levels for c_{66} and c_{11}, but normal T dependence for both of these modes propagated perpendicular to the "c" axis. In contrast, the c_{33} modulus for the longitudinal mode propagated parallel to "c" showed an anomalous T dependence between 4 and 300 K and especially large effects of impurities on the anomaly, as shown in Fig. 35. The value of c_{33} at 0 K decreased by about 6% with increasing purity. The anomaly was, therefore, increased with increasing sample purity. The lower four curves represent the data from a levitation-melted single crystal in its purest form and with intentional additions of Gd, a ferromagnetic impurity. It was also noted that the

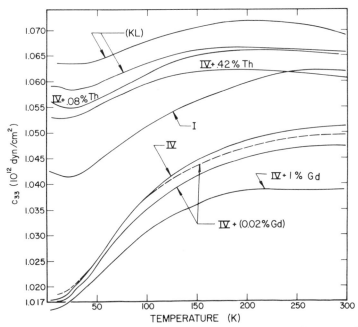

Fig. 35. Temperature dependence of the c_{33} elastic modulus of different scandium lots (Fisher and Dever, 1968). Lot KL, purchased from vendor, lot I from arc-melted high-purity powder, lot IV from levitation melt in vacuum.

paramagnetic susceptibility χ_z along the "c" axis was increased considerably in these crystals. A possible explanation for this unusual sensitivity of a longitudinal elastical modulus to relatively small impurity levels is found in the phonon dispersion studies by Wakabayashi et al. (1971). They noted anomalous "dips," i.e., slope changes in the Δ_1 branch corresponding to the longitudinal modes along the c axis. This change in slope occurs at a relatively low wave vector, as was predicted from neutron diffraction studies of magnetically ordered impurities in Sc. Thus, the degree of kink or dip could have an effect on the velocity of sound, and the addition of magnetic impurities and decrease in nonmagnetic impurity matter would enhance the anomaly. The effect arises due to "nesting" features of the Fermi surface along the c axis. The phonon dispersion measurements in yttrium metal, by the same group, showed no evidence of such an anomaly and this is consistent with the measurements of the c_{33} modulus in Y (Smith and Gjevre, 1960).

E. A-15 Phases

The discussion of electronic contributions to elastic moduli of transition metals would not be complete without reference to the classic examples found in the linear chain A-15 structures. Measurements of the single crystal c_{ij} between 300 and 4 K have been carried out for V_3Si (Testardi et al., 1965) and Nb_3Sn (Keller and Hanak, 1967), both of which are high T_c superconductors, and V_3Ge (Testardi, 1973) with $T_c = 6$ K. In addition, longitudinal and transverse sound velocities of polycrystalline samples have been measured in $Nb_3Al, Nb_3Ga, V_3Ga, Nb_3Ge$, and in the three compounds mentioned above. There is a clear association of the anomalous temperature derivatives of c', c_{44}, and, the transverse sound velocity polycrystalline samples with T_c. For the high T_c compounds V_3Si and Nb_3Sn, both dc'/dT and dc_{44}/dT are anomalous with c', almost disappearing either at the martensitic phase-transition temperatures or at T_c in the nontransformating samples. For V_3Ge, however, the anomalies are relatively small and extend to 80 K instead of to T_c. The data are well summarized by Testardi (1973).

The basic model that has been used to explain the very strong and anomalous dc'/dT in V_3Si and Nb_3Sn assumes a high density of d-band states very near to the Fermi level, as shown in the tight-binding calculations by Labbe and Friedel (1966).

The basic idea for explaining the c' anomalies is exemplified by the RCA Model (Cohen et al., 1967), which uses a step function with a constant density of states and a very low degeneracy temperature in equations similar in form to Eqs. (32) and (35). The model produces an excellent fit to c' versus

T for $T_F \simeq 80$ K and deformation potentials of about 0.3 Ry/unit strain. Measurements of the magnetic susceptibility χ versus T give, however, a T_F of 280 K (Rehwald et al., 1972). Thus the simple density of states versus E models cannot predict the T dependence of both c' and χ. It also does not predict the observed anomalous dc_{44}/dT in both V_3Si and, particularly, Nb_3Sn. It appears clear that a complete understanding of the unusual behavior of c', c_{44}, and their relation to the cubic-to-tetragonal structural transformation will require a detailed band structure calculation and the use of symmetry-dependent deformation potentials, as indicated for the bcc alloys.

F.   Hydrogen Effects on c_{44} of Niobium

It has been known for several years that the dissolution of hydrogen or its isotopes in the group V transition metals causes c_{44} to increase, whereas c' decreases. The latter effect is clearly an anelastic modulus reduction caused by relaxation or movement of the interstitial H and appears to be at least closely related to a Snoek effect (Fisher et al., 1975). The effect on c' is greatest in V and considerably smaller in Nb and Ta, evidently reflecting the inverse effect of host volume on relaxation strength. The increase in c_{44} is, however, independent of the $\Delta c'$ effect and is considerably greater in Nb than in either V or Ta. Recent measurements of these effects in the binary solid solutions of V–Nb and Nb–Ta show that the Δc_{44} effect is very strong in Nb, but decreases very rapidly with additions of either Ta or V as substitutional solutes (Fig. 36). (The strength of the effect is measured as $\Delta c_{44}/c_{44}$ per unit percent concentration of H to metal.) Recalling that c_{44} for pure Nb is anomalously small, the most obvious explanation of this strong variation among the three otherwise similar metals is that H causes changes in the Nb band structure that tend to reduce or eliminate the negative Δc_{el} that leads to the c_{44} anomalies in the 4d, Nb-rich alloys (Fig. 16).

The identification of the large Δc_{44} in Nb as a band structure effect finds confirmation in the most recent calculations of the band energies in several hydrides of V (Switendick, 1975). It should be recalled from Fig. 16 that the anomalous composition dependence of c_{44} and dc_{44}/dT in the Nb-rich alloys of the Nb–Zr system is related qualitatively on a rigid-band model to the high density of states at e/at.~ 4.8. This singularity results from a merging of the ellipsoid and jungle-gym hole surfaces, where the E_F approaches the saddle point on the Σ_1 band (Fig. 24). The very significant differences in the c_{44} values at 300 K between Nb and the other two pure metals indicate that this connectivity of the two hole surfaces still exists in

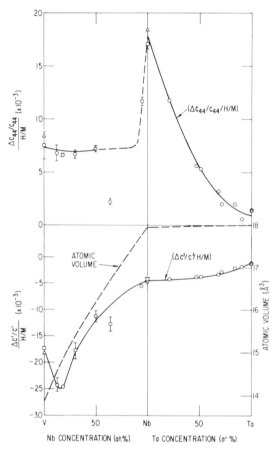

Fig. 36. Hydrogen effects in the c_{44} elastic constant of V–Nb–Ta alloys. □, Fisher *et al.* (1975); △, Magerl *et al.* (1976); ○, Fisher *et al.* (1980).

Nb, i.e., the Σ_1 band is still unoccupied at e/at. = 5, for the 4d alloys. On this basis, the relatively sharp increase in c_{44} by H in Nb suggests that H has the effect of lowering the Σ_1 band in energy so that the saddle point either dips below E_F, thus erasing the merger of the hole surfaces, or brings Σ_1 so close to E_F that some occupation of the band can occur because of temperature elevation. This reduction in Σ_1 is in fact predicted in the process of VH hydride formation, since the N1′ point has s-state symmetry and can, therefore, accept the electron from the H atom. Similarly, in dilute H solutions the Σ_1 states will be those affected by the H, or deuterium, or tritium. Thus, the Δc_{44} effect caused by H appears to be a sensitive monitor of the proximity of the Σ_1 band to E_F. A very primitive interpretation of the data presented in Fig. 36 is that the Σ_1 is shifted only slightly by H in

Ta-rich alloys and also in the range of Nb50V. The reason for this may be that the Σ_1 saddle-point energy is already slight below E_F in the H-free alloy at 300 K and the modification of the $N_{1'}$ state by H has only a slight effect on c_{44}. It may also be noted that a small Δc_{44} is related to the binary Nb–V and Nb–Ta alloys that show very high terminal solubilities for H, i.e., very large terminal solubilities are observed at Nb60V and Nb70Ta (Miller and Westlake, 1979).

V. Concluding Remarks

In this chapter we have presented clear evidence that the elastic properties of a broad range of alloys are extremely sensitive to changes in electron band structure and that electron renormalization effects can strongly influence thermodynamic properties. The understanding of these aspects of cohesion are the basis for the development of material science.

The authors are indebted to the many others in the field with whom they have collaborated on other manuscripts. We are particularly indebted to K. Katahara and M. H. Manghnani at the Hawaii Institute of Geophysics and G. Shannette at the Michigan Technological Institute, F. R. Brotzen at Rice University, M. Peter and his collaborators at the University of Geneva, and M. Posternak and Ch. Pfersich who have permitted inclusion of unpublished data and given precious advice. The authors acknowledge support for their work, which is discussed in this chapter, by the Swiss National Science Foundation and the U. S. Department of Energy.

References

Alers, G. A., and Waldorf, D. L. (1961). *Phys. Rev. Lett.* **6**, 677.
Alers, G. A., and Karbon, J. A. (1968). *J. Appl. Phys.* **39**, 4348.
Armstrong, D. A., and Mordike, B. L. (1970). *J. Less-Common Metals* **22**, 269.
Ashkenazi, J., Dacorogna, M., Peter, M., Talmor, Y., Walker E., and Steinemann, S. (1978). *Phys. Rev. B* **18**, 4120.
Ashkenazi, J., Dacorogna, M., and Peter, M. (1980). Sol. State Comm. **36**, unpublished.
Axe, J. D., Keating, D. T., and Moss, S. C. (1975). *Phys. Rev. Lett.* **35**, 530.
Bardeen, J., and Shockley, W. (1950). *Phys. Rev.* **80**, 72.
Barisic, S. and Labbe, J. (1967). *J. Phys. Chem. Solids* **28**, 2477.
Berghout, C. W. (1962). *Phys. Lett.* **1**, 292.
Bernstein, B. T. (1963). *Phys. Rev.* **132**, 50.
Bernstein, B. T. (1965). *Phys. Rev.* **137**, A1404.
Born, M., and Huang, K. (1954). "Dynamical Theory of Crystal Lattices." Oxford Univ. Press, London.
Boyer, L. L., Papaconstantopoulos, D. A., and Klein, B. M. (1977). *Phys. Rev. B* **15**, 3685.

Bujard, P., Sanjines, R., Walker, E., Ashkenazi, J., and Peter, M. (1980). Unpublished.

Carpenter, M. L., and Shannette, G. W. (1974). Michigan Technological University, unpublished data.

Chen, A. -B., Weisz, G., and Sher, A. (1972). *Phys. Rev. B* **5**, 2897.

Chui, S. T. (1974). *Phys. Rev. B* **9**, 3300.

Cohen, R. W., Cody, G. D., and Halloran, J. J. (1967). *Phys. Rev. Lett.* **19**, 840.

Colavita, E., Franciosi, A., Rosei, R., Sacchetti, F., Giuliano, E. S., Ruggeri, R., and Lynch, D. W. (1979). *Phys. Rev. B* **20**, 4864.

Cowley, R. A. (1967). *Proc. Phys. Soc.* **90**, 1127.

Cowley, R. A. (1976). *Phys. Rev. B* **13**, 4877.

Davidson, D. L., and Brotzen, F. R. (1968). *J. Appl. Phys.* **39**, 5768.

Ducastelle, F. (1970). *J. Phys.* **31**, 1055.

Dutton, D. H., Brockhouse, B. N., and Miller, A. P. (1972). *Can. J. Phys.* **50**, 29.

Elyashar, N., and Koelling, D. D. (1977). *Phys. Rev. B* **15**, 3620.

Endoh (1979). *J. Mag. Magn. Mat.* **10**, 177.

Feynman, R. P. (1939). *Phys. Rev.* **56**, 340.

Fischer, O., Peter, M., and Steinemann, S. (1969). *Helv. Phys. Acta* **42**, 459.

Fisher, E. S., and Dever, D. (1967). *Trans. Metall. Soc. AIME* **239**, 48.

Fisher, E. S., and Dever, D. (1968). *Proc. 7th Rare Earth Conf.*, San Diego, California.

Fisher, E. S., and Dever, D. (1970a). *Acta Metall.* **18**, 265.

Fisher, E. S., and Dever, D. (1970b). "Science Technology and Application of Titanium" (R. Jaffee and N. Promisel, eds.). Pergamon, New York.

Fisher, E. S., and Renken, C. J. (1964). *Phys. Rev.* **135**, A482.

Fisher, E. S., Manghnani, M. H., and Sokolowski, T. J. (1970). *J. Appl. Phys.* **41**, 2991.

Fisher, E. S., Westlake, D. G., and Ockers, S. (1975). *Phys. Status Solidi* **28**, 591.

Fisher, E. S., Miller, J. F., Alberts, H. L., and Westlake, D. G. (1980). Manuscript in preparation.

Fletcher, G. C. (1978). *Physica* **93B**, 149.

Forgacs, R. L. (1960). *I.R.E. Trans. Instrumen.* **I-9**, 359.

Friedel, J. (1964). *Trans. Met. Soc. AIME* **230**, 616.

Friedel, J. (1969). *In* "The Physics of Metals—1. Electrons" (T. M. Ziman, ed.), p. 340. Cambridge Univ. Press, London.

Frey, M. L., Lonnes, J. E., and Shannette, G. W. (1978). *J. Appl. Phys.* **49**, 4406.

Fuchs, K. (1936a). *Proc. Roy. Soc.* **A153**, 622.

Fuchs, K. (1936b). *Proc. Roy. Soc.* **A157**, 444.

Fuchs, K., and Peng. H. W. (1942). *Proc. Roy. Soc.* **A180**, 451.

Gerhardt, V. (1968). *Phys. Rev.* **172**, 651.

Giuliano, E. S., Ruggeri, R., Gyorffy, B. L., and Stocks, G. M. (1978). *In* "Transition Metals— 1977" (M. J. G. Lee, J. M. Perz, and E. Fawcett, eds.), Inst. Phys. Conf. Ser. No. 39, pp. 410–414. Inst. Phys., Bristol, London.

Gladstone, G., Jensen, M. A., and Schrieffer, J. R. (1969). *In* "Superconductivity" (R. D. Parks, ed.), Vol. 2, p. 665. Dekker, New York.

Goasdoue, C., Ho, P. S., and Sass, S. L. (1976). *Acta. Metall.* **20**, 725.

Gray, D. M., and Gray, A. M., (1976). *Phys. Rev. B* **14**, 669.

Gray, D. M., Gray, A. M., and Brown, E. (1975). *Phys. Rev. B* **11**, 1475.

Griessen, R. (1978). In "Transition Metals— 1977" (M. J. G. Lee, J. M. Perz, and E. Fawcett, eds.), Inst. Phys. Conf. Ser. No. 39, pp. 51–60. Inst. Phys., Bristol, London.

Griessen, R., Lee, M. J. G., and Stanley, D. J. (1977). *Phys. Rev. B* **16**, 4385.

Gschneidner, K. A. (1964). *In* "Solid State Physics" (F. Seitz and D. Turnbull, eds.), Vol. 16, p. 275. Academic Press, New York, London.

Hansen, M., and Anderko, K. (1958). "Constitution of Binary Alloys." McGraw-Hill, New York.

Heine, V., and Weaire, D. (1970). *In* "Solid State Physics" (H. Ehrenreich, F. Seitz, and D. Turnbull, eds.), Vol. 24, p. 249. Academic Press, New York, London.

Hellman, H. (1933). *Z. Phys.* **85**, 180.

Herring, C., and Vogt, E. (1956). *Phys. Rev.* **101**, 944.

Hill, R. (1952). *Proc. Phy. Soc.* **A65**, 349.

Hillmann, H., and Pfeiffer, I. (1967). *Z. Metallkd.* **58**, 129.

Hodges, L., Watson, R. E., and Ehrenreich, H. (1972). *Phys. Rev. B* **5**, 3953.

Hubbell, W. C., and Brotzen, F. R. (1972). *J. Appl. Phys.* **43**, 3306.

Huntington, H. B. (1958). *In* "Solid State Physics" (F. Seitz and D. Turnbull, eds.), Vol. 7, p. 213. Academic Press, New York.

Juras, G. E., and Segall, B. (1972). *Phys. Rev. Lett.* **29**, 1246.

Juras, G. E., and Segall, B. (1973). *Surf. Sci.* **37**, 929.

Katahara, K. W., Manghnani, M. H., Ming, L. C., and Fisher, E. S. (1977). *In* "High Pressure Research—Applications to Geophysics" (M. H. Manghnani and S. Akimoto, eds.), pp. 351–366. Academic Press, New York.

Katahara, K., Manghnani, M., and Fisher, E. S. (1979a). *J. Phys. F* **9**, 773.

Katahara, K. W., Nimalendran, M., Manghnani, M. H., and Fisher, E. S. (1979b). *J. Phys. F* **9**, 2167.

Keita, M., and Steinemann, S. (1978). *J. Phys. C* **11**, 4635.

Keller, K. R., and Hanak, J. J. (1967). *Phys. Rev.* **154**, 628.

Keyes, R. W. (1967). *In* "Solid State Physics" (F. Seitz, D. Turnbull, and H. Ehrenreich, eds.), Vol. 20, p. 37. Academic Press, New York.

Kriessman, C., and Callen, H. B. (1954). *Phys. Rev.* **94**, 837.

Labbe, J., and Friedel, J. (1966). *J. Phys. Radium* **27**, 153.

Lakkad, S. C. (1971). *J. Appl. Phys.* **42**, 4277.

Landau, L. D., and Lifshitz, E. M. (1959). "Theory of Elasticity." Addison-Wesley, Reading, Massachusetts.

Landsberg, P. T. (1969). *In* "Solid State Theory—Methods and Applications," p. 41. Wiley-Interscience, London.

Laurent, D. C., Wang, G. S., and Callaway, T. (1978). *Phys. Rev. B* **17**, 455.

Leibfried, G., and Ludwig, W. (1961). *In* "Solid State Physics" (F. Seitz and D. Turnbull, eds.), Vol. 12, p. 275. Academic Press, New York.

Leigh, R. S. (1951). *Philos. Mag.* **42**, 139.

Lenkkeri, J. T. and Lahteenkorva, E. E. (1978). *J. Phys. F* **8**, 1643.

Lifshitz, I. M. (1960). *Sov. Phys. JETP* **11**, 1130.

Lipton, D., and Jacobs, R. L. (1970). *J. Phys. C* **3**, 389.

Löwdin, P. -O. (1963). *J. Mol. Spectrosc.* **10**, 12.

Löwdin, P. -O. (1964). *J. Mol. Spectrosc.* **13**, 326.

Lowrie, R., and Gonas, A. M. (1967). *J. Appl. Phys.* **38**, 4505.

Ludwig, W. (1967). "Recent Developments in Lattice Theory," Springer Tracts in Modern Physics, Vol. 43. Springer Verlag, Berlin–Heidelberg–New York.

MacFarlane, R. E., Rayne, J. E., and Jones, C. K. (1965). *Phys. Lett.* **18**, 91; and Rayne, J. E. (1968). Private communication.

Magerl, A., Berre, B., and Alefeld, G. (1976). *Phys. Status Solidi(a)* **36**, 161.

Mattheiss, L. F. (1970). *Phys. Rev. B* **1**, 373.

McSkimin, H. J. (1964). *In* "Physical Acoustics" (W. Mason, ed.), Vol. 1. Academic Press, New York.

Miller, A. P., and Brockhouse, B. N. (1968). *Phys. Rev. Lett.* **20**, 798.

Miller, J. F., and Westlake, D. G. (1979). Trans. Japan Inst. of Metals, *Proc. Hydrogen in Metals*, Nov. 26, 1979.

Moss, S. C., Keating, D. T., and Axe, J. D. (1973). *Solid State Commun.* **13**, 1465.

Nakagawa, Y., and Woods, A. D. B. (1965). *In* "Lattice Dynamics" (R. F. Wallis, ed.), p. 39. Pergamon Press, Oxford.

Nedyukha, M. M., and Chernyy, V. G. (1966). *Fiz. Metal. Metalloved.* **22**, 114.

Nye, J. F. (1957). "Physical Properties of Crystals." Oxford Univ. Press (Clarendon), London and New York.

Nilsson, P. O., and Sandell, B. (1970). *Solid States Commun.* **8**, 721.

Peter, M., Klose, W., Adam, G., Entel, P., and Kudla, EWA (1974). *Helv. Phys. Acta* **47**, 807.

Pettifor, D. G. (1970). *J. Phys. C* **3**, 366.

Pfersich, C. (1980). Université de Lausanne thesis.

Pfersich, C., Waeber, W. B., and Steinemann, S. G. (1979). *J. Mag. Magn. Mat.* **10**, 172.

Posternak, M. (1976). Université de Lausanne thesis.

Posternak, M., and Steinemann, S. (1970). *Solid State Commun.* **8**, 1373.

Posternak, M., Steinemann, S., and Peter, M. (1968). *Helv. Phys. Acta* **41**, 1296.

Posternak, M., Waeber, W. B., Griessen, R., Joss, W., Van Der Mark, W., and Wejgaard, W. (1975). *J. Low Temp. Phys.* **21**, 47.

Powell, B. M., Martel, P., and Woods, A. D. B. (1968). *Phys. Rev.* **171**, 727.

Powell, B. M., Woods, A. D. B., and Martel, P. (1972). *Proc. Series*, Int. Atomic Energy Agency, Vienna, Neutron Inelastic Scattering Symp., STI/PUB/308, p. 43.

Rayne, J. A. (1960). *Phys. Rev.* **118**, 1545.

Rehwald, W., Rayl, M., Cohen, R. W., and Cody, G. D. (1972). *Phys. Rev. B* **6**, 1452.

Reid, C. N., Routbort, J. L., and Maynard, R. A. (1973). *J. Appl. Phys.* **49**, 1390.

Reuss, A. (1929). *Z. Angew. Math. Mech.* **9**, 49.

Segall, B., Elyashar, N., and Chen, A. B. (1978). *Phys. Rev. B* **18**, 5326.

Shapiro, S. M., Shirane, G., and Axe, J. D. (1975). *Phys. Rev. B* **12**, 4899.

Silva, E. Z., and Ferreira, L. G. (1978). *Rev. Bras. Fis.* **8**, 117.

Slater, J. C. (1939). "Introduction to Chemical Physics." McGraw-Hill, New York and London.

Smith, C. S. (1960). Case-Western Reserve Univ., unpublished data.

Smith, C. S., Schuele, D. E., and Daniels, W. B. (1965). *In* "Physics of Solids at High Pressures" (C. T. Tomizuka and R. M. Emrick, eds.), p. 496. Academic Press, New York.

Smith, J. F., and Gjevre, A. (1960), *J. Appl. Phys.* **31**, 645.

Smith, T. F., and Finlayson, T. R. (1976). *J. Phys. F* **6**, 709.

Stassis, C., Zarestky, J., Arch, D., McMasters, D. D., and Harmon, B. N. (1978). *Phys. Rev. B* **18**, 2632.

Stassis, C., Zarestky, J., and Wakabyashi, N. (1979). *Bull. Am. Phys. Soc.* **24**, 487.

Steinemann, S. (1979). *J. Mag. Magn. Mat.* **12**, 191.

Switendick, A. C. (1975). *In* "Hydrogen Energy," Part B (T. N. Veziragln, ed.), Plenum, New York.

Talmor, Y., Walker, E., and Steinemann, S. (1977). *Solid State Commun.* **23**, 649.

Testardi, L. R. (1973). *In* "Physical Acoustics" (W. P. Mason, ed.), Vol. X, Chapter 4, p. 193. Academic Press, New York.

Testardi, L. R., Bateman, T. B., Reed, W. A., and Chirba, V. E. (1965). *Phys. Rev. Lett.* **15**, 250.

Truell, R., Elbaum, C., and Chick, B. B. (1969). "Ultrasonic Methods in Solid State Physics." Academic Press, New York.

Thurston, R. N., and Brugger, K. (1964). *Phys. Rev.* **133**, A1604.

Tsang, T. W. E., Gschneidner, K. A., and Schmidt, F. A. (1976). *Solid State Commun.* **20**, 737.

Van Torne, L. I., and Thomas, G. (1966). *Acta Metall.* **14**, 621.

Varma, C. M., and Weber, W. (1977). *Phys. Rev. Lett.* **39**, 1094.

Voigt, W. (1889). *Ann. Phys.* **38**, 573.

Wakabayashi, N., Sinha, S. K., and Spedding, F. H. (1971). *Phys. Rev. B* **4**, 2398.

Walker, E. (1978). *Solid State Commun.* **28**, 587.

Walker, E., and Bujard, P. (1980). *Solid State Commun.* **34**, 691.

Walker, E., and Peter M. (1977). *J. Appl. Phys.* **48**, 2820.

Walker, E., Ortelli, J., and Peter, M. (1970). *Phys. Lett. A* **31**, 240.

Wallace, D. C. (1972). "Thermodynamics of Crystals." Wiley, New York, London, Sidney, and Toronto.

Weger, M., and Goldberg, I. B. (1973). *In* "Solid State Physics" (H. Ehrenreich, F. Seitz, and D. Turnbull, eds.), Vol. 28, pp. 1–177. Academic Press, New York and London.

Weinmann, C. (1977). Université de Lausanne thesis.

Weinmann, C., and Steinemann, S. (1974). *Solid State Commun.* **15**, 281.

Weiss, P. (1979). Université de Lausanne thesis.

White, G. K., Collins, J. G., Birch, J. A., Smith, T. F., and Finlayson, T. R. (1978). *In* "Transition Metals—1977 (M. J. G. Lee, J. M. Perz, and E. Fawcett, eds.), Inst. Phys. Conf. Ser. No. 39, Inst. Phys., Bristol, London.

Wilson, A. H. (1965). "The Theory of Metals." Cambridge University Press, London.

Ziman, J. M. (1960). "Electrons and Phonons." Oxford University Press, London.

Keyword List

Elastic constants
Transition metals
Strain response of electronic structure
Softening of elastic constants
Grüneisen gamma—Grüneisen parameter
Pressure derivatives of elastic constants
Temperature derivatives of elastic constants
Lattice stability conditions
Deformation potential—first-order and second-order
Electronic contribution to elastic constants
Electronic anomalies
Electronic band structure and elastic constants
Anharmonic lattice excitation—influence on elastic constants
Elastic anisotropy
Wave propagation
Hydrogen in metals
Electron renormalization of elastic constants
Exchange enhancement in elastic constants
Effective Fermi temperature
Topology changes of Fermi surface
Thermal expansion at low temperatures
Elastic constants from zero sound
Magnetic susceptibility
A-15 phases

5

Electrical Resistivity of Metals

Z. FISK and G. W. WEBB

Institute for Pure and Applied Physical Sciences
University of California
San Diego, La Jolla, California

I. Introduction

The electrical resistivity of a metal is one of its distinguishing properties. The fact that it is also one of the most easily measured properties of a metal encourages one to investigate in detail how much can be learned from this measurement and its temperature dependence. It is noteworthy that there is still very active theoretical research into such heavily investigated topics as the lattice term in the resistivity, just to mention one. Disentangling the many contributions to the scattering of conduction electrons is in many cases not at all straightforward, but in a favorable situation quite detailed information can be obtained, and even qualitative results, such as the shape of the temperature dependence of the resistivity, gives much valuable information.

It is not the purpose of this review to give a thorough survey of the enormous volume of experimental and theoretical work relating to the electrical resistivity of metals. Rather, we wish to provide a syllabus of the variety of behavior that has been observed, along with the thinking on its interpretation, that will be of use to the materials scientist who is either confronted with resistance data which he wishes to interpret or who would like to know whether the measurement of the electrical resistance of his material can help him understand what he has. In general, we have restricted our attention to materials which are macroscopically homogeneous, except for the case of Guinier–Preston zone formation. In cases of particular inhomogeneous systems, the reader should consult the literature (e.g., Proceedings of the First Conference on the Electrical Transport and Optical Properties of Inhomogeneous Media, 1977).

The bibliography is divided into two parts: first the general texts which we have found useful and second the specific references of this review.

II. Scattering Mechanisms

We know from Bloch's theorem that noninteracting electrons in a perfect lattice experience no scattering. In a metal this means that once conduction electrons are accelerated by an electric field they will continue to carry a current which does not decay in time unless they are scattered by imperfections in the lattice potential; for the moment, we neglect electron–electron scattering arising from a weak interaction among the electrons. In this way the study of the electrical resistivity of metals reduces to one of considering the various ways the lattice potential seen by the electrons deviates from perfect long-range order.

We will consider the most important types of disorder which are encountered in real metals. These different types of disorder have been separated according to whether or not they depend on temperature. This is an approximation since those which are thought not to depend on temperature are found under closer scrutiny to have a small temperature dependence.

Another approximation which is very useful is that the scattering frequencies which arise from the different types of disorder are additive. That is, for an electron in a state k, the total scattering frequency ω_k can be written as a sum

$$\omega_k = \sum_s \omega_k^s \tag{1}$$

over the different types of scattering. Essentially this approximation assumes that the scattering events are independent. Systems which show the breakdown of this approximation are treated in Sections II,A,7 and II,B,7. If all the current is assumed to be carried by carriers with the same effective mass, velocity, etc., then Eq. (1) reduces to (Jan, 1957)

$$\rho = \sum_s \rho_s, \tag{2}$$

where ρ_s is the resistivity introduced by the scattering ω_s.

Matthiessen's rule is an important example of Eq. (2) in which

$$\rho = \rho_{\text{impurities}} + \rho_{\text{matrix}}, \quad \text{i.e.,} \quad \rho = \rho_0 + \rho(T), \tag{3}$$

where ρ_0 is the temperature-independent impurity scattering due to such things as dilute amounts of foreign atoms, dislocations, and grain boundaries, and $\rho(T)$ is the temperature-dependent resistivity of the pure host. If, in fact, ρ_0 is not temperature independent, as discussed in Section II,A,7, then Matthiesen's rule will not be valid. Its validity has been reviewed recently by Bass (1972).

If the single-band approximation cannot be made it is still sometimes possible to use a multiband approximation. In doing this, one assumes that there are a set of bands of electrons which carry current in parallel. The total conductivity is then

$$\sigma = \sum_b \sigma_b, \tag{4}$$

where σ_b is the conductivity of an individual band. The resistivity is then

$$\rho = 1/\sum_b \sigma_b = 1/\sum(1/\rho_b) \tag{5}$$

and in general the resistivity of each band is a sum over different types of scattering:

$$\rho_b = \sum_s \rho_b^s. \tag{6}$$

The multiband approximation has been invoked most often to discuss the resistivity of spin-up and spin-down electrons in magnetic materials and in transition metals composed of overlapping sp and d bands.

The above remarks are strictly valid for cubic materials in which the resistivity is independent of crystallographic direction. However, the electrical resistivity is a second-rank tensor and in noncubic materials the electrical resistivity will, in general, be specified by more than one number. For most noncubic elements, this anisotropy is small. The anisotropy is also temperature dependent and a number of papers deal with this temperature dependence for the hexagonal metals Mg and Cd. The rare-earth elements are unusual in that the various hexagonal members are very anisotropic ($\rho_c/\rho_a \sim 2$) at room temperature, with this anisotropy decreasing towards $\rho_c/\rho_a \sim 1$ at high temperatures. There appears to be no ready explanation in terms of the anisotropy of the Fermi surfaces, since easily made estimates of the spin-disorder magnetic terms along the c and a axes do not appear to be anisotropic.

In compounds, extreme values for the anisotropy are sometimes found. The layer type dichalcogenide $1T–TaSe_2$, for example, has $\rho_\perp$ layer$/\rho_\parallel$ layer ≈ 20 (Hambourger and DiSalvo, 1980). This also appears in the quasi–one-dimensional conductors such as TTF–TCNQ and makes measurements on these materials difficult.

A.   Temperature-Independent Scattering

1.   IMPURITY RESISTIVITY

The effect of impurities, here considered to be substitutional or interstitial, is usually considered within the framework of Matthiessen's rule as expressed by Eq. (3). This equation is expected to hold (and in general does) for concentrations small enough not to effect either the lattice vibrations or the electron states of the host significantly. This is often true for concentrations below $\sim 5\%$. Even when the concentration is low there are well-known deviations to be expected when $\rho_0 \approx \rho(T)$ (Bass, 1972), as well as other effects which are the basis of continuing theoretical speculation. In general, these effects are small, typically less than 10% of the effects we consider, and shall not concern us here.

It is generally observed that ρ_0 is proportional to concentration c of impurity, as one expects. In those concentrated alloys for which the end members are similar as to crystal structure, volume, and electron concentration, Nordheim's rule (Mott and Jones, 1936) is approximately obeyed:

$$\rho_0 \propto c(1 - c). \tag{7}$$

Further, for nontransition element solid solutions, it has been found that $\rho_0 \propto (\Delta Z)^2$ for a given host solvent (Mott and Jones, 1936), where ΔZ is the valence difference between impurity and host. In Cu, for example, the proportionality constant is $\sim 0.4\ \mu\Omega\,cm/a_0$ for same-row sp elements.

The actal computation of the resistivity can be done using the Friedel (1954, 1958) sum rule. The charge on an impurity in a metal will be screened at large distances. This leads to a condition between the relative impurity charge and the phase shift $[\eta_L(k_F)]$ of the L angular moment component of the conduction electrons at the Fermi surface:

$$\Delta Z = (2/\pi)\sum_L (2L + 1)\eta_L(k_F). \tag{8}$$

The resistivity due to this scattering is given in terms of these phase shifts by

$$\rho_0 = \left(4\pi n_i / n e^2 k_F\right) \sum_L (2L + 1)\sin^2(\eta_L - \eta_{L+1}), \qquad (9)$$

with n_i the concentration of scatterers. It will often happen that only a few L's are important. Rather good agreement using this expression for impurities in copper has been attained by Friedel and others (Blatt, 1957a,b; de Faget de Casteljau and Friedel, 1956).

The situation for transition-metal impurities is quite different due to the localized character of d electrons. A starting point for the discussion is Friedel's (1958) notion of the virtual bound state. Suppose we have $\Delta Z > 0$, which is generally the case for transition-metal impurities in sp metals. If the impurity potential is strong enough, one of more of the extra electrons may be bound. Consider the case of an impurity with a single d orbital. If the energy of this level lies within the conduction band, however, quantum mechanics tells us that it is not truly a bound state. Anderson (1961) treated this situation theoretically by introducing a conduction-electron d-state interaction V_{kd}. In addition, in his Hamiltonian for the impurity he also includes the d–d Coulomb repulsion U. This tends to keep d electrons with opposite spin off the same impurity. If the position of the virtual level E_d, is such that $E_d + U > E_F$ (the Fermi energy) while $E_d < E_F$, then the d state will be magnetic, provided that the broadening of the level (Δ) is not too great. This case is shown in Fig. 1. Δ is given by the Golden rule

$$\Delta = \pi |\overline{V_{kd}}|^2 \rho(E_F), \qquad (10)$$

where $\rho(E_F)$ is the conduction-electron density of states at the Fermi level for one spin direction and $\overline{V_{kd}}$ an averaged matrix element.

Anderson showed that sufficiently small values of Δ/U and $(E_F - E_d)/U$ lead to magnetic behavior. Subsequent work has shown that the transition between magnetic and nonmagnetic behavior is not sharp, and that a whole range of intermediate behavior characterized by varying spin lifetimes is expected (for review, see Grüner, 1974).

We now turn to the resistivity data for transition-metal impurities in sp metals. Suppose we have N d electrons on the impurity. These are the extra valence electrons $\Delta Z = N$. Hund's rules tell us that the d shell is filled first by five d electrons having parallel spins, the next five going in antiparallel to these. To a first approximation we can assume that the $L = 2$ phase shift is the only important one. This is an intuitively reasonable assumption. In Anderson's nonmagnetic case, then, the up- and down-spin states are unsplit, and the conduction-electron phase shift will be $\eta_2 = N\pi/10$, by the Friedel sum rule. The impurity resistivity then varies with N as $\sin^2 N\pi/10$. This function has a peak at $N = 5$.

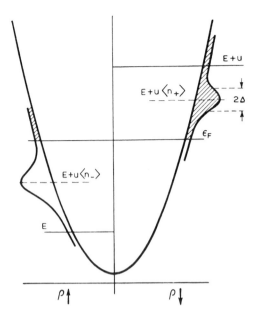

Fig. 1. Density of states for up- and down-spin electrons in the magnetic case. The humps on the free-electron density of states parabola are the virtual d levels. The shaded region is unoccupied. (Anderson, 1961.)

In the magnetic case, where the up- and down-spin virtual bound states are split, the resistivity will be proportional to $\sin^2\eta_{2+} + \sin^2\eta_{2-}$, the sum of the up- and down-spin channels for the now spin split d band. Up-spin conduction electrons are now scattered differently from down-spin conduction electrons. Here, $\eta_{\pm} = N_{\pm}\pi/5$ from the Friedel sum rule, where $\pm$ refers to the spin-up and spin-down states, since $2L + 1 = 5$ for each spin direction. This gives a minimum ($\eta_+ = \pi, \eta_- = 0$) when $N_+ = 5, N_- = 0$: the resistance versus impurity atom plot should have two peaks on either side of the minimum.

We show in Fig. 2 the data of Kedves *et al.* (1972) for first-row transition-metal impurities in Al. At 930 K, there is a double maximum. At 78 K there is a single maximum. The asymmetry of the curve is probably partly due to nonnegligible η_0 and η_1 contributions. Between these temperatures there has been some kind of transition between the nonmagnetic and magnetic cases. This transition is intimately connected with the Kondo effect and for which there is an extensive theoretical literature (for review, see Grüner and Zawadowski, 1974). The temperature characteristic of this smooth change between nonmagnetic and magnetic behavior is the Kondo temperature: $T_K \sim T_F \exp[-1/\rho(E_F)|J|]$, where J is given by the Schrieffer

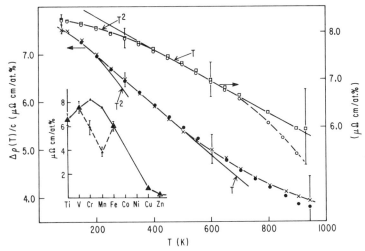

Fig. 2. Temperature dependence of the impurity resistivity of Al–Cr alloys ($\square$, $\bigcirc$) and Al–Mn alloys ($\bullet$, $\times$). The insert shows the impurity resistivity of 3d transition-metal impurities in Al measured at 78 K ($\blacktriangle$) and 930 K ($\triangle$) with estimated maximum errors. (Kedves *et al.*, 1972.)

and Wolff (1966) transformation

$$J \approx |V_{kd}|^2 U/\varepsilon(\varepsilon + U), \quad \text{where} \quad \varepsilon = E_d - E_F. \tag{11}$$

As the temperature is lowered in the magnetic case, the resistivity increases, generally as $\log T$. At lower temperature, the increase is often like T^{-2}, with a leveling off at sufficiently low T. This sort of resistivity behavior represents a gross departure from Mathiessen's rule. The effects are large, as we can see in the figure, and at low temperature there is often a minimum in the resistivity when the lattice term becomes small and the Kondo term then dominates it. Quite generally, when these resistivity effects occur, there are large impurity contributions to the magnetic susceptibility, usually a Curie–Weiss type impurity term.

2. DISLOCATION AND GRAIN BOUNDARY SCATTERING

Dislocations and grain boundaries are of primary importance in many areas of materials research. However, measurements of the electrical resistivity due to dislocations and grain boundaries have not provided much diagnostic help in materials problems in the sense that they have no particular earmark which distinguishes them from other types of crystalline defects. For dislocations in Cu, Ag, and Au, the dislocation resistivity per dislocation length per unit volume (ρ_d/β) is roughly $2 \times 10^{-13}\ \mu\Omega\,\text{cm}^3$. For grain boundaries in these metals, the resistivity due to unit area of boundary per unit volume is (ρ_{gb}/S_v) is roughly $2 \times 10^{-6}\ \mu\Omega\,\text{cm}^2$.

It has been a considerable theoretical problem to calculate the dislocation scattering, most early calculations being one to two orders of magnitude too small. One of the more successful calculations is due to Brown (1967a,b, 1977a). He assumes that there is, in general, resonance scattering by the dislocation core; namely, the scattering phase shift is $\approx \pi/2$. He derives the expression

$$\rho_d/\beta = 8h/3e^2 n_c, \tag{12}$$

where n_c is the number of carriers per primitive cell. This gives reasonable agreement with experiment for a wide range of metals, a fact which is surprising even to Brown in view of the free-electron–like approximations used. We note that Brown (1977b) has provided a comparison of his theory with the equally successful theory of Baskinski et al. (1963).

Brown (1977c) has also used his approach to compute the grain boundary electrical resistivity by considering the scattering from the line defects which make up the boundary. He finds

$$\rho_{gb}/S_v = \left(4h/e^2 n_c\right)\langle 1/D\rangle, \tag{13}$$

where n_c is as above and $\langle 1/D\rangle$ is the average inverse spacing of dislocations in the grain boundary. He again achieves good agreement with experiment for a number of metals.

3. Magnetic Domains and Domain Wall Effects

The measurement of the electrical resistivity of ferromagnetic materials is complicated by the presence of magnetic domains. In general, the electrical resistance measured on the specimen will depend, within each domain, on the relative directions of the magnetization and the current flow. This is due to magnetoresistance effects.

Schindler and LaRoy (1966) have worked out a scheme for determining a residual resistance ratio for ferromagnetic metals which should be indicative of the purity of the material. Their scheme is to apply a longitudinal magnetic field (along the current direction). The effect of this is to both sweep out domain walls, which are one source of resistance, and to align the domains along the current direction. This last is desirable since longitudinal magnetoresistance effects are smaller than transverse effects. By extrapolating the measured resistance back from fields beyond technical saturation to zero field, they arrive at a resistance value which is indicative of the purity of the material.

Their data on iron specimens of various purities show that there is a large initial decrease in the resistivity as the longitudinal magnetic field is applied, followed by a region of monotonic and slow resistance variation with field. This initial change in resistance is found to be as large as $[R(H) - R_0]/R_0 = -0.5$ for rather pure specimens. Here $R(H)$ is the measured sample resistance in a field H, R_0 that at zero field. This value was reached in 200 Oe.

4. GUINIER–PRESTON ZONES

It is sometimes observed that supersaturated solid solutions, formed by quenching from the unsaturated state, have an unusual, time-dependent resistivity: the isothermal resistance at the annealing temperature goes through a pronounced maximum, as a function of time, and then decreases (for review, see Kelly and Nicholson, 1963). For example, this happens in Al–Zn alloys (Fig. 3). This peak is known to be associated with the formation of precipitate of pure solute on the scale of atomic dimensions. At this early stage of precipitation, sometimes called preprecipitation, zones are formed, the so-called Guinier–Preston zones. These are either spherical or plate-like with a typical size of 50–100 Å. The time dependence of the effect is due to growth of the precipitate zones. One plausible explanation

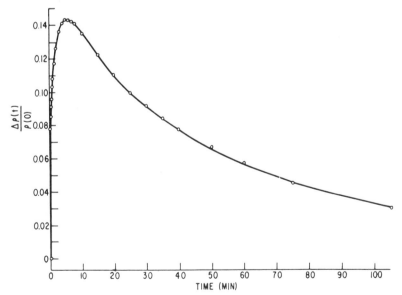

Fig. 3. Isothermal resistivity aging curve at $-45°C$ for an Al-11. 9% Zn alloy homogeneized at $380°C$, then quenched to $-45°C$. The fractional change in the resistivity $\Delta\rho(t)$ from its extrapolated value at time $t = 0$ ($\rho(0)$) is plotted. (Turnbull *et al.*, 1960.)

for the effect is the following (Hillel, 1970; Hillel *et al.*, 1977). Imagine plane-wave electrons from the solid-solution incident on the zone. Certain of these plane waves will be Bragg scattered by the crystal lattice of the precipitate. For small zones, the structure factors for the Bragg scattering will be very broad, so electron states from the whole Fermi surface will be scattered; hence the resistivity increases. As the zones grow, the structure factors become strongly peaked, so that while the scattering is stronger for the appropriate electron states, the number of electrons able to be scattered becomes very small; hence the resistivity decreases.

5. Irradiation and Quenched Thermal Disorder

There is a large and rapidly growing body of literature on the changes in physical properties of materials when they are disordered by a variety of types of radiation. Among the different types of radiation which have been studied are high-energy neutrons (Sweedler *et al.*, 1974, 1978), α particles (Poate *et al.*, 1975), electrons (Wiesman *et al.*, 1977), oxygen ions (Müller *et al.*, 1980), and sulphur ions (Adrian *et al.*, 1978). For a general references, see Billington and Crawford (1961). These high-energy particles disorder the lattice in several ways. The lighter electrons are thought to disorder the lattice by creating one point defect per electron. The heavier charged particles, from neutrons to ions, are thought to transfer energy and momentum to one or more lattice atoms which in turn travel through the lattice creating a cascade of damage. It is found that it takes an ion fluence (numbers of ions per unit area) several orders of magnitude smaller than a given electron fluence to produce equivalent amounts of damage.

The disorder or damage produced in the material has a large effect on many physical properties. Of these the electrical resistivity has been found to be especially interesting because it is a sensitive measure of the total disorder. Figure 4 shows the effects of irradiation by high-energy ($\geq 1 -$ MeV) neutrons on the temperature-dependent resistivity of a single crystal of V_3Si, an A-15 structure compound which is a high T_c superconductor. These data cover a range of fluences, from zero to 4.6×10^{19} neutrons/cm^2. It is evident that there is a monotonic increase in the resistivity at all temperatures. There are several further points of general interest to note about these data. First, the largest effect is in the residual resistivity ρ_0 which increases here by over 100 $\mu\Omega$cm. Simultaneous with the increase in ρ_0, the thermal increment in the resistivity to 300 K, $\rho(300) - \rho_0$, for example, diminishes to values much smaller than in the undamaged state. Thus Matthiessen's rule does not hold here, at least at high fluences. Rather the total resistivity $\rho_0 + \rho(T)$ appears to be approaching an upper limit as a function of fluence.

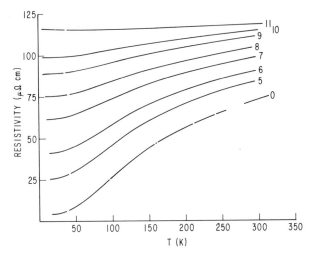

Fig. 4. The temperature-dependent resistivity of a V_3Si single crystal following irradiation by high-energy neutrons. For irradiations 0, 5, 6, 7, 8, 9, 10, and 11 the fluences are (in units of 10^{18} n/cm^2), respectively, 8.2, 11.0, 14.5, 18.0, 21.5, 25.0, 46.1. The last point, 11, was taken on a polycrystalline samples whose behavior at low fluences was the same as the single crystal. (From Caton and Viswanathan, 1978.)

Figure 5 shows the residual resistivity taken from Fig. 4 as a function of fluence. The shape of this curve is not understood in detail, especially at low fluences. However ρ_0 does at high fluences show negative curvature. This could be due to the resistivity saturating at an upper limit or to the amount of disorder reaching an upper limit for a variety of reasons. That the temperature increment of the resistivity has decreased to very small values supports the saturation explanation.

The nature of the defects produced by irradiation in compounds such as V_3Si is still a subject of debate (Testardi *et al.*, 1976). There are several types of defects which have been discussed. The anti-site defect in which an A atom resides on a B site and a B atom is on an A site has been employed to interpret a considerable amount of data (Sweedler *et al.*, 1974, 1978). Another type of a small "bond-bending" distortion along chains of atoms in the A-15 has been suggested and some evidence for its presence inferred from experiments (Testardi *et al.*, 1976). It is expected that vacancies or vacancy interstitial pairs will be produced, but these might require irradiation at low temperature to be retained. The complete loss of crystallinity to an amorphous state has been suggested and in some systems is observed to occur at very high fluences. Finally, an inhomogeneous model of disorder at low fluences has been put forward (Pandé, 1977). This model draws on transmission electron microscopy data which suggest that the initial states

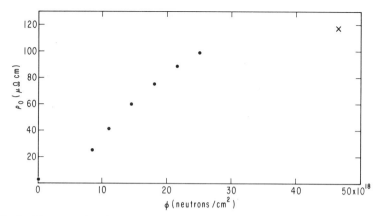

Fig. 5. The extrapolated residual resistivity of irradiated V_3Si as a function of fluence ϕ. The data are taken from Fig. 4.

of irradiation, at least for neutrons, produce heavily damaged local regions ("islands") on a 40-Å scale embedded in an otherwise undamaged host matrix. At higher fluences the damaged regions overlap. Possibly such a model might explain the curvature at low fluences to the residual resistivity in Fig. 5.

At this point it is not clear which is the "correct type" of defect. It is probable that they are all present to some extent. It is also possible that the relative fraction of a particular type of defect depends on the type of irradiating particle and definitely on the temperature of irradiation, a quantity which is difficult to measure with precision in many experiments.

6. Liquid Metals (Faber, 1966, 1972)

The electrical resistivity of liquid metals is outside of the intended scope of this review. However, in some instances the behavior of the electrical resistivity in the liquid state is valuable information in regards to the properties of the solid. In addition, electrical resistivity measurements on liquid metals have recently been made on a number of transition metals (Busch and Güntherodt, 1974), and as more of these data becomes available, it is probable that its importance with regard to the solid state will increase.

In most cases the electrical resistivity of simple metals rises markedly on melting: for alkali metals, roughly by a factor 1.6; for Cu, Ag, and Au, by slightly more than a factor 2. In some cases, the resistivity drops on melting, as with Si and Ge which form metallic liquids. The change of electrical resistivity upon melting of transition metals is often quite small, although the data are still sparse; for example, for Fe and Co the change is 1.01 and 1.09, respectively. Representative data are given in Table I.

TABLE I

LIQUID-STATE RESISTIVITY (ρ_1) AND RATIO OF LIQUID TO SOLID-STATE RESISTIVITY (ρ_1/ρ_s)
FOR SELECTED METALS AT THE MELTING POINT

Metal	ρ ($\mu\Omega$cm)	ρ_1/ρ_s	Metal	ρ ($\mu\Omega$cm)	ρ_1/ρ_s
Li	24.7		La	129.0	
Na	9.6	1.45	Ce	124.0	1.07
K	13.0	1.58	Pr	134.0	
Rb	22.5	1.60	Nd	144.0	
Cs	37.0	1.67	Eu	234.0	
			Gd	188.0	
Cu	21.0	2.1	Tb	186.0	
Ag	17.2	2.1	Dy	202.0	1.05
Au	31.2	2.3	Ho	213.0	1.04
			Er	218.0	
Mg	27.0	1.8	Yb	103.0	
Ca	33.0		Lu	215.0	1.06
Sr	84.8				
Ba	306.0		U	63.0	1.04
Zn	37.0	2.2			
Cd	34.0	2.0	Sb	114.0	0.6
Hg	91.0	3.7–4.9	Bi	128.0	0.35–0.47
Al	24.0	2.2	Mn	175.0	
Ga	26.0	0.45–3.1	Fe	138.0	1.01
In	33.0	2.2	Co	101.0	1.09
Tl	73.0	2.1	Ni	82.0	1.33
Si	80.0				
Ge	66.0				
Sn	48.0	2.1			
Pb	95.0	1.9			

In the liquid state, it is reasonable to suppose that although some short-range order is retained, the atoms have no long-range order, that the Brillouin zone structure has disappeared, and that a nearly free-electron theory, in the pseudopotential sense, is appropriate. This theory was proposed by Ziman(1961, 1967). It computes a scattering time τ which is put into the Drude expression $\rho = m/ne^2\tau$.

The expression for $1/\tau$ is

$$1/\tau = (N/V)mk_F\langle|v|^2a\rangle/\pi h^3. \tag{14}$$

Here (N/V) is the number of ions in volume V, k_F is the Fermi momentum. $\langle|v|^2a\rangle$ is the average of a matrix element of the pseudopotential v for scattering on the Fermi surface times the interference function a. a is related to the pair correlation function $P(r_1,r_2) = P(|r_1 - r_2|)$ by

$$a(K) = 1 + N/V\int_0^\infty 4\pi r^2(P - 1)(\sin Kr/Kr)\,dr. \tag{15}$$

Scattering of neutrons and x rays by liquids gives a measurement of $a(K)$ which can be used in the expression for $(1/\tau)$. What the expression for $(1/\tau)$ says is that the interference function takes into account the effect of short-range order in the liquid. The average of $\langle |v|^2 a \rangle$ is taken over all values of the scattering wavevector (K) between 0 and 2 k_F. The general behavior of $a(K)$ and $v(K)$ are shown in Fig. 6. As k_F varies with the number of electrons per atom, it is clear that $\langle |v|^2 a \rangle$ will vary quite strongly. Qualitatively this accounts for the high resistivity of divalent liquid metals such as Zn and Cd.

In this theory, it is the temperature dependence of the short-range order as contained in a that gives rise to the temperature dependence of the electrical resistivity. In particular the logarithmic derivative

$$\frac{T}{\rho}\left(\frac{\partial\rho}{\partial T}\right)_v = \frac{\langle |v|^2 a(\partial\ln a/\partial\ln T)_v\rangle}{\langle |v|^2 a\rangle}. \tag{16}$$

For $K = 0$, statistical mechanics shows that $a(0) = NkT\beta/V$, where β is the isothermal compressibility. So for small K it follows that

$$\left[\partial(\ln a)/\partial(\ln T)\right]_v = 1. \tag{17}$$

However, as T increases, the effect must be to make $a(K) \to 1$, the gas

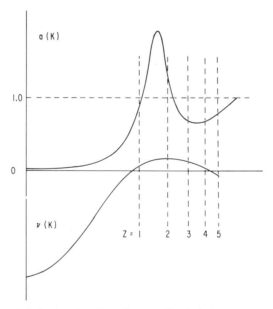

Fig. 6. Schematic behavior of a (K) and v (K). The dashed vertical lines ($Z = 1, 2,$ etc.) show where the limit $K = 2k_F$ occurs for valency Z. (Faber, 1969.)

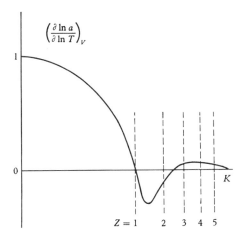

Fig. 7. Schematic behavior of the temperature variation of a (K) at constant volume. The limit $K = 2k_F$ for various valences Z is indicated. (Faber, 1969.)

value. Therefore, we schematically have the behavior for $(\partial \ln a / \partial \ln T)_v$ shown in Fig. 7. Thus in this way the Ziman theory can also account qualitatively for the negative temperature dependence of resistance in divalent metals such as Zn and Cd.

A very speculative application of the nearly free-electron model has been given recently by Delley *et al.* (1978) in connection with the electrical resistivity of liquid rare-earth metals. Going across the rare-earth series, there is found a steady increase in the resistivity of the liquid elements from approximately 130 $\mu\Omega$cm for La to 200 $\mu\Omega$cm for Lu. At the same time, there is a general decrease in the temperature coefficient of electrical resistivity from $\sim +2 \times 10^{-2}$ $\mu\Omega$cm/°C for La to $\sim -1 \times 10^{-2}$ $\mu\Omega$cm/°C for Lu. Their interpretation of this result is that there is a change in the number (n) of d and s electrons across the series. For La, $n_s \sim 0.5$, $n_d \sim 2.5$, while for Lu, $n_s \sim 1.5$, $n_d \sim 1.5$. If the electrical current is assumed to be carried by the s electrons, then k_F increases across the rare-earth series, and it is possible, their calculations suggest it, that the increase in resistivity across the rare-earths series arises much as does the increase on going from mono- to divalent simple metals. And their rough value for Lu indicates that $2k_F$ will be close to the first peak in the interference function, qualitatively explaining the negative temperature coefficient of resistivity. We note parenthetically that a lower liquid resistivity might be expected in Lu if indeed it has more s electrons than La.

There are a number of objections which can be raised regarding the interpretation of this rare-earth data, but their suggestion is an interesting possibility. The application of the nearly free-electron model to transition metals (and rare earths) is also something which has to be justified (if it can be). In fact, liquid transition metals are an area where there has been slow progress, due in part to the difficulty of obtaining the experimental data, as well as due to the theoretical difficulties in dealing with the more complicated electronic structure of transition metals.

7.   RESISTIVITY SATURATION IN ALLOY SYSTEMS

Ioffe and Regel (1960) have noted that there should be a lower bound for the conduction-electron mean free path in a solid of order an interatomic spacing. For the case of a metal, this would apply to electrons at the Fermi level. In the case of semiconductors it would be a lower limit to the mean free path of electrons excited into the conduction band and of holes in the valance band. In general, this lower bound has important ramifications for resistivity as a function of disorder in resistive systems.

Hake (1967) has noted that such a limiting minimum mean free path produces a limiting maximum resistivity as disorder is increased. He also suggested that several concentrated transition-metal alloys with $\rho \approx 150$ $\mu\Omega$ cm appear to approach that limit. Mooij (1973) analyzed the resistive behavior of a large number of concentrated alloy systems, most of which contain at least one transition metal. The systems studied were either concentrated crystalline, e.g., NiCr, substitutional alloys, or amorphous alloys prepared by quenching. In these systems, resistivities in the range 100 to 300 $\mu\Omega$ cm are found. Figure 8 shows an example of the TiAl system. These results are representative of the majority. It is seen that as the

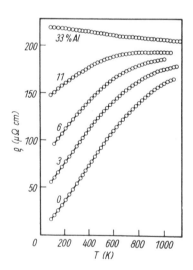

Fig. 8.   Resistivity versus temperature for Ti and TiAl alloys containing 0-, 3-, 6-, 11-, and 33- at.% Al. (Mooij, 1973.)

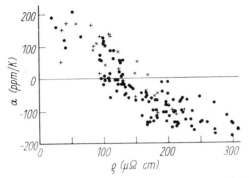

Fig. 9. The temperature coefficient of resistance α versus resistivity. α is defined as $(1/R)dR/dT$ and was measured between 25 and 75°C. Data are from bulk alloys ($+$), thin films ($\bullet$), and amorphous alloys ($\times$). (Mooij, 1973.)

concentration of Al is increased the zero-temperature residual resistivity increases to 100 $\mu\Omega$ cm. At the same time, he notes that the temperature coefficient of resistance $\alpha = 1/R(dR/dT)$ [measured between 25 and 75°C] falls to zero and even becomes negative. Figure 9 shows data from a large number of systems where it has been noted that below 100 $\mu\Omega$ cm, α is positive while above 150 $\mu\Omega$ cm, α is almost always negative. These data led to the suggestion that this behavior is a general property of highly resistive metallic systems. In support of this observation it is found that the correlation between ρ and α is largely independent of whether a material is prepared as a bulk alloy or in thin film form. It is probably independent of crystal structure since several structures show the same behavior. All the systems contain a transition metal although amorphous AuGe and AuCu with $\rho = 100$ $\mu\Omega$ cm and negative α are on the borderline of being exceptions.

Thus it appears from Fig. 10 that the decrease in $d\rho/dT$ is directly related to the increase in ρ. This is especially clear in the TiAl system, as pointed out by Mooij. Here it is evident that the decrease in $d\rho/dT$ is the same whether it is caused by an increase in Al content, i.e., "impurity" scattering due to the random ionic potential, or by an increase in temperature, i.e., thermal disorder due to displacements of atoms from their lattice sites. The more concentrated TiAl of Fig. 8 has a negative $d\rho/dT$ and appears to approach 190 $\mu\Omega$ cm at high temperatures.

These observations above led to the suggestion that the behavior of the resistivity in materials with low α as shown in Fig. 9 or $d\rho/dT$ as shown in Fig. 10 could be understood on the basis of the conduction-electron mean free path approaching a lower limit. In this case, as mean free path becomes shorter as a function of alloy content, the additional scattering due to

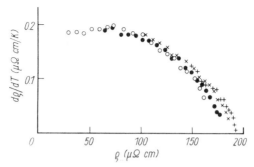

Fig. 10. The temperature derivative of the resistivity $d\rho/dT$ versus the resistivity for Ti(0) and TiAl alloys containing 3-, 6-, and 11-at.% Al ($\bullet$, $\times$, $+$, respectively). (Mooij, 1973.) See Fig. 7.

phonons becomes less effective and both $d\rho/dT$ and α would approach zero. In this model additional assumptions have to be made to explain, for example, a negative α. In the highly resistive materials of Fig. 9 the negative α is an order of magnitude larger than can be accounted for on the basis of thermal expansion alone. At this point, it is not clear what the exact origin of the negative α's is. Liquid metals in a number of uses have negative α, and, while here we are discussing crystalline materials, they may well be a close connection between the disorder of a liquid metal and that of some concentrated solid solutions.

B. Temperature-Dependent Scattering

1. ELECTRON–PHONON INTRABAND SCATTERING

As mentioned above, electrons traveling through a lattice experience scattering only if translational invariance is broken. Therefore a perfect crystal at absolute zero has infinite conductivity or zero resistance. However, the presence of finite temperature brings about an instantaneous displacement of the atoms from their lattice sites. These displacements break translational invariance and thereby scatter electrons. It is most straightforward to treat such scattering processes in the language of phonons. In this description, a scattering event is described by a conduction electron absorbing or creating a phonon such that energy and crystal momentum are conserved.

The calculation of the electrical resistivity at a given temperature involves a knowledge of the electron and phonon distributions as a function of energy and wave vector. Also needed is the wave-vector–dependent electron-ion matrix element which describes their interaction. There are difficult theoretical and computational problems in such calculations. Until recently, the only available calculations were for the simplest elements. Recently, progress has been made in such calculations for some transition elements. In general, the correct application of the theory requires that each material should be considered as a special case, although experimentally, there is some evidence for different classes of behavior. In particular, the temperature dependence of the low-temperature resistivity has been used to group pure metallic materials. Here "pure" is used in the sense that the temperature-independent residual resistivity does not dominate the temperature-dependent part.

It was found that most of the pure nontransition elements have resistivities which vary approximately as T^5 at low temperatures. In the very pure state at very low temperatures there are indications of a T^2 term also, which would be due to electron–electron scattering. However, that term is small and usually overwhelmed by the residual and T^5 terms. An approximate T^5

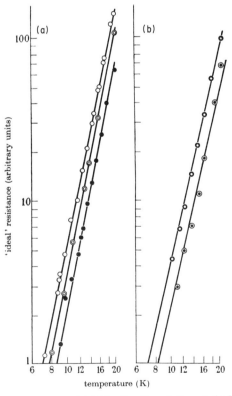

Fig. 11. The "ideal" resistance of (a) Na, ◎, ○, and ●, and (b) Li, ○, and ⊙, versus temperature, computed by subtracting a temperature-independent residual term. The solid lines represent power laws of $T^{4.85}$ for Na and $T^{4.55}$ for Li. (MacDonald and Mendelssohn, 1950.)

has been observed in almost all the monovalent and polyvalent nontransition elements (see, e.g., Aleksandrov and D'Yakov, 1963). Figure 11 gives examples. The original theory, known as the Block–Grüneisen theory, has been developed with monovalent metals in mind whose Fermi surfaces are close to spherical (Wilson, 1956). The phonons are assumed to be adequately described by a Debye model, with a characteristic Debye temperature θ. The result is the well-known Bloch–Grüneisen formula for the resistivity of a metal due to emission and absorption of phonons

$$\rho = 4(T/\theta)^5 \rho_\theta J_5(T) \int_0^{\theta/T} (x/\sinh x/2)^2 x^3 \, dx, \qquad (18)$$

where ρ_θ is the resistivity at the Debye temperature. At low temperature, e.g., $\theta/T \gtrsim 20$, $\rho \propto T^5$, which is observed. The theory has also been applied to experimental results from nontransition elements whose Fermi surfaces intersect Brillouin zone boundaries and are thus nonspherical. It can be argued that the deviations from sphericity introduced by the Brillouin zone boundaries might be small if the bandgaps introduced by the zone boundaries are small. However, the theory should not be taken too literally because of a number of serious deficiencies including the neglect of umklapp processes, problems with scattering by transverse phonons, the assumption of a Debye phonon spectrum, and the above-mentioned assumption of a spherical Fermi surface.

There is a "modernized Bloch–Grüneisen theory" which is free of most of these objections. This theory, which has been developed by Allen, Butler, and co-workers (e.g., Allen and Butler, 1978; Allen, 1971), is closely related to the modern theory of superconductivity. It is described in detail in another chapter of this volume. The main results of the theory are

$$\rho(T) = \left[\tfrac{1}{3} e^2 N(0) \langle v^2 \rangle \, 7\right]^{-1}, \qquad (19)$$

$$1/\tau = (2\pi/\hbar) k_B T \lambda_{\text{tr}}(T), \qquad (20)$$

$$\lambda_{\text{tr}}(T) = 2 \int_0^{\omega_{\max}} (d\omega/\omega) \alpha_{\text{tr}}^2 F(\omega) \left[\frac{x}{2 \sinh x/2}\right], \qquad (21)$$

where $N(0)$ is the density of states for both spins, $\langle v^2 \rangle$ a mean square Fermi velocity, and τ an average scattering time for the electrons. λ_{tr} is a coupling constant closely related to that of superconductivity. In the expression for λ_{tr}, $F(\omega)$ is the actual phonon density of states of the material weighted by the function $\alpha_{\text{tr}}(\omega)$. $\alpha_{\text{tr}}^2(\omega) F(\omega)$ is similar to that of superconductivity theory (McMillan, 1968) but weighted by a factor $1 - \cos\theta$ where θ is the electron scattering angle.

2. ELECTRON–PHONON INTERBAND SCATTERING WITH A DEBYE PHONON DISTRIBUTION

In the 3d, 4d, and 5d transition elements the ground-state free-atom electronic configuration contains both d and s shells partially occupied. In the solid these three states broaden to form conduction bands. The spatial extent of the atomic s states is appreciable compared to observed interatomic spacings in the solid. Therefore the conduction bands formed from overlapping s states are broad and the calculated velocities are high. The same is true of atomic p states which are admixed with the s states in the solid. As the extensive overlap leads to high velocities and low effective masses, the free-electron approximation is often a model for these bands in a transition metal.

The situation for the overlapping d states is more complicated. The more limited spatial extent of the d states produces bands of delocalized states which are narrower in energy. The velocity of these states is lower than that of the bands made of predominantly s and p states. Similarly, the effective masses are, in general, considerably higher.

In a solid made up of overlapping sp and d bands an electrical current will be carried by both types of bands. However, the conductivity of each type of band is proportional to the electron velocity at the Fermi energy. Thus the sp bands are expected to carry a larger share of the current. A major simplification is brought about by making the single-band approximation wherein all the current is assumed to be carried by the s or sp band.

In this approximation there are two types of phonon-induced scattering events leading to electrical resistivity. The first is where an s conduction electron absorbs or emits a phonon and is scattered into another s state. This was discussed in a preceding section. Mott (1936), however, noted that the density of states in the d band is much higher than in the s band so that the higher density of final states for s electrons scattering into electron d states should cause that type of scattering to predominate. The s electron is scattered by either absorbing or emitting a phonon into the low Fermi velocity d band where it is assumed to be completely removed from the conduction process. In other words, the d band acts as a scattering trap. To conserve the number of particles in each band, electrons are also scattered randomly from the d band to the s band.

This type of scattering is usually called s–d scattering. At temperatures higher than the Debye temperature the mechanism leads to a resistance proportional to the first power of the temperature. With one restriction on the electron distribution, to be discussed in a moment, this mechanism will lead to a resistance proportional to the number of quanta of lattice vibrations; thus at very low temperatures, low enough to decompose the lattice vibrations into a system of sound waves, the resistance will vary like the temperature cubed. Between these two limits the behavior of the resistivity is dependent on details of the phonon spectra.

In order to calculate the magnitude and behavior of the resistivity over the full temperature range, Wilson (1938) assumed a Debye phonon distribution. Furthermore, he assumed a spherical conduction band, the s band, and a spherical d band of low Fermi velocity electrons. The phonons cause transitions between these two sheets of the Fermi surface. Wilson observed that for this model, there could be a minimum phonon wave vector q_{min} required to cause transitions between the two bands: this defines a minimum temperature Θ_{min} below which these s–d transitions should exponentially disappear. The resultant Mott–Wilson theory for this contribution to the resistivity is

$$\rho_{sd} = \frac{CT^3}{7.212} \int_{\Theta_{min}/T}^{\Theta/T} \frac{x^3 dx}{(e^x - 1)(1 - e^{-x})}. \tag{22}$$

The constant C contains the electron–phonon interaction energy between s and d states. The constant Θ_{min} is defined by

$$k\Theta_{min} = u_0 |k_d - d_s|, \tag{23}$$

where u_0 is the velocity of sound and $|k_d - k_s|$ defines the minimum phonon wave vector needed to cause s–d transitions. The Θ in the upper limit on the integral can be regarded as an effective Debye temperature for resistivity. Equation (22) leads to a low-temperature $\rho_{sd} \propto T^3$, in contrast with the $\rho \propto T^5$ given by Eq. (18). In this approximation only the longitudinal phonons produce potential fluctuations in the lattice and thus cause transitions. Therefore, we might expect Θ to be somewhat different from the Debye temperature.

For small enough values of θ_{min} we can define three temperature ranges for ρ_{sd}. Let us set $\theta_{min} = \Theta/20$. Then, from high temperatures down to about $T = \frac{1}{2}\theta$, ρ_{sd} is proportional to T. From $T = \theta/5$ down to $T \approx \theta/40$, ρ_{sd} is proportional to T^3. For temperatures below $\theta/40$, ρ_{sd} drops away exponentially with temperature. If, on the other hand, we set $\theta_{min} = 0$, then the T^3 behavior would be observed down to zero temperature.

On the experimental side there are a substantial number of transition elements and compounds whose resistivities vary approximately as T^3 at low temperatures. Figure 12 shows the resistivity of Nb as a function of temperature. At low temperatures it is evident that a T^3 dependence is present. There has been no clear experimental evidence of an exponentially decaying increment to the resistivity of a transition metal indicating that $\theta_{min} = 0$ for the model of Eq. (22).

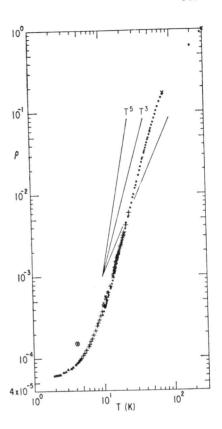

Fig. 12. Resistance data for two samples of pure Nb having residual resistance ratios of 11,000 and 16,500. At low temperatures, the resistance was measured in magnetic field in order to quench superconductivity. Note that the resistance varies as T^3 up to 50 K. Subtracting a term for residual resistance extends the lower-temperature interval over which a T^3 term is observed. (Webb, 1969.)

Nakagawa and Woods (1963) have performed an inelastic neutron scattering study of the phonon system in Nb. These results allow a comparison of the phonon distribution function, derived from experiment, with that of the Debye model. The comparison shows that the Debye model is in good approximation in Nb for phonon frequencies up to 0.5 of the highest found. This lends support to the use of the Debye model in Nb for the calculation of low-temperature properties which are not sensitive to phonons above one-half the phonon frequency.

Fradin (1974) has applied the phenomenological s–d scattering model to the high-temperature resistivity of Pd (the low-temperature resistivity being complicated by the presence of spin fluctuations). Using a realistic electronic band structure, properly broadened by thermal smearing and a Debye model for the phonons, he finds a good one-parameter fit to the data; a small correction for s–s intraband scattering was made using the resistivity of silver.

Recently, Pinski *et al.* (1978) have performed a first-principles calculation of the electron–phonon resistivity of Pd with no adjustable parameters. This calculation uses as input "state of the art" electronic band structure and detailed information about the actual phonons obtained from experiment at 120 K. The agreement with experiment is good over intermediate temperatures in particular up to 600 K. Correcting for the measured temperature dependence of the phonons (anharmonicity) at selected higher temperatures greatly improves the agreement above 600 K. It is especially noteworthy that they find that for Pd at least there is "surprisingly strong support for Mott's s–d model." In particular they find that the Γ centered electron sheet which has only 8% of the density of states carries 80% of the current and that the dominant scattering (81%) of these electrons is to a low Fermi velocity sheet of the Fermi surface which has 89% of the calculated density of states. At low temperatures the calculated resistivity lies considerably below experiment, suggestive of scattering from additional excitations at low temperatures such as spin fluctuations.

Although the phenomenological s–d scattering model has been of utilitarian value, it should not be forgotten that there are a number of serious deficiencies with the theory of Eq. (22). These include the assumption of spherical Fermi surfaces for both bands, the neglect of umklapp processes, the assumption of harmonic (temperature-independent) phonon modes, and a Debye phonon spectrum. In some materials it is known that the Debye model for the phonon spectrum is a poor approximation to the actual one. It turns out that the Debye phonon spectrum in the s–d scattering model can be replaced by a measured spectrum or a better model in a straightforward way. This is discussed in the following section using a specific class of materials as an example.

3. ELECTRON–PHONON INTERBAND SCATTERING WITH A NON-DEBYE PHONON DISTRIBUTION

Equation (22) can be recast in a form emphasizing an arbitrary distribution of phonons $F(\omega)$ as

$$F = A \int \left\{ x / [\sinh(x)]^2 \right\} F(\omega) \, d\omega, \tag{24}$$

where $x = h\omega/k_B T$ and the proportionality constant A contains quantities such as an average transport electron–phonon coupling parameter connecting s and d states, and the s- and d-band densities of states at the Fermi surface (all assumed to be independent of temperature below 50 K). Here s and d are used again to indicate bands of high and low Fermi velocity in the model. The lattice heat capacity in the harmonic approximation in this notation is

$$C_L/3Nk_B = \int (x/\sinh x)^2 F(\omega)\,d\omega, \tag{25}$$

where N is Avogadro's number. With $F(\omega)$ normalized to give the correct number of modes per unit cell there are no adjustable parameters in Eq. (25). We will return to this equation below.

The high superconducting transition temperature (T_c) compounds with the A-15 structure are an important class of materials which has been intensely studied. These compounds exhibit a number of unusual phonon properties (reviewed in Testardi, 1976), including mode softening with decreasing temperature, low-temperature lattice transformations, in some cases anomalously low, or "soft," phonon branches over substantial parts of the Brillouin and unusual low-temperature lattice heat capacities, which in some instances vary as T^2 rather than the usual T^3 behavior. Recent inelastic neutron scattering studies (Schweiss et al., 1976) have shown that most of the materials exhibit phonon densities of states which are quite non-Debye-like in that they have much more strength than a Debye model at low phonon energies.

Another unusual property of these materials is the magnitude and temperature dependence of the normal-state resistivity. Sarachik et al. (1963) and Woodward and Cody (1964) were the first to call attention to the unusual properties of resistivity of some of the highest T_c A-15 materials. For Nb_3Sn, Woodward and Cody found empirically that

$$\rho = \rho_0 + \rho_1 T + \rho_2 e^{-T_0/T}, \tag{26}$$

where ρ_0, ρ_1, ρ_2 are constants independent of temperature T, and $T_0 = 85$ K fits their data closely from $T = 18$ to 800 K. Several theoretical models were considered by them in order to explain Eq. (26) but none was found to be satisfactory. Later it was noted that the low-temperature resistivity of Nb_3Sn can be fit rather well to a simpler T^2 law but over the smaller temperature range from T_c to 50 K (Webb et al., 1977). Similarly, the low-temperature resistivity of the isostructural high T_c compound V_3Si was found to follow a T^2 law, but with a better fit to Eq. (26) over a wider temperature range (Marchenko, 1974).

As the resistivities of the high T_c A-15 structure materials are anomalous over the whole range of measurement temperatures, it is not surprising that a variety of physical ideas have been employed in trying to understand them. Above room temperature, temperature dependence of the resistivity shows strong negative curvature. At high temperature the magnitude of the resistivity rises to about 100 $\mu\Omega$ cm, and appears to be saturating at a value corresponding to an electron mean free path of order one interatomic spacing (Fisk and Webb, 1976; Wiesman *et al.*, 1977). These saturation processes are discussed in more detail in Sections II,A,7 and II,B,7. Over an intermediate temperature interval 50–300 K, the resistivity also shows negative curvature. To account for this, two different treatments of electron–phonon scattering have been invoked. The first is based on a model electronic density of states containing sharp structure near the Fermi energy which gives a rapid Fermi level motion with temperatures of order 100 K (Cohen *et al.*, 1967; Cohen, 1971; Rehwald *et al.*, 1972). The second treatment (Allen *et al.*, 1976) focuses attention on the anharmonic hardening of phonon modes with increasing temperature [opposite to the usual behavior of materials] which has been observed by neutron scattering experiments on some of these materials (Schweiss *et al.*, 1976). Both of these models provide an explanation for the observed negative curvature.

Below 50 K the effect of Fermi level motion is expected to be minimal, based on the deduced Fermi temperatures. Available neutron scattering

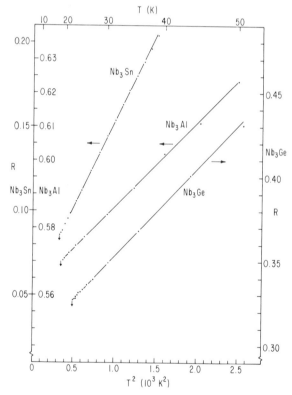

Fig. 13. Low-temperature resistance data for Nb_3Sn, Nb_3Al, and Nb_3Ge plotted versus T^2. These samples have the A-15 structure with superconducting transition temperature $T_c \sim 20$ K. (Webb *et al.*, 1977.)

data (Axe and Shirane, 1973) suggest that phonon mode shifting has largely been arrested, ruling out anharmonic effects. Thus, below 50 K, it is not expected that the resistivity will be influenced by either of these effects or by the effect of the conduction-electron mean free path approaching a lower bound for pure undamaged compounds. However, it has been found that the high T_c compounds ($T_c \simeq 20$ K) Nb_3Sn, Nb_3Al, and Nb_3Ge all have unusual resistivities which vary closely as T^2 in this temperature range (Webb et al., 1977). Some of these data are shown in Fig. 13. This T^2 dependence has been analyzed in terms of the Mott–Wilson s–d scattering model of Eq. (24) but with a more realistic phonon density of states $F(\omega)$. It was also found that the unusual lattice capacities could be reproduced using the same phonon density of states in Eq. (25). However, since the model $F(\omega)$ is related to the data through an integral it is not given a strict test by such a procedure. In fact, another model $F(\omega)$ similar in form but differing in details was found to give an equally good fit to the data (Webb et al., 1977). It is clear that what should be done now is to compute the resistivity using the phonon densities of states as they become available from the inelastic neutron scattering experiments. For additional discussion of the normal-state resistivity of these materials see Gurvitch (1980).

4. ELECTRON–ELECTRON SCATTERING

Within the framework of a single conduction band, electron–electron scattering has been predicted to give a resistivity contribution proportional to T^2. The electrons scatter off each other via a screened Coulomb interaction. Provided there are electrons of different effective masses, a net loss of velocity along the current direction results. This T^2 dependence ought to be experimentally observable at either very low temperatures or at high temperatures roughly twice room temperature. In the transition metals where one has another band of relatively high effective-mass electrons this term ought to be especially large. There is evidence for it in some metals at low temperatures but it seems to be definitely absent at high temperature except in the case of TiS_2 (Thompson, 1975; see also Kukkonen and Maldague, 1976, 1979). The absence of a T^2 term in the high-temperature resistivity of many materials is perhaps not yet fully explained (see, e.g., Appel, 1963; Kukkonen and Maldague, 1979).

5. MAGNETIC EFFECTS

Localized magnetic moments in metallic conductors, in general, strongly interact with the conduction electrons. This makes the measurement of the electrical resistivity of such materials particularly useful for elucidating various magnetic phenomena. In many metals, in fact, it is believed that the principal interaction between local moments is carried by the conduction electrons, the so-called RKKY interaction (Ruderman and Kittel, 1954; Kasuya, 1956; Yosida, 1957). The local-moment–conduction-electron interaction scatters conduction electrons, and this scattering and any temperature dependence it has are easily seen in the electrical resistivity.

a. Rare-Earth Materials. The conceptual picture of the local-moment–conduction-electron interaction which has emerged for the rare earths is in many ways much cleaner than the corresponding case for the transition metals. We discuss the rare earths and their compounds and alloys first.

There is ample experimental evidence in the literature showing that the magnetic behavior of the chemically very similar rare earths (in elemental form and in compounds and alloys) is that expected of the +3 ions with incomplete 4f shells, the 4f ground state being that predicted by Hund's rules. The 4f electrons do not participate in chemical bonding; the electronic structure of a given rare-earth compound is expected to vary little from rare earth to rare earth, with the exception of europium, ytterbium, and cerium when they are not trivalent.

In the paramagnetic state the 4f local moments can, to a good approximation, be treated as independent, even in the pure metals. Therefore we can consider the interaction of one isolated local moment with the conduction-electron sea. It has been found that an interaction of the form

$$H_{\text{int}} = -2J_{\text{sf}}\mathbf{S}\cdot\mathbf{s} \tag{27}$$

can explain in a semiquantitative way a number of phenomena in metals dependent on the interaction between a local spin $\mathbf{S}$ and the conduction electrons $\mathbf{s}$. J_{sf} is the strength of the coupling. This form (27) was originally used by Herring (1958) and Suhl and Matthias (1959) to explain results on the depression of the superconducting transition temperature of La by rare-earth impurities.

For rare earths, the spin–orbit coupling is strong: $\mathbf{S}$ is coupled to $\mathbf{J}$ and should be replaced by its projection on $\mathbf{J}$, i.e., $\mathbf{S} \rightarrow (g - 1)\mathbf{J}$, where g is the Landé g factor. Using this, (27) leads in Born approximation to a magnetic contribution to the electrical resistivity (called the spin-disorder resistivity)

$$\rho_m = (k_{\text{F}}/\pi Z)\left(J_{\text{sf}}V_c m^*/e^2\right)^2 (g - 1)^2 J(J + 1)n. \tag{28}$$

m^* is the effective mass of the conduction electron, k_{F} the Fermi momentum, e the electronic charge, Z the number of free electrons per atom, V_c the atomic volume, and n the concentration of scatterers (see Doniach, 1967).

There are thought to be two principal contributions to J_{sf}. The first, which should vary little across the rare-earth series, is expected from the Coulomb exchange integral and will be positive, according to Hund's rules. The second is expected where the 4f level is a virtual bound state. In this case Schrieffer and Wolff (1966) have shown

$$J_{sf} \approx V_{sf}^2 U / \varepsilon(\varepsilon + U). \tag{29}$$

V_{sf} is a conduction-electron–f-electron mixing matrix element, U is the Coulomb interaction, and ε is the distance of the f level from the Fermi level. For the occupied levels here it is negative. For $U \gg |\varepsilon|$, J_{sf} is negative.

In this last case, Kondo (1964) showed that the Hamiltonian (27) in second Born approximation leads to a term in the electrical resistivity of dilute alloys

$$\rho_k = \rho_m \left[1 + 2 J_{sf} N(E_F) \log(kT/E_F) \right]. \tag{30}$$

ρ_m is given in (28), k is Boltzmann's constant, and $N(E_F)$ is the density of states at the Fermi level E_F. Equation (30) predicts a resistance rise on cooling linear in log T for J_{sf} negative. This has been observed for a number of cerium and ytterbium alloys and compounds and for a few cases involving praeseodymium and europium. At very low temperature, expression (30) is no longer valid. The resistance rise flattens out below a characteristic temperature, the so-called Kondo temperature T_K, approximately given by

$$T_K \sim T_F \exp(-1/N(E_F)|J_{sf}|), \tag{31}$$

where T_F is the Fermi temperature.

At sufficiently low temperature the RKKY interaction, or other competing spin–spin interactions, will lead to the establishment of long-range magnetic order in the material. The spin-disorder resistivity decreases below this temperature, becoming zero when the spins are completely ordered. This loss of resistivity is usually dramatically evident. This magnetic order can introduce new Brillouin zone boundaries, so-called superzone boundaries, corresponding to the periodicity of the spin structure. In many cases these new boundaries differ from the zone boundaries for the crystal lattice, and the magnetic superzones can chop up the Fermi surface in such a way that the resistivity sometimes increases at the ordering temperature, followed by a subsequent decrease due to the loss of the spin-disorder term.

b.   Crystal-Field Effects.   The crystalline field at a rare-earth site can remove some of the (2J + 1)-fold degeneracy of the Hund's rule ground state. In cases of high symmetry it is sometimes possible to make reliable estimates of the crystal-field level scheme. This can be aided by magnetic susceptibility and specific heat measurements which reveal changing crystal-field level populations as the temperature changes. These level repopulations are also reflected in the electrical resistivity. The theory for this has been worked out by Hirst (1967), including both elastic and inelastic scattering.

We illustrate this effect with the electrical resistivity of NdB_6 (Fisk, 1976), a simple cubic metal with one conduction electron per unit cell. Using the notation of Andersen *et al.* (1974), the resistivity due to scattering from the magnetic ions can be written

$$\rho_{mag} = \rho_m \, tr(PQ),\tag{31}$$

where the trace is taken over the $2J + 1$ crystal-field states whose energies are E_i:

$$P_{ij} = \frac{\exp(-E_i/kT)}{\sum_k \exp(-E_k/kT)} \cdot \frac{(E_i - E_j)/kT}{1 - \exp(-[E_i - E_j]/kT)}\tag{32}$$

and

$$Q_{ij} = |\langle i|J_z|j\rangle|^2 + \tfrac{1}{2}|\langle i|J_+|j\rangle|^2 + \tfrac{1}{2}|\langle i|J_-|j\rangle|^2.\tag{33}$$

Q_{ij} is calculated using the crystal-field eigenstates. The lattice term for NdB_6 is taken to be the same as that for isostructural LaB_6, and the difference in their respective resistivities is then the magnetic term ρ_{mag}. The data and fit to expression (31) are shown in Figs. 14 and 15. NdB_6 orders antiferromagnetically at $T_N = 7.5$ K. The loss of spin-disorder resistivity below this temperature is evident in Fig. 15. For this case, $J_{sf} = 0.105$ eV and $\rho_m = 0.277 \, \mu\Omega\,cm$.

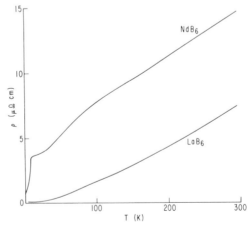

Fig. 14. The temperature dependence of the electrical resistivities of single-crystal NdB_6 and LaB_6. (Fisk, 1976.)

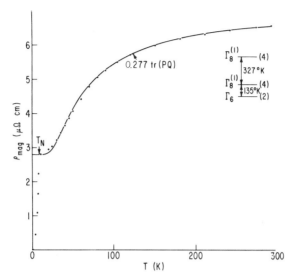

Fig. 15. The experimentally determined temperature-dependent spin-disorder resistivity of NdB$_6$. The solid line is the theoretical fit to the data. (Fisk, 1976.) $\Gamma_8^{(1)} - (4) \leftrightarrow \Gamma_8^{(1)} - (4)$ (327 K) $\leftrightarrow \Gamma_6 - (2)$ (135 K).

c. Aspherical Coulomb Scattering. In addition to the Coulomb exchange interaction, the direct Coulomb interaction can also scatter conduction electrons. Such effects have been reviewed by Fulde and Peschel (1972). The 4f level is not, in general, spherically symmetric (in leading order it has a quadrupole moment) and the disordered quadrupoles can scatter the conduction electrons. Table II lists the quadrupole moments for the fully aligned $J_z = J$ state. Although the magnitude of this scattering is difficult to predict, it cannot be assumed a priori that it is much smaller than the exchange scattering, and this scattering mechanism should be considered in cases for which the quadrupole moment of the fully aligned $J_z = J$ state is large. As with exchange scattering, a temperature-dependent contribution will arise from this source due to crystal-field splittings. We illustrate this with the electrical resistivity of PrB$_6$ (Fisk and Johnston, 1977), isostructural with NdB$_6$ considered above.

In addition to the exchange scattering, there is now an additional term

$$\rho = \rho_A \operatorname{tr}(PQ^A), \tag{34}$$

with P given by (32) and

$$\sum_{M=-2}^{+2} |\langle i | y_2^M | j \rangle|^2. \tag{35}$$

TABLE II

ELECTRIC QUADRUPOLE MOMENTS FOR
THE FULLY ALIGNED 4f SHELL[a]

Ion	J	Q
Ce (4f)^1	$\frac{5}{2}$	$-\frac{2}{7}$
Pr (4f)^2	4	$-\frac{728}{2475}$
Nd (4f)^3	$\frac{9}{2}$	$-\frac{14}{121}$
Pm (4f)^4	4	$\frac{106}{1815}$
Sm (4f)^5	$\frac{5}{2}$	$\frac{13}{63}$
Eu (4f)^6	0	0
Gd (4f)^7	$\frac{7}{2}$	0
Tb (4f)^8	6	$-\frac{1}{3}$
Dy (4f)^9	$\frac{15}{2}$	$-\frac{1}{3}$
Ho (4f)^{10}	8	$-\frac{2}{15}$
Er (4f)^{11}	$\frac{15}{2}$	$\frac{2}{15}$
Tm (4f)^{12}	6	$\frac{1}{3}$
Yb (4f)^{13}	$\frac{7}{2}$	$\frac{1}{3}$

[a] From Stevens (1952).

ρ_A is a constant depending on the interaction strength and the y_2^M are operator equivalents for $L = 2$.

To fit the resistivity of PrB$_6$, the ratio ρ_m/ρ_A was varied. The data are shown in Fig. 16, the fit in Fig. 17. It is found that the aspherical Coulomb scattering is equally important with the exchange scattering in this case. In the case of NdB$_6$, the aspherical Coulomb scattering can be estimated to be only 6% of the exchange term, due to the much smaller quadrupole moment of Nd. This accounts for the reasonable fit in the case of NdB$_6$ in spite of the neglect of the aspherical coulomb term.

d. Transition-Metal Elements. The situation for the d transition metals appears to be rather different than that for the rare earths. For the magnetic transition elements it appears that a mixture of an itinerant picture and a localized-spin picture is appropriate. Above the magnetic ordering temperature there are no magnetic moments present. This means that upon ordering, the change in electrical resistivity will not simply be a loss of the spin-disorder scattering, it will also involve effects from the splitting of the up-spin and down-spin d bands and changes in the s–d scattering.

The upshot of this is that resistance measurements for the elements are more diagnostic than quantitative. Namely, a kink in the resistivity can be

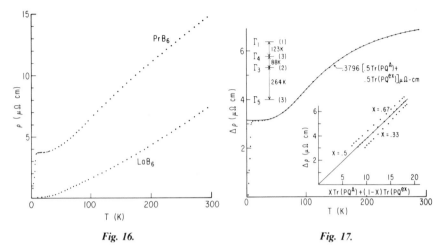

Fig. 16.   Fig. 17.

Fig. 16. The temperature dependence of the electrical resistivities of single-crystal PrB_6 and LaB_6. (Fisk and Johnston, 1977.)

Fig. 17. Excess resistivity of PrB_6 over lattice and residual term as a function of tempera-ture. Solid line gives fit to theory using both exchange and aspherical Coulomb scattering. Inset shows sensitivity of fit to relative amount of these two scattering mechanisms. (Fisk and Johnston, 1977.) $\Gamma_1 - (1) \leftrightarrow \Gamma_4 - (3)$ (123 K) $\leftrightarrow \Gamma_3 - (2)$ (88 K) $\leftrightarrow \Gamma_5 - (3)$ (264 K).

used to locate a Curie or Neel point and to follow how it changes with alloying, or some other variable, e.g., pressure. In a number of compounds, such as Sc_3In and $ZrZn_2$, which are thought to be itinerant ferromagnets, the resistance changes at T_c are slight precluding the extraction of quantita-tive scattering information.

In the case of ferromagnets, T_c can be pinpointed much more sensitively following a scheme due to Sales (1979). His resistance data for Ni near the Curie point at various frequencies are shown in Fig. 18. At 1000 Hz the anomaly is enormous. The effect here is due to the changing skin depth δ. We know that $\delta \propto (\mu f \sigma)^{-1/2}$, where μ is the magnetic permeability, f the frequency of the current, and σ the conductivity. For copper at 300 K and 60 Hz this number is 0.85 cm. What one sees for Ni is that the very large initial permeability of Ni exactly at T_c makes the skin depth small. With a proper choice of frequency and sample dimension, the skin depth at T_c will change from a value larger than a sample dimension to a value smaller than a sample dimension above T_c to a value much smaller than a sample dimension below T_c with a corresponding change in sample resistance. It is clear that this technique will have an application to the study of ferromag-netic materials.

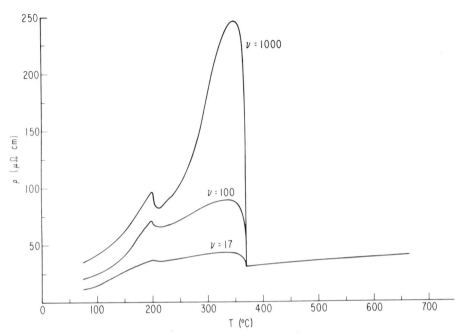

Fig. 18. Temperature-dependent electrical resistivity of Ni measured at three different frequencies ν. The anomaly near 200°C is a real effect, not an experimental artifact. (Sales, 1979.)

e. Two Current Conduction in Ferromagnets. Some very interesting resistivity work has been done by Campbell and co-workers (1967, 1970; Campbell, 1970) on the question of the splitting of the spin-up and spin-down d bands in ferromagnetic iron and nickel. These bands conduct in parallel, and if impurities scatter spin-up and spin-down d electrons differently, then gross deviations from Matthiessen's rule are to be expected, both at low temperature where incremental impurity resistivities from two different impurities will be nonadditive, and as the temperature increases, where spin-flip scattering and the electron–phonon interaction cause deviations. These effects turn out to be pronounced as can be seen in Figs. 19 and 20. The effects are understandable on the parallel conduction model, Eqs. (4) and (5), when coupled with Friedel's interpretation of impurity resistivities using the notion of virtual bound states [Eq. (9)]. Campbell's papers should be consulted for the details.

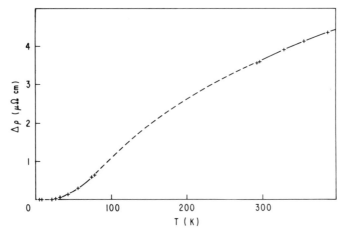

Fig. 19. Temperature-dependent deviation of the electrical resistivity for Fe—3.3 wt./%
Os. The deviation is given by the difference between the measured resistivity of the alloy, and
that of its residual term added to the temperature dependence of pure Fe. (Campbell *et al.*,
1967.)

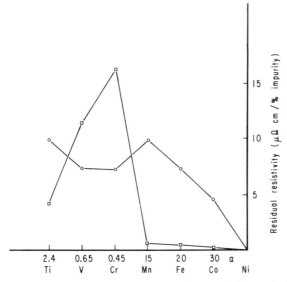

Fig. 20. Experimental values of the up-spin band (□) and down-spin band (○) residual
resistivities due to first-row transition-metal impurities in Ni. (Campbell *et al.*, 1970.)

6. TEMPERATURE DEPENDENCE OF SATURATION

When the electronic mean free path becomes short, on the order of a few interatomic spacings, the scattering cross section will no longer be linear in the scattering perturbation. In particular, if the electron–phonon interaction is the dominant temperature-dependent scattering mechanism, the resistance will no longer be proportional to the mean square atomic displacement, which is proportional to T for a harmonic potential. Instead, the resistance will rise less rapidly than linearly in T and exhibit a negative curvature $(d^2\rho/dT^2 < 0)$. This effect is seen in a majority of the transition-metal superconductors (Fisk and Lawson, 1973; Fisk and Webb, 1976).

It is an unsolved theoretical problem as to the exact behavior to be expected as the mean free path approaches one interatomic spacing. Wiesman et al. (1977) have used an empirical expression

$$1/\rho(T) = (1/\rho_{\text{sat}}) + (1/\rho_l), \tag{36}$$

where $\rho_l = \rho_0 + \rho_{\text{ideal}}(T)$, ρ_0 is the residual resistivity, and $\rho_{\text{ideal}}(T)$ is the temperature-dependent lattice term in the absence of saturation effects. ρ_{sat} is the limiting high-temperature resistivity.

A rough derivation for this formula can be given if it is assumed that the usual electron mean free path l is replaced by $l + a$ (a being the interatomic spacing). This expression is reasonable, since infinitely strong scattering can only reduce the electron mean free path to a.

We then have in the Drude expression for the electrical conductivity

$$\sigma = \frac{ne^2\tau}{m} = \frac{ne^2(l + a)}{mv_f} = \frac{ne^2a}{mv_f}\left(1 + \frac{l}{a}\right) \equiv \sigma_{\text{sat}} + \sigma_{\text{lattice}}(T). \tag{37}$$

Calculation of ρ_{ideal} for Nb_3Al and Nb_3Ge (Allen et al., 1978) give mean free paths of ~ 5 Å at 300 K, lending support to the notion of thermally induced saturation in these materials.

Chakraborty and Allen (1979) have made a detailed general investigation of the effect of strong scattering within the framework of the Boltzmann transport equation. They find that the interband scattering opens up new "nonclassical conduction channels" and that this leads to the parallel resistor expression above [Eq. (36)]. Qualitatively, one might expect such effects for very strong scattering. However, the theory has not as yet been developed to the point where a detailed quantitative comparison with experiment is possible.

7. PHASE CHANGES

The electrical resistivity, as pointed out earlier, is an extremely useful tool for locating phase transitions, especially as a function of some variable such as composition, pressure, or temperature. We briefly mention here some of the types of resistance behavior observed at phase transitions.

(i) First order; pre- and postmelting. The earmark of a first-order phase transition is a distinct, discontinuous change in the electrical resistivity. In a number of cases, however, it is found that there are definitive premonitory effects on both sides of the transition. For example, McDonald (1953) has found that the electrical resistivities of Li, Na, and K increase more than expected just below the melting point. This he attributes to vacancy formation, and from the data obtains energies of vacancy formation in reasonable agreement with self-diffusion activation energies.

(ii) Second-order; critical exponents. The situation with second-order transitions is different. Here there are in general very definite power laws obeyed by various physical quantities, as a function of the quantity $t = (T - T_p)/T_p$, where T_p is the temperature of the phase transition. The so-called critical exponents of t are different above and below T_p, as well as different for different physical quantities. There is much current effort in the theory of critical phenomena to calculate these exponents. The interested reader is referred to the monographs of Stanley (1971) and Ma (1976).

We mention two particular cases. Although the case of magnetic transitions is dealt with elsewhere in this review, we comment here about the electrical resistivity near the magnetic transition temperature. Fisher and Langer (1968) give general arguments to show that the dominant part of the magnetic resistivity at the transition temperature comes from short-range spin fluctuations and that this implies that $d\rho_{mag}/dT$ should be proportional to the magnetic specific heat.

Spectacular effects occur at a special type of second-order transitions known as order–disorder transitions (for reviews, see Muto and Takagi, 1955; Guttmann, 1956). In these there is atomic ordering (i.e., different types of atoms occupy different sublattices) which happens below the transition temperature (T_c), whereas above this temperature, various atomic sites in the crystal can be occupied by several different atoms. Large resistance effects are to be expected here; above T_c the solid is highly disordered and the electrons will be strongly scattered, while below T_c order sets in and this scattering will decrease as the order parameter increases.

Figure 21 shows data for both quenched and annealed Cu–Au alloys (Johansson and Linde, 1936). The quenched alloys are random substitutional face-centered-cubic alloys and the resistivity has the parabolic shape

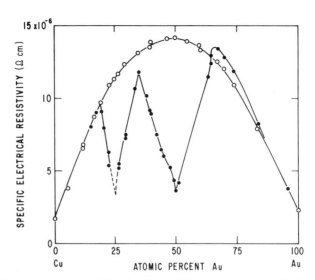

Fig. 21. Electrical resistivity of Cu–Au alloys. (Johansson and Linde, 1936.) ○, quenched from 650°C; ●, annealed at 200°C.

expected from Nordheim's rule. At Cu_3Au and CuAu, however, annealing produces much lower resistivities. The transition to these ordered states is, in fact, second order, and for Cu_3Au occurs at 400°C, with a rapid loss of resistivity below this temperature due to the atomic ordering of the Cu atoms to the face-center sites and the Au atoms to the corner positions giving the $L1_2$ structure. In CuAu there are actually two related ordered structures, orthorhombic CuAu II stable below 380°C and tetragonal CuAu I stable between 380 and 420°C, the order–disorder critical point. CuAu II is known to have a long period (~ 40 Å) superlattice. The reader can find details of this in Sato and Toth (1965).

It can happen that the ordered alloy has a higher resistivity than the disordered alloy at a given temperature, although the residual resistivity will, in general, be lower for the ordered alloy. This happens in Au_3Cu (Hirabayashi, 1952), as well as CoPt (Newkirk *et al.*, 1951) and Fe_3Al (Bennett, 1952). This effect can be due, for example, to change in the number of conduction electrons as well as changes in the phonon spectrum.

(iii) *Crystallographic distortions.* Generally, it is true that both the electrical and lattice vibrational properties of a material change when a distortion of the crystal structure occurs. Such distortions are observed to be second-order changes in some cases and first-order in others. It very often happens that the lattice distortion is coupled to another phase change taking place. This is particularly true in many magnetic phase transitions where the magnetoelastic coupling drives the lattice distortion. In these cases, it is difficult to pick out the part of the electrical resistance change due to the distortion.

A different situation is that of the martensitic transformation in A-15 superconductors slightly above the superconducting transition temperature. Here it is observed that the cubic structure undergoes a transition, possibly weakly first order, to a low-temperature tetragonal phase with $(c/a - 1) = 2.5 \times 10^{-3}$, in the case of V_3Si (Batterman and Barrett, 1964, 1966) and in the case of Nb_3Sn, $(c/a - 1) = -6 \times 10^{-3}$ (Mailfert *et al.*, 1969; Vieland and Wicklund, 1971). This distortion has been described by Labbé and Friedel (1966) as a Jahn–Teller type distortion which lifts the high electron degeneracy of the Fermi level arising from the transition-metal atom chains. The resistance change is very subtle at this transition. This is seen in Fig. 22 for Nb_3Sn.

(iv) Superconductivity. The most prominent property of a superconductor is that the electrical resistivity falls abruptly to zero below the superconducting transition temperature. In inhomogeneous materials, superconducting regions will short circuit normal (nonsuperconducting) regions, and it is not possible to determine from a resistance measurement whether one is measuring a bulk property of a material. This type of short circuiting

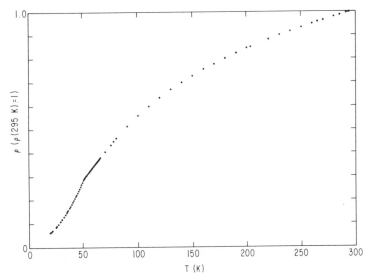

Fig. 22. Temperature dependence of the resistance of pure Nb_3Sn between 300 and $T_c = 18$ K. The resistance ratio $R(300)/R(18\ K) = 18$. Extrapolating the normal-state resistivity to 0 K gives $R(300\ K)/R(0\ K) \sim 50$. At 51 K there is a kink in the resistivity due to the martensitic transformation. (Webb, unpublished observations.)

is the reason for resistively determined transition temperatures generally being somewhat higher than those determined inductively or by heat capacity on the same sample.

In very pure materials, it is possible to observe fluctuation effects very near T_c. These are governed by a critical exponent $(d-4)/2$, where d is the dimensionality of the material. These effects are difficult to observe experimentally because in homogeneous materials the phase transition is very sharp. This sharpness is due to the superconducting coherence length being much larger than the interatomic distance. (See Tinkham, 1975, for a discussion of this effect.)

(v) Charge and spin density waves. Under a very special set of circumstances, the electron gas is unstable with respect to the formation of either spin density or charge density. It is clear that this will have a large effect on the electrical resistivity.

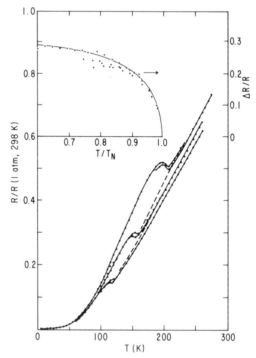

Fig. 23. Electrical resistivity of two different samples of Cr at 26, 46, and 65 kbar descending from the top of the figure. The dashed curves give the extrapolated paramagnetic resistance. The inset gives the difference between resistance in the spin-wave state and this extrapolated paramagnetic resistance. The solid curve in the inset is calculated from the theory of McWhan and Rice (1967).

Briefly, the best known case of a spin density wave occurs in Cr (see discussions in Arrott, 1966; Barker and Ditzenberger, 1970). Resistance data are shown in Fig. 23. The qualitative physical data behind this phenomenon are that a single wave vector connects a large number of electron and hole states at the Fermi level, and these electron–hole pairs condense via their Coulomb interaction into a spin density wave with this

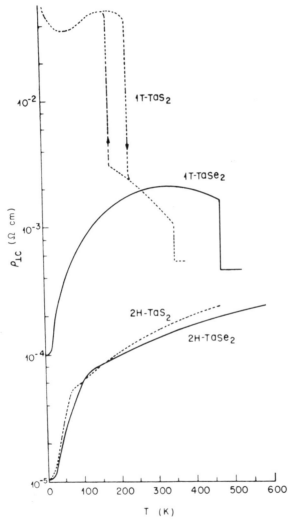

Fig. 24. Basal plane resistivities for two different stacking polytypes of TaS$_2$ and TaSe$_2$. (Wilson *et al.*, 1975.)

wave vector. In the case of Cr, this wave vector is not commensurate with the lattice, and this leads initially to a rise in the electrical resistivity at the transition point.

There are not many candidates for itinerant magnetism known. The other suspected examples known at present are $ZrZn_2$ (Matthias and Bozorth, 1958), Sc_3In (Matthias et al., 1961), and $TiBe_2$ (Matthias et al., 1978). For these last three, the resistance anomalies at the magnetic transition temperature are very small.

The physics behind the charge density wave is similar, but the resistance effects are much more spectacular. We show data for several different forms of TaS_2, Figs. 24 and 25 as examples. The effect of the various charge density waves of the electrical resistivity is large, and it is important to note that it is the strength of electron–phonon interaction relative to the Coulomb interaction which determines whether a charge or spin density wave forms, the charge density wave being favored in the strong electron–phonon coupling case. It is often the case that the charge density wave is at high temperature incommensurate with the lattice, and as it builds in amplitude

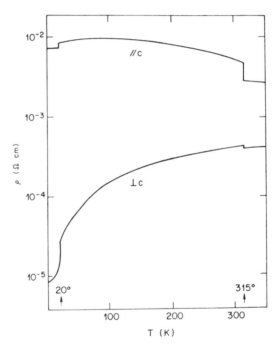

Fig. 25. Resistivity parallel and perpendicular to the c axis for the 4 Hb-polytype of TaS_2. (DiSalvo et al., 1973.)

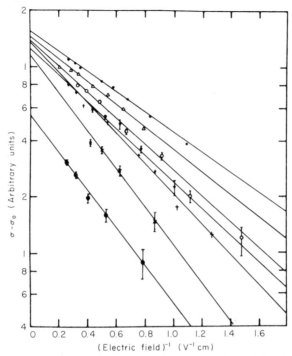

Fig. 26. Semilog plot of the change in resistivity against reciprocal electric field for $NbSe_3$ at various temperatures between the two resistance anomalies observed at 145 and 59 K. Both these anomalies have been tentatively associated with the formation of charge density waves. (Ong and Monceau, 1977.) ⏀ = 135.5 K, ⏀ = 129.2 K, △ = 123.4 K, ● = 115.8 K, ⏀ = 91.0 K, + = 81.8 K, and ⏀ = 72.1 K.

as the temperature decreases, there is often an incommensurate–commensurate transition. (For an extensive review, see Wilson *et al.*, 1975.)

The charge and spin density waves are favored in structures with lower dimensionality. Some materials have quite anisotropic bonding. The so-called layer compounds with very weak van der Waals bonding between layers and strong intralayer bonding approximate two-dimensional metals. TaS_2 is quasi–two dimensional. In the quasi–one-dimensional case, the electrical resistance is highly singular, as the charge density wave buildup often leads to a lattice distortion (The Peierls distortion) which in one dimension is insulating.

Another effect in the charge density wave systems is non-Ohmic behavior (Ong and Monceau, 1977). This is because it is possible to apply electric fields large enough, near the transition, to excite electrons out of the charge density state. Some data for such a case are shown in Fig. 26.

III.   Measurement Techniques

A.   Standard Four-Probe Method

The standard measurement technique for electrical resistivity employs long thin specimens of uniform cross-sectional area with four attached leads as shown in Fig. 27. Use of thin specimens ensures a uniform current distribution between the voltage contacts. The voltage leads should be attached well inside the current contacts where the current distribution is uniform across the sample to avoid complications due to the current "spreading" resistance. If the sample resistance R is measured then the resistivity ρ can be calculated as $\rho = RA/L$, where A is the cross-sectional area and L the distance between voltage leads along a principal direction. When the input impedance of the voltmeter is much greater than the impedance of the sample plus voltage leads then the voltage measurement is independent of that impedance.

The measurement can be made with either direct or alternating current. With direct current care must be taken to eliminate the effects of thermo-electric voltages in the voltage circuit by frequently reversing the sample current. The sample voltage then reverses with the current but the thermal voltages do not. These complications are eliminated with the use of alternating current, even in the low audio-frequency range. With ac, however, care must be taken to minimize the inductive coupling between the driving current and voltage circuits. This is accomplished by using twisted pairs for the current and voltages which are separated from each other and using a low frequency; the inductive voltage is directly proportional to frequency. As the inductive signal is 90° out of phase with the sample resistive signal it can be suppressed further by the use of phase sensitive detection. For an accurate measurement of the sample resistivity, it is additionally necessary to use a low enough measurement frequency that the skin depth is much larger than the maximum cross-sectional dimension. A convenient expression for the skin depth δ in centimeters is

$$\delta = (1/2\pi)\sqrt{\rho/f} \quad \text{cm,} \tag{38}$$

where ρ is in $\mu\Omega$cm and f is in kHz.

The circuit shown in Fig. 28 is fairly typical for ac resistance measurements. It was developed and used for measuring low resistance very pure metals as a function of temperature. By tuning the potentiometer, one, in effect, moves the ground along the length of the sample. This grounding arrangement provides much less noise than single-ended operation. With care taken in laying out the input leads and experimenting with ground arrangements, this circuit provides a noise level of less than 5×10^{-10} V in the low audio-frequency range. This noise figure refers to a time constant of about 3 s. A constant-amplitude current source provided a sinusoidal sample current in the low audio range from 17–270 Hz. Root-mean-square sample current densities between 3 and 60 A/cm^2 were used; the corresponding total sample currents were in the range 1–100 mA. The sample voltage, proportional to its resistance, was measured by a voltmeter whose impedance was high compared to the input circuit.

Sample temperatures below 40 K can be easily measured with a calibrated germanium resistance thermometer in good thermal contact with the sample. Temperatures above 40 K are easily measured by platinum resistance thermometry. In practice, point-by-point sample resistance measurements are taken as the temperature was slowly swept both up and down. The rate of change of temperature should be kept slow enough so that the measurements taken on heating and cooling coincide.

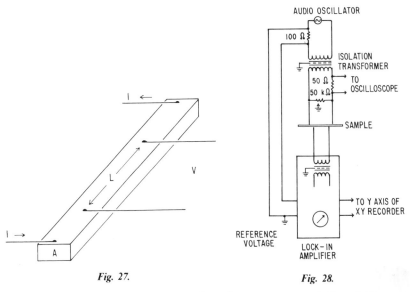

Fig. 27.   *Fig. 28.*

Fig. 27.   Four lead arrangement for resistance measurement. $R = \rho L / A$.

Fig. 28.   A sample resistance measuring circuit. The 50-kΩ grounding potentiometer can be adjusted for minimum noise. The input impedance of the voltmeter should be much less than that of the grounding potentiometer and much greater than that of the sample. (Webb, 1969.)

The main part of the sample holder used with the circuit of Fig. 28 is a quartz single crystal of semicircular cross section about $1\frac{1}{2}$ in. long and 0.3 in. wide across the flat portion. A single crystal of quartz was chosen since it is an electrical insulator with high thermal conductivity at cryogenic temperatures. It is also convenient to use insulated copper as a construction material. The sample was pressed against the quartz with four pure-indium-tipped Be–Cu electrical contacts. The pressure contacts did not deform the sample to an extent that was observable under a 40-power microscope. Other applications permit the use of spot welded or soldered leads.

Figure 29 shows a schematic of a stainless steel dewar assembly which has been used for resistance measurements in the temperature range 1.5 to 350 K. The heater winding on the outside of the dewar is glued and taped to the tube. A convenient resistance for the heater winding is of order 100 Ω. The

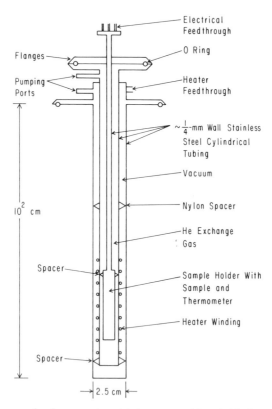

Fig. 29. An example of an experimental dewar assembly suitable for resistance measurements from room temperature to the liquid-helium range.

amount of He exchange gas in the central space is not critical above a few Torr of pressure. A slight amount of exchange gas is introduced into annular vacuum space to provide cooling. It is found that reproducible data spanning the range from room temperature to 4 K can be recorded over a period of order of 2 h.

B. The van der Pauw (1958) Method

This technique permits the measurement of the resistivity of an isotropic sample of uniform flat thickness but with arbitrary shape if it meets the criteria that

(1) the contacts are at the periphery of the sample,
(2) the contacts are sufficiently small,
(3) the sample is uniform in thickness,
(4) the sample is simply connected, i.e., it does not contain holes, and
(5) the sample is homogeneous.

The measurement relies on a theorem proved for a sample of arbitrary shape with leads shown as in Fig. 30. The resistance $R_{AB,CD}$ is defined as the potential difference $V_D - V_C$ between contacts C and D when unit current is put through contacts A and B, i.e.,

$$R_{AB,CD} \equiv |V_C - V_D|/|I_{AB}|. \tag{39}$$

It is proved that the implicit relation holds that

$$\exp(-\pi R_{AB,CD}d/\rho) + \exp(-\pi R_{BC,DA}d/\rho) = 1, \tag{40}$$

where d is the thickness of the sample and R is the resistivity. A consistent set of units is, for example, d expressed in cm, R in Ω and ρ in Ω cm.

The resistivity can be written

$$\rho = \frac{\pi d}{\ln 2} \frac{(R_{AB,CD} + R_{BC,DA})}{2} F\left(\frac{R_{AB,CD}}{R_{BC,DA}}\right). \tag{41}$$

Here F is a function of the ratio of resistance such that

$$\frac{R_{AB,CD} - R_{BC,DA}}{R_{AB,CD} + R_{BC,DA}} = F \operatorname{arccosh}\left[\frac{\exp(\ln 2/F)}{2}\right]. \tag{42}$$

van der Pauw has given a graphical representation for F, but this has been supplanted by the modern electronic calculator.

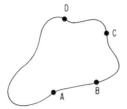

Fig. 30. A sample of arbitrary shape with four small contacts at arbitrary places along the circumference (van der Pauw, 1958).

In the special case of samples and contacts which are invariant under rotation of 90°C, the sheet resistance $R_s \equiv \rho/d$ reduces to simple form

$$R_s = \frac{\pi}{\ln 2} \frac{V}{I} = \left(\frac{\pi}{\ln 2}\right) R_{AB,CD} = \left(\frac{\pi}{\ln 2}\right) R_{AC,BD}. \qquad (43)$$

Specimens with such symmetry are often easily prepared. Here V is the voltage between two voltage contacts and I is the current flowing between the two current contacts. Versnel (1978) has extended this equation to include the case of finite length contacts on structures which are invariant under 90° rotations.

Logan *et al.* (1971) have analyzed the special case of determining the resistance of an isotropic rectangular prism with four contacts attached to the corners of one face. Montgomery (1971) has analyzed in detail the geometry of a rectangular prism having contacts in four corners which can be strongly anisotropic. This was accomplished by finding a mapping which carries the anisotropic case over to the isotropic case of Logan *et al.* It is important to take into account anisotropy, especially if temperature dependent, as has been noted by Schafer *et al.* (1974) for the case of TTF–TCNQ.

C. Contactless Methods of Measurement

There have been essentially two general contactless methods of measurement of resistivity. Both of these methods employ induced eddy currents in the sample by magnetic field changing in time. One of the methods applies a step function or single square wave magnetic field to the sample and then monitors the decay of the induced eddy currents with an oscilloscope. This is accomplished by having the sample in a coil and watching the voltage output of the coil with an oscilloscope. The voltage is characterized by an infinite series of exponential time constants which is a defect of the method for some applications. One then examines the time dependence of the system in the long time limit where the longest time constant is dominant. For details the reader is referred to the literature (Bean *et al.*, 1959).

The other method is a steady-state induction method in which the sample is placed in a coil. The coil is usually driven with a sinusoidal alternating current. The magnetic field of the coil excitation current induces eddy currents in the sample. The induced eddy currents in the sample have the effect of changing both the in phase, reactive component of the impedance, and the out of phase, loss component of the coil.

Zimmerman (1961) has treated these effects on the impedance of a coil for samples in two convenient geometries, namely an infinite circular cylinder and a sphere. The following are those results. The impedance between the terminals of the coil is the sum of a resistive and reactive component. It is written as $R + i\omega L$ where ω is the angular frequency. R, the resistance component, is defined as P/I^2, where P is the power dissipated in the coil plus sample and I the rms current driving the coil. As P can be represented with two components, one the power dissipated in the coil and the second the power dissipated in the sample, he writes

$$R = R_0 + R' \tag{44}$$

where R is the total resistive component, R_0 is the resistive component without a sample in the coil, and R' is the increase due to the sample.

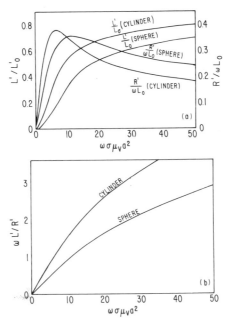

Fig. 31. (a) Relative reactance and resistance variation for an infinite circular cylinder of radius a and for a sphere of radius a of conducting material of conductivity σ, in a uniform applied ac field of frequency ω. $\mu_v = 4\pi \times 10^{-7}$ h/M. (b) Ratio of reactance to resistance variation. (Zimmerman, 1961.)

Similarly, the inductance is written as

$$L = L_0 - L', \tag{45}$$

where L is the total measured inductance, L_0 that of the empty coil, and $-L'$ the negative change in the inductance due to the insertion of the sample.

Zimmerman has found closed expressions for L' and R' for the special cases of an infinite circular cylinder of radius a and of a sphere or radius a, both in a uniform applied field. These results are shown in Fig. 28. Here σ is the conductivity of the material in the coil and $\mu_v = 4\pi \times 10^{-7}$ h/M. α is a filling factor which is difficult to calculate except in the case of the infinite cylinder where it is the ratio of the cross-sectional area of the coil to that of the cylinder. For the sphere he finds that if L_0 is taken as the inductance per unit length of the coil then $\alpha = A^2/2a^3$ where A is the radius of the coil.

Several points have been noted about Fig. 31a. One point was that the curves for the sphere and the cylinder have similar shapes, differing only by about a factor of 2 in the horizontal scale; this indicates that specimens with less than ideal shapes can be treated by just rescaling the horizontal axis. The second point made was that the ratio $\omega L'/R'$ for the two geometries does not depend on L_0 and α so that no properties of the coil are involved. The significance of this fact is obvious.

Zimmerman gives several examples of coils and bridge circuits used in the measurement. Comparisons of the technique with known materials were rather good except in perhaps one case. In general, the strongest advantage of the technique would be to monitor the resistivity at fixed experimental conditions of a large number of samples of fixed shape and size.

Acknowledgments

It is a great pleasure to thank Dr. M. B. Maple for several useful discussions and Annetta Whiteman for untiring labor in typing several versions of this chapter. Support by the National Science foundation during the course of this work is also gratefully acknowledged.

General References

Blatt, F. J. (1968). "Physics of Electronic Conduction in Solids." McGraw-Hill, New York.

Meaden, G. T. (1965). "Electrical Resistance of Metals." Plenum, New York.

Mott, N. F., and Jones, H. (1958). "The Theory of the Properties of Metals and Alloys." Dover, New York.

Olsen, J. L. (1962). "Electron Transport in Metals." Wiley (Interscience), New York.

Peierls, R. E. (1965). "Quantum Theory of Solids." Oxford Univ. Press, London and New York.

Wilson, A. H. (1965). "The Theory of Metals," 2nd ed. Cambridge Univ. Press, London and New York.

Ziman, J. M. (1963). "Electrons and Phonons." Oxford Univ. Press, London and New York.
Ziman, J. M. (1964). "Principles of the Theory of Solids." Cambridge Univ. Press, London and New York.
Ziman, J. M., ed. (1969). "The Physics of Metals. 1. Electrons." Cambridge Univ. Press, London and New York.

References

Adrian, H., Ischenko, G., Lehman, M., Müller, P., Braun, H., and Linker, G. (1978). *J. Less-Common Met.* **62**, 99.
Aleksandrov, B. N., and D'Yakov, I. G. (1963). *Sov. Phys.—JETP* **16**, 603.
Allen, P. B. (1971). *Phys. Rev. B***3**, 305.
Allen, P. B., and Butler, W. H. (1978). *Phys. Today*, December, p. 44.
Allen, P. B., and Hui, J. C. K., Pickett, W. E., Varma, C. M., and Fisk, Z. (1976). *Solid State Commun.* **18**, 1157.
Allen, P. B., Pickett, W. E., Ho, K. M., and Cohen, M. L. (1978). *Phys. Rev. Lett.* **40**, 1532.
Andersen, H. H., Gregers-Hansen, P. E., Holm, E., Smith, H., and Vogt, O. (1974). *Phys. Rev. Lett.* **32**, 1321.
Anderson, P. W. (1961). *Phys. Rev.* **124**, 41.
Appel, J. (1963). *Philos. Mag.* **8**, 1071.
Arrott, A. (1966). *In* "Magnetism" (G. T. Rado and H. Suhl, eds), Vol. 2B, p. 296. Academic Press, New York.
Axe, J. D., and Shirane, G. (1973). *Phys. Rev. B***8**, 1965.
Barker, A. S., Jr., and Ditzenberger, J. A. (1970). *Phys. Rev. B***1**, 4378.
Bansinski, Z. S., Dugdale, J. S., and Howie, A. (1963). *Philos. Mag.* **8**, 1989.
Bass, J. (1972). *Adv. Phys.* **21**, 431.
Batterman, B. W., and Barrett, C. S. (1964). *Phys. Rev. Lett.* **13**, 390.
Batterman, B. W., and Barrett, C. S. (1966). *Phys. Rev.* **145**, 296.
Bean, C. P., DeBlois, R. W., and Nesbitt, L. B. (1959). *J. Appl. Phys.* **30**, 1976.
Bennett, W. D. (1952). *J. Iron Steel Inst., London* **171**, 372.
Billington, D. S., and Crawford, J. H. (1961). "Radiation Damage in Solids." Princeton Univ. Press, Princeton, New Jersey.
Blatt, F. J. (1957a). *Phys. Rev.* **108**, 285.
Blatt, F. J. (1957b). *Phys. Rev.* **108**, 1204.
Brown, R. A. (1967a). *Phys. Rev.* **156**, 692.
Brown, R. A. (1967b). *Phys. Rev.* **156**, 889.
Brown, R. A. (1977a). *J. Phys. F.* **7**, 1269.
Brown, R. A. (1977b). *J. Phys. F.* **7**, L297.
Brown, R. A. (1977c). *J. Phys. F.* **7**, 1477.
Busch, G., and Güntherodt, H.-J. (1974). *Solid State Phys.* **29**, 235.
Campbell, I. A. (1970). *Phys. Rev. Lett.* **24**, 269.
Campbell, I. A., Fert, A., and Pomeroy, A. R. (1967). *Philos. Mag.* **15**, 977.
Campbell, I. A., Fert, A., and Jaoul, O. (1970). *J. Phys. C* **1**, Suppl., S95.
Caton, R., and Viswanathan, R. (1978). *J. Phys. (Paris), Colloq.* **39**, 385.
Chakraborty, B., and Allen, P. B. (1979). *Phys. Rev. Lett.* **38**, 782.
Cohen, R. W. (1971). *Proc. Conf. Superconduct. d- f-Band Met., Rochester, N.Y.* p. 144.
Cohen, R. W., Cody, G. D., and Halloran, J. J. (1967). *Phys. Rev. Lett.* **19**, 840.
de Faget de Casteljau, and Friedel, J. (1956). *J. Phys. Radium* **17**, 27.
Delley, B., Beck, H., Kunzi, H. U., and Güntherodt, H.-J. (1978). *Phys. Rev. Lett.* **40**, 193.
DiSalvo, F. J., Bagley, B. G., Voorhoeve, J. M., and Waszczah, J. V. (1973) *J. Phys. Chem. Solids* **34**, 1357.

Doniach, S. (1967). *In* "Theory of Magnetism in Transition Metals" (W. Marshall, ed.), p. 255. Academic Press, New York.

Faber, T. E. (1966). *Adv. Phys.* **15**, 547.

Faber, T. E. (1969). *In* "Physics of Metals", Part 1, (J. M. Ziman, ed.) p. 282. Cambridge Univ. Press, London and New York.

Faber, T. E. (1972). "Introduction to the Theory of Liquid Metals." Cambridge Univ. Press, London and New York.

Fisher, M. E., and Langer, J. S. (1968). *Phys. Rev. Lett.* **20**, 665.

Fisk, Z. (1976). *Solid State Commun.* **18**, 221.

Fisk, Z., and Johnston, D. C. (1977). *Solid State Commun.* **22**, 359.

Fisk, Z., and Lawson, A. C. (1973). *Solid State Commun.* **13**, 277.

Fisk, Z., and Webb, G. W. (1976). *Phys. Rev. Lett.* **36**, 1084.

Fradin, F. Y. (1974). *Phys. Rev. Lett.* **33**, 158.

Friedel, J. (1954). *Adv. Phys.* **3**, 446.

Friedel, J. (1958). *Nuovo Cimento, Suppl.* **7**, 287.

Fulde, P., and Peschel, I. (1972). *Adv. Phys.* **21**, 1.

Grüner, G. (1974). *Adv. Phys.* **23**, 941.

Grüner, G., and Zawadowski, A. (1974). *Rep. Prog. Phys.* **37**, 1497.

Gurvitch, M. (1980). *Proc. Conf. Superconduct. d-f-Band Met., 3rd*, La Jolla, p. 317. (to be published).

Guttmann, L. (1956). *Solid State Phys.* **3**, 146.

Hake, R. R. (1967). *Appl. Phys. Lett.* **10**, 189.

Hambourger, P. D., and DiSalvo, F. J. (1980). *Physica (Utrecht)* **99B**, 173.

Herring, C. (1958). *Physica* **24**, S184.

Hillel, A. J. (1970). *Acta Metall.* **18**, 253.

Hillel, A. J., Edwards, J. T., and Wilkes, P. (1977). *Philos. Mag.* **35**, 829.

Hirabayashi, M. (1952). *Nippon Kinzoku Gakkaishi* **16**, 67.

Hirst, L. L. (1967). *Solid State Commun.* **5**, 751.

Ioffe, A. F., and Regel, A. R. (1960). *Prog. Semicond.* **4**, 237.

Jan, J. P. (1957). *Solid State Phys.* **5**, 1.

Johansson, C. H., and Linde, J. O. (1936). *Ann. Phys. (Leipzig)* **25**, 1.

Kasuya, T. (1956). *Prog. Theor. Phys.* **16**, 45.

Kedves, F. J., Hordos, M., and Gergely, L. (1972). *Solid State Commun.* **11**, 1067.

Kelly, A., and Nicholson, R. B. (1963). *Prog. Mater. Sci.* **10**, 149.

Kondo, J. (1964). *Prog. Theor. Phys.* **32**, 37.

Kukkonen, C. A., and Maldague, P. M. (1976). *Phys. Rev. Lett.* **37**, 782.

Kukkonen, C. A., and Maldague, P. M. (1979). *Phys. Rev. B* **19**, 2394.

Labbé, J., and Friedel, J. (1966). *J. Phys. (Paris)* **27**, 153, 303.

Logan, B. F., Rice, S. O., and Wick, R. F. (1971). *J. Appl. Phys.* **42**, 2975.

Ma, S. (1976). "Modern Theory of Critical Phenomena." Benjamin, New York.

MacDonald, D. K. C. (1953). *J. Chem. Phys.* **21**, 177.

MacDonald, D. K. C., and Mendelssohn, K. (1950). *Proc. Roy. Soc. (London)* **A202**, 103.

McMillan, W. L. (1968). *Phys. Rev.* **167**, 331.

McWhan, D. B., and Rice, T. M. (1967). *Phys. Rev. Lett.* **19**, 846.

Mailfert, R., Batterman, B. W., and Hanak, J. J. (1969). *Phys. Status Solidi* **32**, K67.

Marchenko, V. A. (1974). *Sov. Phys.—Solid State* **15**, 1261.

Matthias, B. T., and Bozorth, R. M. (1958). *Phys. Rev.* **109**, 604.

Matthias, B. T., Clogston, A. M., Williams, H. J., Corenzwit, E., and Sherwood, R. C. (1961). *Phys. Rev. Lett.* **7**, 7.

Matthias, B. T., Giorgi, A., Struebing, V. O., and Smith, J. L. (1978). *Phys. Lett.* **A69**, 221.

Montgomery, H. C. (1971). *J. Appl. Phys.* **42**, 2971.

Mooij, J. H. (1973). *Phys. Status Solidi A* **17**, 521.

Mott, N. F. (1936). *Proc. R. Soc. London, Ser. A* **153**, 699.

Mott, N. F., and Jones, H. (1936). "The Theory of the Properties of Metals and Alloys." Oxford Univ. Press, London and New York.

Müller, P., Ischenko, G., Adrian, H., Bieger, J., Lehman, M., and Haase, E. L. (1980). *Proc. Conf. Superconduct. d-f-Band Met., 3rd*, La Jolla, p. 369 (to be published).

Muto, T., and Takagi, Y. (1955). *Solid State Phys.* **1**, 194.

Nakagawa, Y., and Woods, A. D. B. (1963). *Phys. Rev. Lett.* **11**, 271.

Newkirk, J. B., Smoluchowski, R., Geisler, A. H., and Martin, D. L. (1951). *J. Appl. Phys.* **22**, 950.

Ong, N. P., and Monceau, D. (1977). *Phys. Rev. B* **16**, 3443.

Pandé, C. S. (1977). *Solid State Commun.* **24**, 241.

Pinski, F. J., Allen, P. B., and Butler, W. H. (1978). *Phys. Rev. Lett.* **41**, 431.

Poate, J. M., Testardi, L. R., Storm, A. R., and Augustyniak, W. M. (1975). *Phys. Rev. Lett.* **35**, 1290.

Rehwald, W., Rayl, M., Cohen, R. W., and Cody, G. D. (1972). *Phys. Rev. B* **6**, 363.

Ruderman, M. A., and Kittel, C. (1954). *Phys. Rev.* **96**, 99.

Sales, B. (1979). Personal communication.

Sarachik, M. P., Smith, G. E., and Wernick, J. H. (1963). *Can. J. Phys.* **41**, 1542.

Sato, H., and Toth, R. S. (1965). *In* "Alloying Behavior and Effects in Concentrated Solid Solutions" (T. B. Massalski, ed.), p. 295. Gordon & Breach, New York.

Schafer, D. E., Wudl, F., Thomas, G. A., Ferraris, J. P., and Cowan, D. O. (1974). *Solid State Commun.* **14**, 347.

Schindler, A. C., and LaRoy, B. C. (1966). *J. Appl. Phys.* **37**, 3610.

Schrïeffer, J. R., and Wolff, P. A. (1966). *Phys. Rev.* **149**, 491.

Schweiss, B. P., Renker, B., Schneider, E., and Reichardt, W. (1976). *Proc. Conf. Superconduct. d- f-Band Met., 2nd, Rochester, N. Y.* p. 189.

Stanley, H. E. (1971). "Introduction to Phase Transitions and Critical Phenomena." Oxford Univ. Press, London and New York.

Stevens, K. W. H. (1952). *Proc. Phys. Soc., London, Sect. A* **65**, 209.

Suhl, H., and Matthias, B. T. (1959). *Phys. Rev.* **114**, 977.

Sweedler, A. R., Schweitzer, D. G., and Webb, G. W. (1974). *Phys. Rev. Lett.* **33**, 168.

Sweedler, A. R., Cox, D. E., and Moehlecke, S. (1978). *J. Nucl. Mater.* **72**, 50.

Testardi, L. R., Poate, J. M., and Levinstein, H. J. (1976). *Phys. Rev. Lett.* **37**, 637.

Thompson, A. H. (1975). *Phys. Rev. Lett.* **35**, 1786.

Tinkham, M. (1975). "Introduction to Superconductivity." McGraw-Hill, New York.

Turnbull, D., Rosenbaum, H. S., and Treaftis, H. N. (1960) *Acta Met.* **8**, 277.

van der Pauw, L. J. (1958). *Philips Res. Rep.* **13**, 1.

Versnel, W. (1978). *Solid-State Electron.* **21**, 1261.

Vieland, L. J., and Wicklund, A. (1971). *Phys. Lett. A* **34**, 43.

Webb, G. W. (1969). *Phys. Rev.* **181**, 1127.

Webb, G. W., Fisk, Z., Engelhardt, J. J., and Bader, S. D. (1977). *Phys. Rev. B* **15**, 2624.

Wiesman, H., Gurvitch, M., Lutz, H., Ghosh, A., Schwartz, B., Strongin, M., Allen, P. B., and Halley, J. W. (1977). *Phys. Rev. Lett.* **38**, 782.

Wilson, A. H. (1938). *Proc. R. Soc. London, Ser. A* **167**, 580.

Wilson, A. H. (1956). "The Theory of Metals, 2nd ed. Cambridge Univ. Press, London and New York.

Wilson, J. A., DiSalvo, F. J., and Mahajan, S. (1975). *Adv. Phys.* **24**, 117.

Woodward, D. W., and Cody, G. D. (1964). *Phys. Rev.* **136**, 166A.

Yosida, K. (1957). *Phys. Rev.* **106**, 893.

Ziman, J. M. (1961). *Philos. Mag.* **6**, 1013.

Ziman, J. M. (1967). *Adv. Phys.* **16**, 551.

Zimmerman, J. E. (1961). *Rev. Sci. Instrum.* **32**, 402.

6

Electronic Structure of Point Defects in Metals

P. JENA

Physics Department, Michigan Technological University
Houghton, Michigan

I. Introduction

The study of imperfect crystalline solids has been a topic of great interest to physicists, chemists, metallurgists, and material scientists for many years. Such an interest stems not only from the fact that no solid in nature exists in a 100% pure state, but the effect of defects on the mechanical, thermodynamic, and electronic properties of crystalline materials has important fundamental and technological significance. In general, imperfections in solids include point defects and their clusters, dislocations, grain boundaries, interphase boundaries, stacking faults, and surfaces. This constitutes an enormous area of research, and it is impossible to discuss all facets of imperfections in solids in this chapter. We shall only be concerned with the discussion of the electronic structure of point defects in metals.

The point defects in solids may constitute intrinsic defects (vacancies, self-interstitials), and impurity atoms (substitutional or interstitial). There are several recent reviews (Flynn, 1972; Caglioti, 1976; Stoneham, 1975; Crawford and Slifkin, 1977; Gehlen *et al.*, 1972; Hannay, 1975) which treat the properties of point defects on an atomic scale. In this chapter we shall only deal with two of the most elementary form of point defects—intrinsic defects such as vacancies and extrinsic defects such as positrons, hydrogen, and its isotopes. A vacancy which is normally located at a lattice site corresponds to the removal of an atom from the bulk of the solid and can be created by thermal processes and/or electron and neutron irradiation. Impurities such as hydrogen, on the other hand, enter the interstitial sites of the host. Apart from the academic interest in the study of vacancies and hydrogen in solids, there are practical considerations that warrant a fundamental understanding of these types of point defects. For example, the vacancies cluster to form voids that become important in fast nuclear reactor technology, the embrittlement of metals due to dissolved hydrogen (Kolachev, 1968; Birnbaum, 1978) and hydrogen storage (Reilly, 1978) are important problems in energy-related technology. Obviously, one would like to understand how, why, and to what extent the properties of the host materials are affected on an electronic scale due to these point defects.

A metallic host is characterized by a lattice of positive ions and a sea of itinerant conduction electrons. The lattice ions are almost entirely screened by these conduction electrons within the Wigner–Seitz volume. The introduction of a point defect into this system provides an external perturbation. The conduction electrons in their effort to screen the charge on the point defect get redistributed. This redistributed electron charge, of course, approaches the ambient or unperturbed electron-charge density at distances far from the defect. The nature of the spatial distribution of this screening charge is the key to our understanding of many electronic properties of the imperfect crystalline metal.

A rigorous theoretical treatment of the electronic structure of point defects is hampered because of the loss of lattice periodicity. As a consequence, Bloch's theorem is no longer applicable. In spite of this difficulty numerous theoretical attempts and progress have been made in this area. Some of the more common methods used to study the impurity problem will be reviewed in the following section. Sections III and IV will be devoted to a study of light interstitials (hydrogen, deuterium, and positive muon), and monovacancies and microvoids (although voids cannot be classified under point defects) in metals, respectively. In each of these sections, the discussions will center around electron distribution around the point defect, changes in the ground-state energy of the system due to imperfections, and interactions of the electron charge and spin density distributions with the host and impurity nuclei. Particular care will be taken to compare the data obtained from a variety of experimental techniques, such as x-ray photoelectron spectroscopy, Compton scattering, Mössbauer spectroscopy, nuclear magnetic and quadrupole resonance, electrical conductivity measurements, and positron annihilation techniques with theoretical studies. It is hoped that such a comparison will provide a "unified" picture of the electron behavior in imperfect crystalline metals.

II. Theoretical Formulation

In calculating the electron response to a metal defect, one is always faced with the problem of finding a suitable choice for the electron-defect potential. For a weak external potential, the perturbation on the electronic system is small. Consequently, simple theories based on statistical (Thomas, 1927; Fermi, 1928; March, 1957) or perturbation methods (Harrison, 1966) may be adequate in calculating the electronic properties of the imperfect system. In a metal, the potential experienced by an electron due to the ion may be simplified by realizing that the orthogonality of the conduction electrons to the ion-core orbitals leads to a weak potential. The resulting pseudopotentials have been very successful (Harrison, 1966; Cohen and Heine, 1970, Cohen, 1979) in the interpretation of a variety of solid-state properties. Naturally, one would expect the effective potential due to a monovacancy to be weak since it corresponds to the absence of an ionic pseudopotential. Even in this case, one always wonders if the pseudopotential is indeed weak enough to justify a linear response of the electrons to this defect potential. In the case of hydrogen impurity in metals, the problem is somewhat more clear. Due to the absence of any conventional electronic core structure around a proton, the coulomb interaction between a proton and an. electron is singular at the proton site. Thus the interaction potential is much stronger than any pseudopotential and warrants a more sophisticated treatment of the impurity problem.

An extreme limit of this pseudopotential concept is to neglect entirely the effect of lattice ions on the behavior of host electrons. Thus the charges on the lattice ions are smeared uniformly with a constant density

$$n_0 = Z_v/\Omega_0,\tag{1}$$

where Z_v is the charge on each ion and Ω_0 is the volume occupied by an atom in the solid. Superimposed on this positive background is a homogeneous electron gas of density exactly equal to n_0 to maintain charge neutrality. Conventionally the density of the homogeneous electron gas is expressed in terms of a parameter r_s such that $(4\pi/3)r_s^3 a_0^3 = 1/n_0$. Here a_0 is the Bohr radius for the hydrogen atom. For most metals, $2 \lesssim r_s \lesssim 6$. Physically r_s represents the average distance between two electrons in the metal. Thus large (small) r_s corresponds to a low- (high-) density metal. This so called jellium model is shown schematically in Fig. 1a for a solid in two dimensions for the sake of convenience. The left-hand side of Fig. 1 corresponds to a real situation whereas the right-hand side represents an equivalent jellium description. Figures 1b–d correspond, respectively, to an

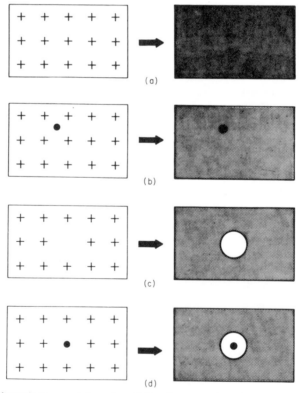

Fig. 1. Schematic representation of a solid in discrete and continuum (jellium) model. (a) Perfect solid. (b) Interstitial impurity. (c) Monovacancy. (d) Substitutional impurity.

interstitial impurity, monovacancy, and substitutional impurity. The radius of the hole a created in the positive background must correspond to the Wigner–Seitz radius R_{WS} in order that it may represent a monovacancy. The external perturbation $n_{ext}(\mathbf{r})$ on the electron gas may be given mathematically as

$$n_{ext}(\mathbf{r}) = Z_I \delta(\mathbf{r}) + n_0 \theta(\mathbf{r} - \mathbf{a}), \tag{2}$$

where Z_I is the impurity charge and equals unity for a proton. The Heavidise function θ is defined such that it is zero for $r < a$ and is unity for $r \geq a$. Thus, Figs. 1a–d correspond, respectively, to $Z_I = 0$, $a = 0$; $Z_I = 1$, $a = 0$; $Z_I = 0$, $a = R_{WS}$; and $Z_I = 1$, $a = R_{WS}$. Although in principle no metal exists in nature for which the jellium model for a solid is strictly valid, in practice this model describes quite well many electronic properties of both perfect and imperfect systems. We shall discuss in this chapter results in both jellium and more sophisticated models.

In this section our objective is to outline briefly the common methods normally employed in the study of point defects with emphasis on the interrelationship, if any, between them. The various approximations and limitations of the methods will be discussed. Comparisons of the results between these methods will be delegated to the Sections III and IV.

A. Statistical Models

In the Thomas–Fermi statistical model (Thomas, 1927; Fermi, 1928; March, 1957), the potential $V(r)$ in which the electrons move is assumed to be varying slowly over an electron wavelength such that many electrons can be localized within a volume over which the potential changes by a small fraction of itself. The electrons obey Fermi–Dirac statistics and can be treated by statistical mechanics. Consequently, the electrons fill states in order of increasing energy. The wave vector of the highest occupied state is then related to the electron density $n(r)$ through the relation

$$k_F^3(r)/3\pi^2 = n(r). \tag{3}$$

For a perfect host, the electron density n_0 is uniform and corresponds to a Fermi wave vector k_F^0

$$\left(k_F^0\right)^3/3\pi^2 = n_0. \tag{4}$$

According to the Thomas–Fermi model, the total energy of the electron at any point in space around the defect is equal to the host chemical potential, i.e.,

$$k_F^2(r) + V(r) = \left(k_F^0\right)^2. \tag{5}$$

Substituting Eqs. (3) and (4) in (5), we have

$$n^{2/3}(r) - n_0^{2/3} = -V(r)/(3\pi^2)^{2/3}. \qquad (6)$$

Here, and throughout this chapter, we shall use atomic units ($\hbar = 1$, $m = \frac{1}{2}$, $e^2 = 2$). The energy is in Rydbergs (1 Ry = 13.6 eV) and length in Bohr radius, a_0 for hydrogen ($1a_0 = 0.5292$ Å). The electrostatic potential energy $V(r)$ is related to the charge distribution through the Poisson's equation

$$\nabla^2 V(r) = -8\pi[n_{ext}(r) - n(r)]. \qquad (7)$$

Since both $n(r)$ and $n_{ext}(r)$ approach n_0 for $r \to \infty$, the screened potential $V(r)$ in Eq. (7) is of limited range. Simultaneous solutions of Eqs. (6) and (7) result in the Thomas–Fermi charge distribution $n(r)$ around the impurity.

There are mainly three shortcomings in the Thomas–Fermi model for a point defect:

(1) In the form discussed, exchange and correlation to the effective potential are neglected.

(2) The electron density $n(r)$ at large distances from the defect does not oscillate. These oscillations, usually known as Friedel oscillations (Kittel, 1963), are characteristic of a sharply defined Fermi surface of the host metal.

(3) For a point impurity, i.e., $Z_I \neq 0$ in Eq. (2), the Thomas–Fermi model leads to a divergent electron density at the nuclear (origin) site.

In the Thomas–Fermi–Dirac (TFD) approximation, the local exchange energy can be included in the total energy. The inhomogeneity corrections to the total energy can be taken into account (Kirzhnits, 1957; Weizsäcker, 1935; Jones and Young, 1971) by including density gradients. The later modification does not significantly improve the results based upon the TF approximation as far as electron distribution around vacancies are concerned (Nieminen et al., 1975).

B. Response Theories

For a point defect interacting only weakly with the host electron system, the electron distribution can be given by a simple expression (Harrison, 1970), namely,

$$\delta n(q) = \{[1/\varepsilon(q)] - 1\}(q^2/4\pi e^2)\Delta w(q), \qquad (8)$$

where $\varepsilon(q)$ is the dielectric constant of the medium and $\Delta w(q)$ is the difference between the bare pseudopotentials of the host $w_0(q)$ and the point defect $w_i(q)$,

$$\Delta w(q) = w_i(q) - w_0(q), \tag{9}$$

and $\delta n(q)$ is the Fourier transform of the perturbed electron charge density $\delta n(r)$,

$$\delta n(r) = n(r) - n_0. \tag{10}$$

For a weak pseudopotential, higher-order corrections can be neglected and $\delta n(q)$ in Eq. (8) is linear in the external perturbation. However, there has always been the nagging question of whether or not the often-used linear response theory is, in fact, adequate.

Sjölander and Stott (1972) have considered the problem of a strong interaction between the electrons and the defect. They have derived an integral equation for the Fourier transform of the polarization charge around an impurity in an electron gas by extending the theory of electron correlations in a homogeneous system by Singwi et al. (1968). Recently, Gupta et al. (1978) have generalized the Sjölander–Stott theory for a two-component plasma consisting of a static impurity and the electron gas by introducing a density derivative term with an unknown coefficient a_{12} exactly in the same way as is done in the Vashishta–Singwi theory (1972) of a homogeneous electron gas. The preceding formulations treat the response of the electrons to all orders in the defect potential. The steps of the theory are rather complicated and the reader is referred to the original papers (Sjölander and Stott, 1972; Gupta et al., 1978) for details. In the following, we only present the final result of the generalized Sjölander–Stott theory (Gupta et al., 1978),

$$\delta n(q) = \left(\frac{1}{\varepsilon(q) - 1} \right) \frac{q^2}{4\pi e^2} \left[\phi(\mathbf{q}) + \left(1 + a_{12} \frac{d}{dn_0} \right) \right.$$

$$\left. \times \left(\frac{1}{n_0} \int \frac{d^3k}{(2\pi)^3} \frac{\mathbf{k} \cdot \mathbf{q}}{q^2} \phi(\mathbf{k}) \delta n(\mathbf{k} - \mathbf{q}) \right) \right]. \tag{11}$$

Note that the linear-response theory result is obtained by keeping only the first term within the bracket. By setting $a_{12} = 0$ in Eq. (11), one recovers the result of the Sjölander–Stott (1972) theory. In Sections III and IV we shall compare the results of the mean-field formulation with other methods.

C. Density Functional Formalism

Hohenberg, Kohn, and Sham (HKS) (Hohenberg and Kohn, 1964; Kohn and Sham, 1965) have developed a theory for studying the properties of interacting inhomogeneous systems. In this theory the ground-state energy of an interacting electron gas in a static external potential, $V_{ext}(r)$ is written as a universal function of the particle density $n(r)$ as

$$E[n(\mathbf{r})] = T_0[n(\mathbf{r})] + \int d\mathbf{r}\, n(\mathbf{r})V_{ext}(\mathbf{r})$$

$$+ \frac{1}{2} \int \int \frac{n(\mathbf{r})n(\mathbf{r}')}{|\mathbf{r} - \mathbf{r}'|}\, d\mathbf{r}\, d\mathbf{r}' + E_{xc}[n(\mathbf{r})], \qquad (12)$$

where $T_0[n(\mathbf{r})]$ is the kinetic energy of a system of noninteracting electrons with the same density, $n(\mathbf{r})$ and $E_{xc}[n(\mathbf{r})]$ is the exchange correlation energy of the interacting system. The true ground-state density is that $n(\mathbf{r})$, which minimizes $E[n(\mathbf{r})]$; this minimum being the true ground-state energy.

This variational principle leads to the following set of self-consistent equations:

$$\{-\nabla^2 + V_{eff}[n(\mathbf{r}), r]\}\psi_i(\mathbf{r}) = \varepsilon_i \psi_i(\mathbf{r}), \qquad (13)$$

where
$$n(\mathbf{r}) = \sum_i^{occ} |\psi_i(\mathbf{r})|^2 \qquad (14)$$

and
$$V_{eff}[n(\mathbf{r}), r] = \Phi(\mathbf{r}) + V_{xc}[n(\mathbf{r})]. \qquad (15)$$

The summation in Eq. (14) is carried over all occupied states of the electrons. The electrostatic potential $\Phi(\mathbf{r})$ is related to the ground-state density, and external perturbation through the Poisson's equation

$$\nabla^2 \Phi = -8\pi[n_{ext}(\mathbf{r}) - n(\mathbf{r})]. \qquad (16)$$

The exchange-correlation potential $V_{xc}[n(\mathbf{r})]$ is given by

$$V_{xc}[n(\mathbf{r})] = \partial E_{xc}[n(\mathbf{r})]/\partial n(\mathbf{r}). \qquad (17)$$

In most applications, the local density approximation for $E_{xc}[n(\mathbf{r})]$ is used, i.e.,

$$E_{xc}[n(\mathbf{r})] \cong \int d\mathbf{r}\, n(\mathbf{r})\varepsilon_{xc}[n(\mathbf{r})], \qquad (18)$$

where $\varepsilon_{xc}(n)$ is the exchange correlation energy per particle for a homogeneous electron gas of density $n(\mathbf{r})$. Several formulations of $\varepsilon_{xc}(n)$ are available in the literature. We quote only one of those forms, due to Vashishta and Singwi (1972), for the exchange-correlation energy per particle ε_{xc},

$$\varepsilon_{xc}(r_s) = -\frac{0.9163}{r_s} - 0.112 + 0.0335 \ln r_s - \frac{0.02}{0.1 + r_s}. \qquad (19)$$

In this approximation, we then have

$$V_{xc}[n(\mathbf{r})] = \frac{d}{dn}[n\varepsilon_{xc}(n)] - \mu_{xc}(n_0), \qquad (20)$$

where $V_{xc}[n]$ has been defined with respect to the exchange correlation potential μ_{xc} for the average density n_0 of the host metal. As $r \rightarrow \infty$, both $\phi(\mathbf{r})$ and $V_{xc}(n)$ tend to zero. The local density approximation is certainly valid for systems where the particle density varies slowly over a typical electron wavelength. Higher-order gradient corrections (Rasolt and Geldart, 1975) to $V_{xc}(n)$ for systems where the density variation is large, e.g., surface profile and electron pileup around a singular point charge (proton), are found to be usually small (Gupta and Singwi, 1977; Jena and Singwi, 1978a; Perdew *et al.*, 1977; Gunnarson, 1978). Thus, for most practical purposes, it is sufficient to solve Eqs. (13)–(16) self-consistently to determine the inhomogeneous particle density $n(\mathbf{r})$.

To treat the problem of a point defect in a ferromagnetic background in the above formalism, one has to introduce a spin component to the exchange-correlation potential, conduction-electron wave function, and particle density. The generalizations of the HKS formalism to a spin-polarized system are available in the literature (von Barth and Hedin, 1972; Rajagopal and Callaway, 1973) and the reader is referred to these papers for details. In essence one has to solve two coupled spin-dependent Schrödinger-type equations (13) to compute the spin density, $n\uparrow(r)-n\downarrow(r)$.

Like the previous two methods, the density functional formalism has been used to study the electronic structure of point defects based upon the jellium model. Lang and Kohn (1970) have introduced the effect of the underlying discrete lattice, assuming it is weak, on the electron distribution through first-order perturbation theory. Perdew and Monnier (1976) have proposed a variational generalization of the jellium-based density functional method to incorporate the effect of lattice ions. Almbladh and von Barth (1976), on the other hand, have proposed a spherical solid model where the potentials from the sites surrounding the defect are approximated by pseudopotentials. Retaining only the spherical average of these potentials, separate self-consistent calculations for the ground state are made. In Sections III and IV, we shall compare the results obtained using these models.

D. Molecular Cluster Approach

In this model, an imperfect crystalline solid is mimicked by a molecular cluster consisting of the impurity and a small number of host atoms. It is assumed that the electronic properties of the point defect are dictated mainly by its local environment. Such an assumption, it is hoped, may not be entirely unreasonable since, in a metallic environment, the potentials associated with both host and impurity ions are short ranged due to an efficient screening of the ionic charge. Consequently, a molecular cluster model may provide meaningful results for the electronic structure of impurities in non–free-electron–like systems.

There are several different procedures (Ellis and Painter, 1970; Averill and Ellis, 1973; Myles and Dow, 1979; Johnson et al., 1979) available in the literature to carry out the molecular cluster calculations. These methods are based on some variations of the Hartree–Fock scheme (Slater, 1974). In the following, we outline only one such method (Averill and Ellis, 1973; Rosén and Ellis, 1976; Ellis et al., 1978) based upon a general variational principle for calculating energy bands and charge densities in a solid.

The basic problem is finding the eigenvalues and eigenfunctions of a one-electron Hamiltonian,

$$H(r) = -\nabla^2 + V_{\text{Coul}}(r) + V_{\text{ex}}(r), \tag{21}$$

where the first two terms are the kinetic and Coulomb potential. The exchange potential is taken in the usual $x\alpha$ form,

$$V_{\text{ex}}(r) = -6\alpha[3n(r)/8\pi]^{1/3}. \tag{22}$$

$\alpha = 1$ corresponds to the well-known Slater $\rho^{1/3}$ exchange potential, whereas $\alpha = \frac{2}{3}$ in the HKS theory (Hohenberg and Kohn, 1964; Kohn and Sham, 1965). There exists more elaborate local density exchange and correlation potentials (Wang and Callaway, 1977), which are found to lead to small differences in self-consistent energy levels and charge densities for transition metals. In the discrete variational method (Ellis and Painter, 1970), one seeks approximate solutions expressible as

$$\psi_n(\mathbf{r}) = \sum_j \chi_j(\mathbf{r}) C_{jn}, \tag{23}$$

where the functions $\chi_j(\mathbf{r})$ are symmetrized linear combinations of single-site orbitals, i.e.,

$$\chi_j(\mathbf{r}) = \sum_{\nu,m,l} a^{jl}_{\nu m} u_{nl}(r_\nu) Y_{lm}(r_\nu). \tag{24}$$

$u_{nl}(r_\nu) Y_{lm}(r_\nu)$ is the single-site orbital centered on the νth nucleus. The coefficients $a^{jl}_{\nu m}$ are the symmetrization coefficients obtained by group theory with j-enumerating basis functions for irreducible representations of the molecular point group, and r_ν is the distance from the νth nucleus to the point r.

Application of a linear-discrete variational procedure leads to the secular matrix equation

$$HC = ESC, \tag{25}$$

where the matrix elements of H and S are

$$H_{ij} = \langle \chi_i | H | \chi_j \rangle, \qquad S_{ij} = \langle \chi_i | \chi_j \rangle. \tag{26}$$

The variational coefficients $\{C_{jn}\}$ are obtained by solving the secular equation (25). The single-particle equation

$$H\psi_i = E_i\psi_i \tag{27}$$

is solved approximately by minimizing certain error moments on a sampling grid in $\mathbf{r}$. The electron charge density $n(\mathbf{r})$ is computed by summing over all occupied states $|\psi_i(\mathbf{r})|^2$, where the different states are filled according to the prescription of the Mulliken population analysis (Mulliken, 1955a, b).

To facilitate comparison with band structure results, it is possible to calculate the partial density of states (PDOS). The cluster PDOS is found as a sum of Lorentzian lines of width γ centered at the molecular orbital energies,

$$D_n(E) = \sum_i f_{nj}(\pi/\gamma)/\left[(E - E_j)^2 + \gamma^2\right]. \tag{28}$$

The choice of the width γ should be consistent with the discrete level structure of the cluster and should reflect the uncertainty in cluster levels due to basis set limitations; f_{nj} are taken to be atomic populations obtained from a Mulliken population analysis. The advantage of the above cluster approach is that it can be used in association with any reasonable set of basis functions.

An obvious limitation of the cluster approach for the study of point defects is that one is often restricted to use only small clusters because of large computation time involved. As a result, cluster calculations seldom exhibit the Friedel oscillations of electron density at large distances from the impurity. Second, the matching of the boundary conditions at the cluster surface with the perfect solid is not always unambiguous.

E. Band Structure Calculations

As noted earlier, the conventional band structure approach for studying the electronic structure of point defects is not very useful due to the loss of periodicity of the lattice. However, several successful attempts (Freeman, 1981; Gupta and Siegel, 1977; Zunger and Freeman, 1977; Louie et al., 1976) have recently been made to circumvent this problem in metals, ionic solids, and semiconductors. The technique known as "supercell" method is based on a bulk-crystal calculation with periodically repeated large unit cell containing one defect per cell. The calculation, of course, can be repeated to achieve self-consistency in energy bands, potential, and charge density. In principle this method can exactly reproduce the properties of a point defect if the radius of the supercell is infinitely large. In practice, however, this is not possible even on the most advanced computers. Consequently, one has to settle for a finite supercell size that corresponds to a rather significant

impurity concentration. One, then hopes that the impurity–impurity interactions are weak so that the artificially introduced large impurity concentration in the calculation does not influence the theoretical predictions. The
small size of a supercell, just as in "small" cluster calculations, does not
reproduce the well-known Friedel oscillations.

Other approaches (Baraff and Schlüter, 1979; Kane and Schlüter, 1979;
Zeller and Dederichs, 1979; Katayama et al., 1979a, b; Terakura, 1976a, b;
Dupree, 1961; Beeby, 1967; Johnson, 1968; Harris, 1970; Lasseter and
Soven, 1973; Coleridge et al., 1974) based on Green's function technique
have been used in the study of point defects. These methods rely on the
availability of accurate first principles calculation of the electronic structure
of the perfect host. The advantage in using the Green's function method is
that one needs to consider the problem only in that region of space for
which the impurity potential exists. Since such a region is typically much
smaller than the region over which the wave functions extend, the above
scheme becomes computationally convenient. In the following we shall
provide a comparative analysis of the results obtained in the various models
discussed in this section.

III. Hydrogenlike Impurities in Metals

In this section we discuss the electronic structure of light impurities such
as hydrogen and its isotopes that occupy interstitial sites in metals. Hydrogen upon entering into a metal is believed to dissociate into a proton and an
electron. A proton with no conventional core electrons is the simplest kind
of impurity that can be implanted into a solid. However, it is the absence of
core electrons that causes the proton to interact rather strongly with the host
electrons (the electron–proton potential has a Coulomb singularity at the
origin). As a result, the properties of the host electrons are strongly
influenced by the presence of hydrogen. In addition, there are other reasons
for which the study of metal–hydrogen systems is so very interesting. For
example,

(i) The diffusivity of hydrogen in metals is extremely high, 15–20
orders of magnitude higher than that of oxygen or nitrogen. At low
temperatures hydrogen is known to exhibit quantum tunneling (Flynn and
Stoneham, 1970; Sussmann, 1971; Stoneham, 1972; Kagan and Klinger,
1974). Interestingly, the electron response time is several orders of magnitude higher than the hydrogen jump time. Consequently, the electrons
always "see" a static proton in spite of the large diffusivity of hydrogen.

(ii) No other systems are better suited to study the isotope effect on the
electronic structure than the hydrogen–metal systems since the lightest

isotope of hydrogen, the positive muon has $\frac{1}{9}$ of the proton mass and the heaviest isotope, tritium is three times heavier than the proton. The different amplitude of quantum vibration of these particles gives rise to a variety of interesting phenomena that can be studied.

(iii)   The onset of superconductivity (Satterthwaite and Toepke, 1970; Skoskiewicz, 1972) on hydrogenation of metals, the reverse isotope effect (Stritzker and Buckel, 1972) on the superconducting transition temperature, and the technological importance of hydrogen embrittlement (Kolachev, 1968) and hydrogen storage (Reilly, 1977) have resulted in considerable research activity (Alefeld and Völkl, 1978; Mueller *et al.*, 1968; Lewis, 1967; Westlake *et al.*, 1978) in recent years. In the following we discuss some of the interesting properties of metal–hydrogen systems.

A.   *Electron-charge Density Distribution around a Proton in Metals*

There have been, in the past, mainly four schools of thought to characterize the state of hydrogen in metals. These are

(i)   the anionic model in which a hydrogen atom is assumed to accept an electron from the metal atom; this is exemplified by the alkali hydrides such as LiH;

(ii)   the covalent model which assumes the metal–hydrogen bond as covalent;

(iii)   the screened proton model in which the hydrogen enters the metal as a proton and contributes its electron to the unfilled metal bands according to a rigid-band model;

(iv)   the atomic model in which the hydrogen is dissolved as neutral hydrogen atoms.

To see if any of these simple pictures are valid, it is necessary to obtain accurate information regarding the redistribution of electrons around hydrogen.

We shall first discuss the electron distribution around a proton in a paramagnetic environment. In the absence of any applied magnetic field the electrons of either spin are equally perturbed. Thus one is only interested in the electron-charge density distribution. In Fig. 2 we compare the electron distribution around a proton in Al obtained in the jellium model using linear response theory, Thomas–Fermi model and density functional formalism (see preceding section). Note that the linear response theory underestimates the strength of the electron–proton interaction. Consequently, it gives rise to a much smaller pileup of electrons on the proton. The nonlinear response of the electrons to the effective electron–proton interaction has

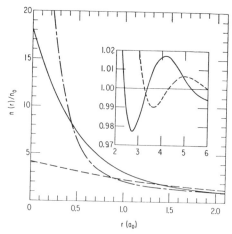

Fig. 2. Normalized electron charge density distribution $n(r)/n_0$ around a proton in Al in the jellium model. The solid, dashed, and dot-dashed curves correspond, respectively, to self-consistent density functional, linear response theory, and Thomas–Fermi results. The inset exhibits the Friedel oscillations on a magnified scale.

been realized recently by many authors (Friedel, 1952; Bhattacharya and Singwi, 1972a; Popović *et al.*, 1976; Almbladh *et al.*, 1976; Zaremba *et al.*, 1977; Jena and Singwi, 1978a; Jena, 1979). There also exists substantial difference in the amplitude and phase of the Friedel oscillations in the charge density at large distances from the proton. The Thomas–Fermi model, on the other hand, predicts an infinitely large charge pileup at the proton site. This is a reflection of the singularity at the proton site in the Thomas–Fermi potential. Furthermore, it does not exhibit charge density oscillation—an essential feature in the screening of impurities in metals. Even the generalized nonlinear response theory for a two-component plasma (Gupta, *et al.*, 1978) provides a poor description of electron density around the proton. It is recognized that it is important to treat the kinetic energy of the noninteracting electrons properly for a system where the impurity interacts rather strongly with its environment. Thus response theories or statistical methods cannot provide reliable results for metal–hydrogen systems.

The electron distribution around a proton embedded at the octahedral interstitial site in Pd metal obtained from the pseudojellium model is compared with the molecular cluster (Pd_6H) result in Fig. 3. The electron densities at the proton site in these two calculations (Jena *et al.*, 1979) differ from each other by about 17%, while the discrepancy gets narrower as one goes farther away from the proton. The charge distribution remains isotropic within a sphere of one Bohr radius around the proton. This result along with the agreement between pseudojellium and molecular cluster models may, at first, be surprising. An analysis of the different angular momentum components of the charge density reveals that the s electrons play the dominant role in screening the proton. This result is consistent with the angular momentum–resolved partial density of states inside the hydrogen sphere (see below) as well as with the predominant s-wave scattering from the hydrogen determined from de Haas–van Alphen experiments (Wampler and Lengeler, 1977) in copper containing dilute amounts of hydrogen. It is interesting that the charge density at the proton site in the pseudojellium model is higher than that obtained in the molecular cluster calculation. This result is consistent with one's physical intuition that in the molecular cluster model, a fraction of the electrons around hydrogen will be pulled away to screen the Pd atoms as well. We shall discuss this point further in this section.

In Fig. 4 we compare the effect of the background electron density on proton screening. As the electron density of the host metal decreases (i.e., r_s increases), the perturbation caused by the proton on the electron gas increases sharply due to a less-efficient screening of the point charge. This results in electron density profiles $n(r)/n_0$ that correspond to larger pileup of charge at the proton site with decreasing background density. The different amplitudes and phases of the charge density oscillations also exhibit the r_s dependence. However, the radius within which the proton

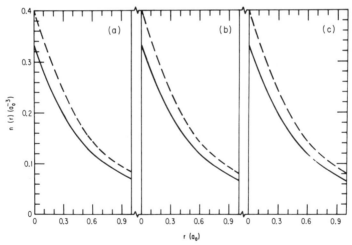

Fig. 3. Comparison between the electron charge densities along (a) (100), (b) (110), and (c) (111) direction around an octahedrally located hydrogen atom calculated self-consistently in the molecular cluster (solid curve) and pseudojellium (dashed curve) models. (From Jena *et al.*, 1979.)

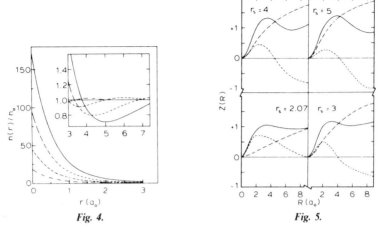

Fig. 4.   **Fig. 5.**

Fig. 4. Normalized electron density distribution $n(r)/n_0$ around a proton in electron gas. The curves $\cdots\cdots$, $---$, $-\cdot-\cdot-$, and —— correspond, respectively, to background electron density parameter $r_s = 2.07$, 3, 4, and 5. The inset shows Friedel oscillations on a magnified scale. (From Jena and Singwi, 1978a.)

Fig. 5. Screening charge within a sphere of radius R around a proton due to bound ($-\cdots-$) and scattering ($----$) states. The solid line represents the spatial dependence of the total screening charge. (From Jena and Singwi, 1978a.)

charge is almost entirely screened is not found to depend on the background density r_s. This is studied by computing the integral

$$Z(R) = \int_0^R d^3r \, \delta n(r). \tag{29}$$

Here $Z(R)$ corresponds to the number of electrons contained in a sphere of radius R around the proton. It is evident that $Z(\infty) = 1$ to satisfy charge neutrality. In Fig. 5 we have plotted $Z(R)$ for four different background densities (only note the solid curves. We shall discuss the other curves in Fig. 5 later in this subsection). Note that $Z(R) = 1$ for $R \sim 2a_0$ for all these metallic densities, i.e., the screening radius of the proton is relatively independent of its environment. This is simply because the potential energy is dominated by the Coulomb attraction between electrons and a proton which is independent of r_s. Infinitely large pileup of the electrons on the proton is avoided due to accompanying high kinetic energy costs. Because of the dominant role of Coulomb interaction, the electronic structure around a proton in metals does not differ substantially from that in hydrogen atom.

B. Electronic Structure of Hydrogen in Metals

In order to gain a deeper understanding of the electronic environment of a proton in a metal in terms of conventional simple models, we recall that because of the strong effective electron–proton potential, electron bound states appear in the solution of Kohn–Sham equation (13) throughout the metallic density range ($2 \lesssim r_s < 6$). The bound states that contain two electrons are rather weak even for low-density metals ($r_s \sim 6$, i.e., Cs). The wave functions extend to distances several times the lattice constant. This is clear from Fig. 5 where the bound-state contribution to electron number

$$Z_b(R) = \int_0^R d^3r \, \delta n_b(r) \tag{30}$$

is plotted. Again, for $R \to \infty$, $Z_b(R) = 2$. Naturally the scattering state contribution

$$Z_{sc}(R) = \int_0^R d^3r \, \delta n_{sc}(r) \tag{31}$$

is such that $Z_{sc}(\infty) = -1$. Thus, one may be tempted to describe the electronic structure of hydrogen in metals as an extended H^- ion with an equally extended hole in the continuum. However, such a picture is not valid for the following reasons. The single-particle eigenvalues (viz., bound states) have no fundamental meaning in the Kohn–Sham theory (even though the whole of band theory based on HKS formalism rests on their interpretation). The calculated lifetime broadening (Hedin, 1965) of these states due to electron–electron interaction is large compared to their binding energies making the physical significance of the latter questionable. Finally, the interaction of the lattice ions with the extended wave function would make the bound states unstable.

An alternative way (Jena et al., 1979) of viewing this problem is to compare the calculated $Z(R)$ for a proton in a metal with that of a free hydrogen atom. The result for $r_s = 2.7$ (corresponding to hydrogen tetrahedrally bonded to Zr atoms) is shown in Fig. 6. Of course, in both cases $Z(\infty) = 1$. It is interesting to note that for distances as large as $3a_0$ from the proton, the number of electrons around the proton in a metal is larger than that of a free hydrogen atom. The result of this excess screening (a common feature at all metallic densities) makes hydrogen in metals appear to be in a "slightly anionic" form. This has to result from a charge transfer from metal ion to hydrogen.

In order to substantiate this point, we discuss the result of a cluster calculation (Jena et al., 1979) involving 6 Pd atoms located on the face

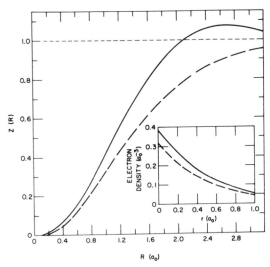

Fig. 6. The number of electrons contained in a sphere of radius R around a proton in a jellium of $r_s = 2.7$ (corresponding to hydrogen tetrahedrally bonded to Zr atoms) (solid curve) vs. that around a proton in free-hydrogen atom (dashed curve). $Z(R) = 4\pi \int_0^R r^2 \, \delta n(r) \, dr$. (From Jena *et al.*, 1979.)

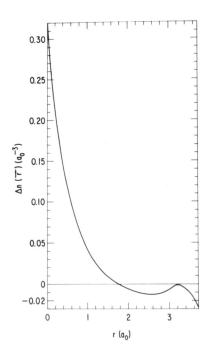

Fig. 7. The difference in the electron charge density in Pd_6H and Pd_6 clusters. The negative region indicates the zone from which metal charge has been transferred due to hydrogenation. $\Delta n(\mathbf{r}) = n_{Pd_6H}(\mathbf{r}) - n_{Pd_6}(\mathbf{r})$. (From Jena *et al.*, 1979.)

center of a Pd matrix and 6 Pd atoms with one body-centered hydrogen atom. The difference in the electron density around the octahedral site between Pd_6H and Pd_6 cluster

$$\Delta n(\mathbf{r}) = n_{Pd_6H}(\mathbf{r}) - n_{Pd_6}(\mathbf{r}) \tag{32}$$

is shown in Fig. 7 along the (100) direction. The negative region of electron density for distances beyond $\sim 2a_0$ from the proton suggests that the charge from the vicinity of the metal ion has been transferred to the hydrogen sphere. Similar conclusions have also been reached recently using band structure calculations (Gupta and Freeman, 1978; Gelatt et al., 1978; Peterman et al., 1979). This viewpoint is supported by the following experiments.

1. X-Ray Photoelectron Spectroscopy

A direct consequence of the electron transfer from the vicinity of a metal ion is to reduce the strength of Coulomb repulsion between core and conduction electrons. Consequently, one would expect an increase in the core level binding energy. In Table I we compare the energies of the 4s, 4p, and 3d core levels of the Pd atom in the free-atom, Pd_6, and Pd_6H cluster configuration. In the Pd_6 cluster, some charge from each atom is donated to the conduction sea resulting in an increase of about 0.4 eV in ion core level binding. The addition of hydrogen accentuates this trend. The Pd–H bonding charge is being drawn from the vicinity of the metal ion core (Fig. 7) leaving core levels still more tightly bound.

One of the experimental techniques that is sensitive to core level energy shifts involves x-ray photoelectron spectroscopy (XPS) (Siegbahn et al., 1967). Recently, Veal et al. (1979) have measured the core level structure of Zr and $ZrH_{1.65}$. The shift of about 1 eV towards higher binding of the Zr 4p level in $ZrH_{1.65}$ compared to that in Zr has been attributed to the charge transfer from the metal atom to hydrogen. However, observed shifts are

TABLE I

COMPARISON OF PD CORE LEVEL ENERGIES (IN EV) RELATIVE TO FERMI ENERGY OF ATOM AND CLUSTER IN NONRELATIVISTIC SELF-CONSISTENT LOCAL DENSITY MODEL

Level	Atom	Pd_6	Pd_6H
4p	46.3	46.7	47.0
4s	75.5	75.9	76.2
3d	328.3	329.2	329.5

difficult to interpret quantitatively because of several other effects (Kowalczyk *et al.*, 1974; Williams and Lang, 1978; Watson *et al.*, 1976; Lang and Williams, 1977) that can influence the result. The reader is referred to the work of Veal *et al.* (1979) and Veal (1981) for a thorough discussion of this problem.

2. Mössbauer Studies

The sign of the charge on hydrogen in metals can also be deduced from Mössbauer experiments on rare-earth metal hydrides (Shenoy *et al.*, 1976; Suits *et al.*, 1977; Vicarro *et al.*, 1979) since the nature of the electronic ground state of a rare-earth ion in a hydride depends on the local symmetry of the ion and the crystalline electric field produced by hydrogen and rare-earth neighbors. Following the crystal field diagram of Lea *et al.* (1962), Shenoy and coworkers have measured the hyperfine splitting in an external magnetic field and have shown that the crystal field ground state is a Γ_6 Kramers doublet. From symmetry considerations this ground state corresponds to a net negative charge on hydrogen and thus supports the charge transfer picture.

3. Compton Scattering Studies

Another experimental technique well suited for the determination of the electronic structure of hydrogen in metals is the Compton scattering of high-energy photons (Williams, 1977). The Compton profiles $J(q)$, which measure the electron momentum distribution, are usually obtained for the same sample before and after hydrogen loading. The difference in the profiles $\Delta J(q)$ can then be analyzed to study the influence of hydrogen on the electronic distribution since the contribution of core electrons can be easily separated from that of the valence electrons. The latter are the only ones affected by the presence of hydrogen.

The difference profiles

$$\Delta J(q) = J_{\text{hydride}}(q) - J_{\text{host}}(q) \tag{33}$$

are plotted in Fig. 8 for crystalline $PdH_{0.72}$. The experimental profiles indicated by asterisks are compared with the anionic, atomic, and protonic models (Lässer and Lengeler, 1978). The apparent disagreement between simple theoretical models and experiment is a clear indication that none of these simple pictures can adequately describe the electronic structure of hydrogen in metals.

A proper understanding of the difference Compton profiles requires an analysis of the wave functions obtained from more sophisticated calcula-

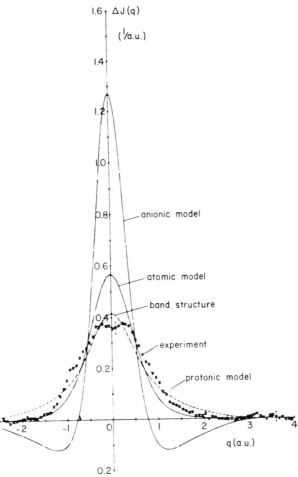

Fig. 8. Difference Compton profiles for polycrystalline $PdH_{0.72}$ and Pd compared with the profiles for four different model wave functions describing hydrogen in the metal. The model profiles are convoluted with the energy-resolution function and the experimental data are deconvoluted. (From Lässer and Lengeler, 1978.)

tions described in the previous section. The recent band structure (Gupta and Freeman, 1978; Gelatt *et al.*, 1978; Peterman *et al.*, 1979; Switendick, 1972; Papaconstantopoulos and Klein, 1975; Klein *et al.*, 1977; Papaconstantopoulos *et al.*, 1978; Faulkner, 1976) as well as cluster (Jena *et al.*, 1979; Adachi *et al.*, 1978; Simpson *et al.*, 1978) calculations of metal hydrides show that a hydrogen–metal bonding state appears several volts below the Fermi level. A typical partial density of states plot (obtained in cluster approach) centered about Pd and hydrogen spheres that exhibits the

bonding level is shown in Fig. 9. This plot is very similar to the partial
density of states obtained from APW calculations of PdH (Gupta and
Freeman, 1978) and thus testifies to the importance of local environment.
These bonding states have been observed in several photoemission experiments (Eastman *et al.*, 1971; Antonageli, 1975; Fukai *et al.*, 1976). In PdH,
this state can accommodate 0.5 electrons per hydrogen atom. An analysis of
the angular momentum–resolved partial density of states reveal that the

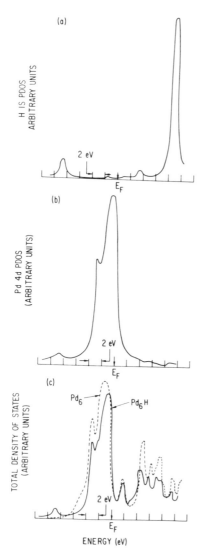

Fig. 9. Partial density of states (PDOS) in arbitrary units for (a) hydrogen 1s, (b) Pd 4d states, and (c) total density of states for Pd_6H (solid line) and Pd_6 (dotted line) clusters. (From Jena *et al.*, 1979.)

bonding state is primarily composed of s electrons. A similar result is also obtained from an analysis of the different angular momentum components of the charge density based on the pseudojellium model (Jena *et al.*, 1979). Taking these features into account, Lässer and Lengler have computed $\Delta J(q)$. The profile termed as "band structure" in Fig. 8 compares very well with the experimental data.

Thus, the band structure as well as cluster models give adequate description of the electronic structure of hydrogen. These models predict a charge transfer from metal ion to hydrogen and the formation of a metal hydrogen bond. The electrons around the proton have predominantly s character, which is consistent with the predominant s-wave scattering from hydrogen determined from the de Haas–van Alphen experiments (Wampler and Lengeler, 1977; Lengeler, 1977) in copper containing dilute amounts of hydrogen.

C. Heat of Solution of Hydrogen

Heat of solution is a basic quantity of interest in the study of metal–hydrogen systems. It is equal to the change in energy of a hydrogen atom dissolved in the metal compared with the energy per hydrogen atom in a hydrogen molecule. Experimentally the heat of solution ΔH is obtained as the slope of the logarithmic plot of the solubility versus $1/T$ at constant hydrogen pressure. It can be calculated from the expression (Popović *et al.*, 1976)

$$\Delta H = 15.86 + \Delta H_{\text{el-p}} \quad \text{eV}, \tag{34}$$

where $\Delta H_{\text{el-p}}$ is the heat of solution for a free electron and proton from vacuum. The first term in the above equation is equal to the sum of the ionization energy of a hydrogen atom (13.6 eV) and the dissociation energy of a hydrogen molecule (2.26 eV). For simple metals, the electron contribution to ΔH can be calculated with reasonable confidence using pseudopotential theory, whereas the contribution of the proton has to be corrected using the results of the nonlinear screening of the proton previously described.

The total energy, E^{T} of N pseudoions immersed in an electron gas calculated to second order in perturbation theory can be given by (Harrison, 1966)

$$E^{\text{T}} = N[Z_V V_{\text{el}} + V_{\text{e}} + V_{\text{b}}],$$

where V_{el} is the energy per electron for the uniform, interacting electron gas plus the average value of the electron–ion interaction. This term will be modified owing to the introduction of an additional electron in the hydrogenation process. V_e is the electrostatic ion–ion interaction energy and can be evaluated by Ewald's method for any given configuration of lattice ions. This term will change because of the addition of a proton and its evaluation assumes a particular equilibrium configuration for the proton. The last term, V_b is the band structure energy, which changes owing to the introduction of the proton and the additional electron.

Adding up all the changes in the energy, the expression for the heat of solution of ionized hydrogen is (Popović *et al.*, 1976)

$$\Delta H_{\text{el-p}} = \Delta H_1 + \Delta H_2, \tag{35}$$

where ΔH_1 depends only on the properties of the perfect crystal and ΔH_2 depends on the nonlinear response of the electrons to the presence of the proton and is given by

$$\Delta H_2 = \sum_{q \neq 0} S(\mathbf{q}) W(\mathbf{q}) \, \delta n(\mathbf{q}) + E_{\text{corr}}, \tag{36}$$

where the Fourier transform of the displaced charge $\delta n(\mathbf{r})$ around the proton is

$$\delta n(\mathbf{q}) = \int_0^\infty dr \, 4\pi r^2 \, \delta n(\mathbf{r}) \left[\sin(qr)/qr \right]. \tag{37}$$

$S(\mathbf{q})$ is the lattice structure factor and $W(\mathbf{q})$ is the bare-ion pseudopotential of the host lattice. The first term in Eq. (36) can be viewed as the interaction energy between the electron screening charge around the proton and the lattice ions. Thus the magnitude and phase of the Friedel oscillations in the displaced charge density play an important role in the determination of this contribution. The second term in Eq. (36) is the electron–proton correlation energy and can be calculated using the Feynman–Hellman theorem

$$E_{\text{corr}} = \int_0^1 \frac{dZ}{Z} V_{\text{int}}(Z), \tag{38}$$

with $V_{\text{int}}(Z)$ as the interaction energy between the electron and a fictitious point charge Ze (the value of which lies between 0 and 1),

$$V_{\text{int}}(Z) = -2 \int d^3 r (Z/r) \, \delta n(Z, r). \tag{39}$$

$\delta n(Z, r)$ is the displaced nonlinear charge around a heavy point impurity carrying a charge Ze.

Using various models for the displaced charge density around the hydrogen atom, Popović *et al.* (1976) and Jena and Singwi (1978b) have calculated the heat of solution of hydrogen in Al and Mg. Although their results are in fair agreement with experiment (Eichenauer, 1968), it should be pointed out that the heat of solution is very sensitive to the electron-charge distribution around the proton. For example, the heats of solution computed by taking the charge densities of the last two successive iterations of a self-consistent calculation differ in the second place of decimal. The difference of 0.28 and 0.08 eV in E_{corr} for Al and Mg, respectively, between the values of Zaremba et al. (1977) (who performed a fully self-consistent calculation) and Popović et al. (1976) [who calculated $\delta n(r)$ approximately self-consistently] further testifies to the sensitiveness of ΔH on calculated $\Delta n(r)$. (Both authors used the exchange-correlation energy given by Hedin and Lundquist.) The results are also sensitive to the approximations in the exchange-correlation energy functional. For example, E_{corr} calculated self-consistently by Jena and

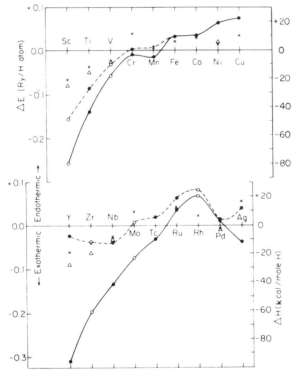

Fig. 10. Calculated heats of formation per hydrogen atom for stoichiometric hydrides without (solid line) and with (dashed line) Coulomb energy corrections. Gelatt et al., (1978). The experimental point represent enthalpies of formation of dilute (after Mueller et al., 1970) and nondilute (after Mueller et al., 1968) 3d and 4d hydrides. ●—○, theory; ●---○, Coulomb corrected theory; ×, experiment (α phase); △, experiment (hydrides).

Singwi (1978b) (who used Vashishta–Singwi form) and Zaremba *et al.* (1976) (who used Hedin–Lundquist form) differ from each other by 0.09 eV for Al and 0.17 eV for Mg. (The experimental heat of solution of hydrogen in Al and Mg are 0.65 and 0.25, respectively.) Similarly, the inclusion of first gradient correction in the exchange-correlation energy functional yields heats of solution that differs significantly from that in the local density approximation (Jena and Singwi, 1978b).

The calculation of heat of formation of hydrogen in transition-metal hydrides is an even more difficult problem since in this case pseudopotential theories for the perfect host are inadequate. Gelatt *et al.* (1978) have computed complex energy bands and heats of formation for 3d and 4d metal hydrides using an extension of the average *t*-matrix formulation (Ehrenreich and Swartz, 1976) for interstitial alloys. Their results for the heats of formation of the stoichiometric 3d and 4d monohydrides are shown in Fig. 10. The gross features of the heat of formation curve reflect the general experimental situation; namely, that with the exception of Pd and Ni, stable concentrated hydrides form only on the left-hand side of the period.

In order to assess the accuracy of the calculated heat of solution ΔH, it should be pointed out that ΔH is not expected to depend on the change in volume of the sample. It also does not depend on the state of the metal surface, since the hydrogen atom to be dissolved is electrically neutral and does not react with the surface dipole (created by spilling out of electrons from the surface of a metal) as it passes through the surface. Quantitative calculations of the heat of solution should take into account the lattice relaxation around hydrogen.

D. Nuclear Quadrupole Interaction

The interaction between the quadrupole moment of the nucleus and electric field gradient caused by the ions and conduction electrons of a noncubic crystal provide useful insight into the nature of electron-charge density distribution. In cubic systems, this interaction is zero since the electric field gradient in a perfect cubic crystal vanishes due to symmetry considerations (Slichter, 1963; Abragam, 1961). When impurities are introduced into cubic metals, this cubic symmetry around a host metal atom is, however, broken and there is a finite distribution of electric field gradients at near-neighbor host nuclear sites. This results from the screening of the impurity charge by host conduction electrons. Blandin and Friedel (1960) and Kohn and Vosko (1960) have formulated a theory of the origin of electric field gradient in cubic metal alloys that successfully explained the rapid loss (Rowland, 1960) of the intensity of the nuclear resonance signal in Cu when small quantities of other elements were alloyed with it.

According to the above theory, the electric field gradient (efg) at *large* distances from the impurity is given by

$$q(\mathbf{r}) = (8\pi/3)\alpha(\mathbf{k}_F)\,\delta n(\mathbf{r}), \qquad (40)$$

where $\delta n(\mathbf{r})$ is the perturbation in the bulk electron density and $\alpha(\mathbf{k}_F)$ is the so-called Bloch enhancement factor

$$\alpha(\mathbf{k}_F) = \frac{\int d^3 r'\ \Psi_{\mathbf{k}_F}^2(\mathbf{r}')P_2(\cos\theta')/r'^3}{\int d^3 r'\ \exp(2i\mathbf{k}_F\cdot\mathbf{r}')P_2(\cos\theta')/r'^3}. \qquad (41)$$

Here θ' is the angle between $\mathbf{r}'$ and $\mathbf{k}_F$, and $\psi_{\mathbf{k}_F}(\mathbf{r}')$ is the wave function of the Bloch electron in the unperturbed lattice. Thus $\alpha(\mathbf{k}_F)$ is a property of the perfect host and $\delta n(\mathbf{r})$ depends on the impurity–host interacting potential. The calculation of $q(\mathbf{r})$ then requires an accurate knowledge of both $\delta n(\mathbf{r})$ and $\alpha(\mathbf{k}_F)$.

The normal practice (Kohn and Vosko, 1960) in the past has been to compute $\alpha(\mathbf{k}_F)$ by approximating $\psi_{\mathbf{k}_F}(\mathbf{r})$ by a single orthogonalized plane wave (OPW), thus ignoring any band structure effects. The perturbed charged density, $\delta n(\mathbf{r})$ is often calculated from the asymptotic expression:

$$\delta n(\mathbf{r}) = A\cos(2k_F r + \theta)/r^3, \qquad (42)$$

where the amplitude, A and phase θ are obtained (Kohn and Vosko, 1960) from scattering phase shifts. The expression in Eq. (42) is valid only at *large* distances from the impurity. In a typical experiment, electric field gradients at first four or five near neighbors are measured and the validity of Eq. (42) in that region is not certain. Attempts (Fukai and Watanabe, 1970; 1974; Jena *et al.*, 1978b) have been made to calculate $\delta n(\mathbf{r})$ in the linear response theory using pseudopotentials for host and impurity atoms. Not only is the linear response theory inadequate even for pseudopotentials, but $\delta n(\mathbf{r})$ is sensitive to the choice of pseudopotential (Jena *et al.*, 1978b). Since the choice of a pseudopotential is arbitrary to a certain extent, any agreement or disagreement between Eq. (40) and experiment does not throw much light on the limitations of the theory of Kohn and Vosko (1960) and Blandin and Friedel (1960).

The above ambiguity does not arise in the calculation of electric field gradient at near-neighbor sites due to a proton or a positive muon, since no pseudopotentials are involved in the calculation of the perturbed electron density $\delta n(\mathbf{r})$. Recently Camani *et al.* (1977) have measured the electric field gradient at the nearest-neighbor Cu nuclei due to an interstitial positive muon by studying the depolarization of muons in single crystals as a function of applied magnetic field and crystal orientation. The analysis is based on a modified version (Hartmann, 1977) of Van Vleck's (1948) theory of dipolar broadening of magnetic resonance lines. From the orientational dependence of the dipolar width due to a combined electric and magnetic interaction, Camani et al. (1977) find the nearest-neighbor field gradient to be 0.27 Å^{-3}.

Jena et al. (1978a) have studied this problem theoretically by computing $\delta n(\mathbf{r})$ self-consistently in the density functional formalism (Hohenberg and Kohn, 1964; Kohn and Sham, 1965) and $\alpha(\mathbf{k}_F)$ from a band structure calculation based on linear combination of atomic orbitals (LCAO). They have found that $\alpha(\mathbf{k}_F)$ in the LCAO calculation is not only strongly anisotropic (it is isotropic in the single-OPW calculation), but differs from the single-OPW estimate by as much as a factor of 4 along the first nearest-neighbor direction. With the LCAO value for $\alpha(\mathbf{k}_F)$ and self-consistent value of $\delta n(\mathbf{r}_1)$ at the first near-neighbor site (0.00488 Å^{-3}), Jena et al. (1978a) computed the efg due to μ^+ to be 0.26 Å^{-3}, in good agreement with experiment. To appreciate the sensitivity of $\delta n(\mathbf{r}_1)$ to model calculations, we point out that the model calculation of Meier (1975) and Thomas–Fermi model predict the magnitude of $\delta n(\mathbf{r}_1)$ to be 0.015 Å^{-3} and 0.008 Å^{-3}, respectively. While the electron density at the muon site calculated by Meier (1975) is in good agreement with the self-consistent result (Jena et al., 1978a), the discrepancy at large r is due to correlation effects neglected in Meier's calculation. Manninen and Nieminen (1979a) have calculated the efg due to μ^+ in Cu using spherical solid model and found about 20% change from the jellium result. They also found that efg at nearest-neighbor Cu nuclei does not depend sensitively on the site preference of the muon.

Thus based on the above calculations, it could be concluded that the Kohn–Vosko theory provides an adequate understanding of the nuclear quadrupole interaction in cubic metal alloys. However, this does not seem to be entirely true when one compares the efg due to μ^+ in Al between theory (Jena et al., 1978a; Manninen and Nieminen, 1979a) and experiment (Hartmann et al., 1978). The spherical solid model predicts the efg to be 0.07 and 0.04 Å^{-3} for octahedral and tetrahedral configurations of the muon, respectively. The corresponding numbers in the jellium model are 0.16 and -0.05 Å^{-3}. The experimental efg for tetrahedral configuration is 0.18 Å^{-3}. This apparent disagreement in Al suggests further evaluation of the theory. In all of the above calculations, the contributions to efg due to lattice strain and lattice relaxation around the impurities are ignored. Recently Sagalyn and Alexander (1977) have pointed out that such effects could be important and are necessary to explain the asymmetry in the electric field gradient tensor in cubic metal alloys. A more detailed study of this problem is certainly necessary.

E. Impurity Resistivity due to Hydrogen

The resistivity of a metal containing impurity atoms may be written in the form

$$\rho = \Delta\rho + \rho_0, \qquad (43)$$

where ρ_0 is the resistivity caused by thermal motion of the lattice, and $\Delta\rho$ is the resistivity caused by scattering of the electron waves by impurity atoms that disturb the periodicity of the lattice. For $T \to 0$, ρ_0 vanishes. Thus the residual resistance $\Delta\rho$ is equal to the extrapolated resistivity at 0 K. For a system of free electrons scattering off an impurity, $\Delta\rho$ is given by (Kittel, 1963)

$$\Delta\rho = \frac{4\pi n_i \hbar m}{n_0 e^2 k_F} \sum_{l=0}^{\infty} (l+1)\sin^2[\delta_l(E_F) - \delta_{l+1}(E_F)], \qquad (44)$$

where n_i is the impurity concentration and $\delta_l(E_F)$ is the scattering phase shift for the lth partial wave on the Fermi surface. An expression similar to Eq. (44) has been recently derived by Gupta and Benedek (1979) for the scattering of Bloch electrons.

Using the density functional formalism, Manninen and Jena (unpublished) have evaluated the scattering phase shifts $\delta_l(E_F)$ for hydrogen in the metallic density range for $0 \le l \le 7$ by matching the radial solution of the Schrödinger equation (13) to the asymptotic form. Using Eq. (44) they have calculated $\Delta\rho$ to be 3.4, 3.6, 4.2, 4.2, and 3.8 ($\mu\Omega$cm/at.%) for Cu, V, Nb, Ta, and Pd, respectively. The experimental result for these metals, however, lies in the range of 0.6–1.0 $\mu\Omega$cm/at.%. (Peterson and Jensen, 1978; MacLachlan et al., 1975; Verdini, 1976, Westlake and Ockers, 1975; Pryde and Tsong, 1971; Haywood and Verdini, 1968; Westlake, 1967). The poor agreement between theory and experiment may be attributed to (i) the inaccuracy in the determination of n_0 at the interstitial site, (ii) approximating the electrons involved in the scattering process by free, instead of Bloch, electrons, and (iii) neglecting the effect of lattice relaxation. To our knowledge no theoretical calculations of $\Delta\rho$ in metal hydrogen systems based on more sophisticated models are available at present. Such calculations are eagerly awaited.

F. Magnetic Interaction of Hydrogen with Electrons

In ferromagnetic metals or paramagnetic systems under the influence of an external magnetic field, the conduction electrons can be spin-polarized via exchange interaction. The dominant interaction between the spin-polarized electrons and the magnetic moment of the proton is given by the

Fermi contact Hamiltonian,

$$H_{en} = (8\pi/3)\gamma_e\gamma_n\hbar^2 \sum_l \mathbf{I}\cdot\mathbf{S}_l\delta(\mathbf{r}_l). \tag{45}$$

γ_e and γ_n are, respectively, the electron and nuclear (proton) gyromagnetic ratios, $\mathbf{I}$ is the spin of the proton that defines the origin of the coordinate system, and $\mathbf{S}_l$ is the spin of the ith electron with radius vector $\mathbf{r}_l$. The additional shift of the nuclear levels caused by the interaction in Eq. (45) is known as the Knight shift and is given (Slichter, 1963) by the equation

$$K = (8\pi/3)\chi_s\Omega\langle|\Psi_{\mathbf{k}_F}(0)|^2\rangle, \tag{46}$$

where χ_s is the Pauli paramagnetic susceptibility and Ω is the volume over which the conduction-electron wave functions $\Psi_k(\mathbf{r})$ are normalized. The notation $\langle\ \rangle$ implies that the wave-function density, $|\Psi_{\mathbf{k}_F}(0)|^2$ has to be averaged over the Fermi surface. The expression in Eq. (46) ignores the distortions in the wave functions induced by the magnetic field. A more complete theory of the Knight shift based on the Fermi contact interaction has been recently given by Munjal and Petzinger (1978). The modified expression for the Knight shift is

$$K = (8\pi/3)\chi_s\rho_s(0), \tag{47}$$

where $\rho_s(0)$ is the electron spin density at the nuclear site and is given by

$$\rho_s(0) = \frac{n^\uparrow(0) - n^\downarrow(0)}{n_0^\uparrow - n_0^\downarrow}. \tag{48}$$

$n^\sigma(0)$ is the density of electrons of spin σ at the proton site and $(n^\uparrow 0 - n_0^\downarrow)$ is the ambient spin-polarization density caused by the external magnetic field on the electron gas of the perfect host. In ferromagnetic metals the conduction electrons are spin-polarized through exchange interaction mediated by the ever-present internal field. The interaction of the impurity with the conduction electron spins can be measured through the hyperfine field

$$B_{hf} = -(8\pi/3)\mu_B[n^\uparrow(0) - n^\downarrow(0)]. \tag{49}$$

The sign convention used here is such that a spin-up ($\uparrow$) electron has a magnetic moment of $-\mu_B$ and hyperfine fields are positive when parallel to the magnetization in the positive z direction.

Another related quantity is the nuclear spin-lattice relaxation rate $1/T_1$, which proceeds via the Fermi contact interaction, and is given by

$$1/T_1T = 4\pi\hbar\gamma_n[N(E_F)B_{hf}]^2. \tag{50}$$

Here $N(E_F)$ is the density of states at the Fermi energy for one direction of spin. Thus Eqs. (47), (49), and (50) give a measure of the effect of an impurity on the electron spin distribution of the host.

1. ELECTRON SPIN DISTRIBUTION AROUND A PROTON

Analogous to our previous discussion on the electron charge distribution around a proton, we concentrate on the effect of a proton on the homogeneous distribution of the conduction-electron spins. The simpler models such as linear response theory and statistical approach fail to describe the electron spin response to the proton accurately (just as the charge distribution described earlier) and are no longer discussed. We shall present results based on the self-consistent density-functional approach, molecular cluster, and band structure models.

Recently several self-consistent calculations (Munjal and Petzinger, 1978; Petzinger and Munjal, 1977; Jena *et al.*, 1978c; Manninen and Nieminen, 1979a) of the electron spin density distribution around a fixed point charge (proton) based on the generalized density-functional formalism (von Barth and Hedin, 1972; Rajagopal and Callaway, 1973) have appeared in the

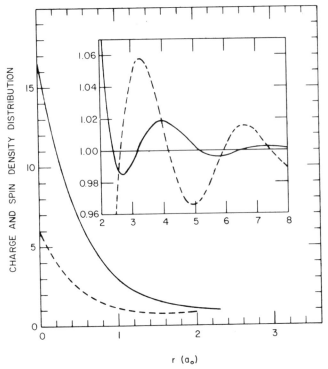

Fig. 11. Charge- and spin-density distribution around a proton in a spin-polarized electron gas with $r_s = 2$ and $\zeta_0 = (n_0^\uparrow - n_0^\downarrow)/n_0 = 0.17$. The solid and the dashed curves correspond, respectively, to normalized charge density $n(r)/n_0$ and normalized spin density $n(r)\zeta(r)/n_0\zeta_0$. (From Jena *et al.*, 1978c.)

literature using the exchange-correlation energy of Gunnarson *et al.* (1974) throughout the metallic density range ($1 < r_s < 6$). In Fig. 11 we have plotted typical results for the normalized charge and spin density distributions for $r_s = 2$. We note a strong buildup of charge at the impurity just as in the paramagnetic case [$\zeta_0 = (n_0^\uparrow - n_0^\downarrow)/n_0 = 0$]. As the bulk ζ_0 is increased from zero, the charge density at the origin $n(0)$ changes little from its paramagnetic value (less than 4% up to background polarization $\zeta_0 = 0.6$) with a drop of about 10–15% in the purely ferromagnetic case ($\zeta_0 = 1$). Thus the contact charge density is largely fixed by the Coulomb singularity in the external potential, and the main effect of increasing bulk polarization on the density profiles is to make them more spread out in accord with the increased screening length in a polarized electron gas (Gunnarson *et al.*, 1974).

At distances far from the proton, the spin densities, just like the charge densities, exhibit the Friedel oscillations. Note that not only the spin density at the origin in enhanced over the ambient polarization to a much lesser degree than the charge density, but also the Friedel oscillations differ in phase by about $\pi/2$. The amplitudes of the charge and spin oscillations increase with increasing r_s (decreasing ambient electron density). This results from a larger enhancement of spin and charge density at the proton site as r_s increases.

In Fig. 12 we have compared the electron-charge density enhancement at the proton site with that of the spin density enhancement. Note that the charge density enhancement is not only much larger than the spin density enhancement, but it increases more steadily with r_s than the spin density enhancement. The former arises from the fact that in a spin-polarized medium further pile up of electrons of a given spin costs additional kinetic energy and is, therefore, avoided. Increasingly less-efficient screening of the point charge gives rise to the later effect. It is because of this reason that the electron density at a proton site in spin-polarized electron gas of a given density is somewhat less than that in a paramagnetic electron gas of the *same* density.

In Fig. 12 we have also plotted (open circles) the quantity $V\langle|\Psi_{k_F}(0)|^2\rangle$, an analog of spin density [see Eqs. (46) and (47)] for various r_s. The deviation between $\rho_s(0)$ and $V\langle|\Psi_{k_F}(0)|^2\rangle$ increases with increasing r_s. Its origin lies in the relative importance of scattering and bound states that appear, as in the paramagnetic system, for all $r_s \gtrsim 1.9$. The minority state is always more tightly bound than the majority spin state that results in a negative bound-state contribution to the total spin density $\rho_s(0)$ near the origin. The scattering state density is always larger and positive. Increasing bound-state contribution at larger r_s is responsible for the above discrepancy that can be linked to familiar core-polarization effects (Gaspari *et al.*, 1964).

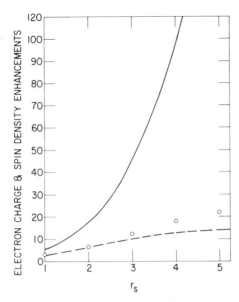

Fig. 12. Charge density enhancement $n(0)/n_0$ (solid line) and spin density enhancement $n(0)\zeta(0)/n_0\zeta_0$ at the proton site vs. electron density parameter r_s. The open circles correspond to spin density, $V\langle|\Psi_{k_F}(0)|^2\rangle$ calculated for paramagnetic electron gas.

It follows from the fact that even for a spatially uniform field, the exchange and correlation potentials for up ($\uparrow$) and down ($\downarrow$) spin directions are different and, more importantly, spatially nonuniform. A static field induces an inhomogeneous spin polarization $\zeta(r)$ in the electron gas. This phenomenon is present not only for impurities but also for pure metals, and its effect is expected to be significant for impurities with well-defined core-electronic structure.

2. Knight Shifts of Protons and Positive Muons

Using the calculated exchange-enhanced Pauli spin susceptibility χ_s and the spin density discussed above, the Knight shift at a proton site can be computed via Eqs. (46) and (47). The results can be compared with experimental data obtained from proton nuclear magnetic resonance. For the purpose of comparison, one can also use the information obtained from muon spin rotation (μSR) experiments. This is a relatively new technique (Schenck, 1976; Brewer *et al.*, 1975; Seeger, 1978) and has been very successful in furthering our understanding of the behavior of hydrogen in metals. The positive muon μ^+, like a proton, carries one unit of positive charge and has $\frac{1}{9}$ of the proton mass. Since μ^+ is still about 200 times heavier than the electron, to a first approximation it can be regarded, like a proton, as a static point charge. Within this approximation the μ^+ and proton are identical particles. (We shall discuss the effect of their masses on the electronic structure later in this section.) The positive muon has a lifetime of 2.2 μsec and decays into a positron (e^+), a neutrino (ν), and antineutrino ($\bar{\nu}$).

In a typical μSR experiment a beam of nearly 100% polarized muons is directed into the sample under study. One measures an oscillatory time-dependent positron-counting rate as the muon decays while precessing around an applied magnetic field transverse to the muon polarization. This time histogram contains two major pieces of information. First, the Larmor precession frequency ω_μ of the μ^+ is related to the internal field at the μ^+ site through the relation $\omega_\mu = \gamma_\mu B_\mu$, where γ_μ is the muon gyromagnetic ratio. The Knight shift is obtained from a careful measurement of the magnetic field B_μ at the implanted muon and the external field B_{ext}:

$$B_\mu = \left\{ 1 + \left[(4\pi/3) - N \right] \chi_t + K \right\} B_{\text{ext}}. \qquad (51)$$

N is the demagnetization factor and χ_t is the total magnetic susceptibility of the target material. For spherical targets, $N = 4\pi/3$.

In Fig. 13 we compare the most recent experimental (Camani et al., 1979) results of muon Knight shift with various theoretical calculations (Jena, 1979; Munjal and Petzinger, 1978; Manninen and Nieminen, 1979a; Keller and Schenk, 1979). Note that self-consistent calculations based on the jellium model using Eq. (47) (curve 2) provide a consistently better understanding of the nature of Knight shift variation than that obtained using Eq. (46) (curve 1). This indicates the importance of magnetic field–induced distortions of the conduction-electron wave functions. Calculations of Manninen and Nieminen (1979a) (curve 4) who went a step further to include the effect of the periodic lattice on the muon Knight shift through the spherical solid model (Almbladh and von Barth, 1976) are in much better agreement with the experimental value in Cu, Na, and Al. None of these calculations is able, however, to predict a negative Knight shift. Keller and Schenck (1979) (curve 5) through a first-principles calculation on the electronic structure of μ^+ in Be metal using the cluster multiple-scattering technique (Keller, 1973) have suggested that diamagnetic shielding analogous to a chemical shift in molecules may simulate a negative total Knight shift. Such an effect is to be expected if bound states or certain bonding states do exist.

Experiments on proton Knight shifts are not available for many systems since proton nmr experiments normally require significant hydrogen concentration and not many metals easily absorb hydrogen. However, some

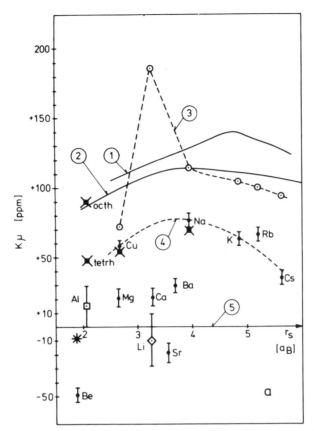

Fig. 13. Theoretical and experimental Knight shifts at interstitial muon sites in metals versus electron density parameter r_s. Curves labeled 1, 4, and 5 are from the theoretical work of Jena (1979), Manninen and Nieminen (1979), and Keller and Schenck (1979), respectively. Curves labeled 2 and 3 are from the theory of Munjal and Petzinger (1978). The dashed lines are intended to guide the eye only. (From Camani *et al.*, 1979.)

experimental data (Kazama and Fukai, 1976, 1977) are available on the transition metal–hydrogen system. Interestingly enough, the observed Knight shifts are negative. It is believed that this is due to negative spin density at the interstitial site in transition metal systems.

3. PROTON SPIN–LATTICE RELAXATION

When the perturbation responsible for inducing transitions in a nuclear spin system is switched off, the spin system relaxes to its ground state by transferring its excess energy to the lattice in time T_1, the spin–lattice relaxation time. The nuclear spin–lattice relaxation can be viewed as a spin–flip scattering experiment of the conduction electrons due to the magnetic hyperfine interaction with the nuclear spin (Abragam, 1961).

In a rigorous theoretical calculation of $1/T_1$ [see Eq. (50)], the hyperfine field in the unfactorized form is given by (Gupta and Freeman, 1978)

$$N(E_F)B_{hf} = \frac{\Omega}{(2\pi)^3} \int_{FS} \frac{dS_k}{|\nabla_k E_k|} |\Psi_k(0)|^2, \tag{52}$$

where Ω is the atomic volume, $\nabla_k E_k$ is the gradient of the energy band, dS_k is a surface element on the Fermi surface, and the integration is carried over the Fermi surface. $|\Psi_k(0)|^2$ is the wave-function density at the proton site. Gupta and Freeman (1978) have calculated the above quantity for hydrogen in stoichiometric PdH using augmented plane-wave (APW) band structure method. Their value of

$$[N_s(E_F)B_{hf}]_{theo.}^{PdH} = 4.4 \times 10^{15} \text{ G/erg} \tag{53}$$

agrees rather well with the experimental result (Jena et al., 1978d; Wiley and Fradin, 1978),

$$[N_s(E_F)B_{hf}]_{expt}^{PdH} = 4.00(4) \times 10^{15} \text{ G/erg}. \tag{54}$$

While this theoretical result is only 7% higher than the experiment, the corresponding quantity calculated in the pseudojellium model (Jena et al., 1979) discussed earlier (based on self-consistent generalized density-functional formalism) is about 57% higher. Keeping in mind that the result in Eq. (53) is based on a non–self-consistent calculation, the apparent disagreement between the two theories can be attributed largely due to the neglect of the periodic array of metal ions in the pseudojellium model.

The good agreement between Eqs. (53) and (54) should, however, be treated with caution. The additional terms in the hyperfine interaction, such as core polarization, the spin dipolar, and the orbital contributions would add to the result of Eq. (53), thus increasing further the gap between theory and experiment. The results can also be affected by changes in $N_s(E_F)$. For example, the value of $N_s(E_F)$ calculated by Papaconstantopoulos and Klein (1975) is 20% higher than that obtained by Gupta and Freeman (1978). Thus, in view of the uncertainties in the potential due to Slater local-exchange approximation and the lack of self-consistency, it is reasonable to state that the theory provides only a semiquantitative understanding of the proton spin–lattice relaxation rate in metal hydrides.

4. Hyperfine Fields at μ^+ Site in Ferromagnetic Metals

One of the interesting applications of the μSR technique is to probe the interstitial magnetization of ferromagnetic materials. The internal field B_μ is usually a sum of three contributions,

$$B_\mu = B_d + (4\pi/3)M + B_{hf}, \tag{55}$$

where B_d is the field due to the local dipole moments inside a sphere centered on the muon site, and $(4\pi/3)M$ is the so-called Lorentz field due to induced magnetic charges on the surface of the sphere. These two terms can be easily calculated from a knowledge of the crystal geometry and equilibrium configuration of the muon in the lattice. Thus from the sign and magnitude of the measured B_μ, B_{hf} can be computed. Since B_{hf} is proportional to the conduction-electron density at the muon site [see Eq. (49)] and, as has been pointed out earlier, the muon only enhances the ambient spin density around it without changing the sign of polarization, one can obtain information regarding the interstitial magnetization density and how it is perturbed by an external agent. These are some of the features of great interest to our understanding of the origin of ferromagnetism.

In Table II we compare the experimental B_{hf} (Foy et al., 1973; Gurevich et al., 1976a, b; Nishida et al., 1977; Graf et al., 1976, 1977) in Fe, Co, Ni, Gd, and Dy with various theoretical models and with the ambient magnetization density obtained from polarized neutron-scattering experiments (Moon, 1964; Mook, 1966; Shull and Mook, 1966; Moon et al., 1972; Shull, 1967). It is encouraging that all theoretical calculations (Katayama et al., 1979a; Jena, 1979; Keller, 1979; Jepsen et al., 1980) (which differ drastically from each other in terms of their approximations) predict the same negative sign for B_{hf} that is in agreement with experiment. Furthermore, the sign of the hyperfine fields is the same as that of the ambient polarization obtained

TABLE II

Comparison between Experimental and Theoretical Hyperfine Fields at an Interstitial μ^+ Site in Ferromagnetic Fe, Co, Ni, Gd, and Dy

Host	Ambient spin density $n_0^\downarrow - n_0^\uparrow (\mu_B/\text{Å}^{-3})$	Experimental	B_{hf} (kOe) Jellium	Molecular cluster	Layer	KKR
Fe	− 0.0136	− 11.1	− 9.4	− 10.1		
Co	− 0.0204	− 6.2	− 13.4	− 5.7		
Ni	− 0.0085	− 0.64	− 6.9	− 0.59	− 0.46	− 0.72
Gd	− 0.037	− 7.5	− 24.4	− 8.7		

from neutron scattering. This suggests that μSR can be used as a complementary tool in determining the sign of interstitial magnetization of the perfect host. The magnitudes of B_{hf} predicted by various models are, however, very different from each other. In the following we discuss separately the limitations of these calculations.

In the jellium model (Jena, 1979; Petzinger and Munjal, 1977; Jena et al., 1978c; Patterson and Falicov, 1974; Meier, 1975) for the ferromagnets the following assumptions have been made:

(i)   the electronic structure of the ferromagnetic host is modeled by a localized magnetic moment at the nuclear site and a uniformly spin-polarized conduction band composed of free electrons;
(ii)   the density of the conduction electrons is obtained by estimating in some ad hoc manner the number of free electrons per magnetic ion;
(iii)   the magnitude of the ambient interstitial spin polarization is taken from neutron-scattering experiments.

Thus any error in the neutron-scattering analysis can directly influence the calculated B_{hf}. This is not unlikely since the Fourier transform of the neutron-scattering form factors, which yields the spin polarization in real space, is carried out for finite reciprocal lattice vectors (Mook, 1966). Rath et al. (1979) have examined this point by calculating the magnetization density around the octahedral site in Ni using the self-consistent band structure scheme of Wang and Callaway (1977). Their result of the ambient spin density $(n_0^\uparrow - n_0^\downarrow)$ is a factor of two smaller than the experiment, although the signs are in agreement. In view of the accuracy with which the band structure results have explained the neutron-scattering form factors and other electronic properties, the above disagreement between theory and experiment could be attributed to the errors in the Fourier transform technique in the neutron-scattering analysis of the interstitial region. In this regard, it may be worth pointing out that large discrepancies in the interstitial spin density in ferromagnetic Gd exist between band theory and experiment (Harmon and Freeman, 1972).

A more serious shortcoming of the above jellium model is the assumption that the quasilocalized d electrons respond to an external perturbation as if they were free. This assumption is most likely the reason for the large discrepancy between jellium-based theory and experiments in Table II. The inadequacy of this model in treating the screening of impurities in ferromagnetic systems is even more apparent when one tries to explain the hyperfine field systematics. Using Eq. (2) and taking Z as the nuclear charge, Manninen and Jena (1980a) have calculated the hyperfine field at interstitial impurities from the second row of the periodic table (H, He, Be, B, C, N, O, F, and Ne) using the generalized self-consistent density functional formalism (von Barth and Hedin, 1972; Rajagopal and Callaway, 1973). Their results failed to account for the experimental data (Murnick *et al.*, 1976; Hass *et al.*, 1976; Hamagaki *et al.*, 1976), although the model is more successful in describing the Knight shifts in paramagnetic hosts (see Fig. 13). The failure of the jellium-based theory in accounting for the impurity hyperfine field systematics is due to the fact that the internal magnetic field in a ferromagnet is spatially inhomogeneous, whereas in Knight shift measurements, the external magnetic field is homogeneous in space. Following the models of earlier authors (Daniel and Friedel, 1963; Daniel 1967; Jena and Geldart, 1978; Jena, 1976a), Manninen and Jena (1980a) have included the effect of quasilocalized 3d electrons into the density-functional theory and successfully accounted for the hyperfine field systematics in a large class of ferromagnetic materials.

Recently several first-principles calculations (Katayama *et al.*, 1979a; Keller, 1979; Jepsen *et al.*, 1980) of muon hyperfine fields have been carried out. The results of spin-polarized cluster calculations (Keller, 1979) are in good agreement with the experiment (see Table II). These calculations involve only small clusters (the muon and first near-neighbor atoms). The effect of cluster size and the imposed boundary condition on the electron spin density at the μ^+ site are not known. Also unavailable is the ambient electron spin polarization in the interstitial region of the pure metal cluster that can be compared with band structure calculations. Until this information is available, it may be premature to judge the potential of the finite size molecular cluster calculation in interpreting muon hyperfine fields.

Katayama *et al.* (1979a) have carried out an ab initio calculation of the electronic structure around μ^+ by using the Korringa–Kohn–Rostoker (KKR) formalism developed by Soven (1970) for a muffin-tin potential model of an impurity atom. In this model the perturbing potential is assumed to be confined within the muffin-tin sphere of the impurity. The calculated muon hyperfine field of -0.72 kOe in Ni agrees well with the experimental data. The authors were also able to explain the observed temperature dependence of the hyperfine field to be due predominantly to single-particle excitations. The model also describes the electronic properties of pure Ni rather well, and thus shows promise as a quantitative tool for understanding impurity hyperfine fields.

Jepsen *et al.* (1980) have also carried out self-consistent calculation of the muon hyperfine field in Ni using the linear augmented plane-wave method for slab geometry. They have introduced a central layer of muons to a five-layer-thick Ni film and computed the hyperfine field to be -0.46 kOe.

The limitations on the quantitative nature of their results are twofold:

(i) The concentration of muons is too high, just as in the cluster model, while in μSR experiments one is concerned with a single muon at a time in the sample.

(ii) The results of five-layer Ni film (such as the variation of magnetic moment from layer to layer) are at variance with the self-consistent nine-layer calculation of Wang and Freeman (1979).

Thus the muon hyperfine field may be model dependent. In spite of all these criticisms, it is encouraging that these models yield semiquantitative results for the magnetic interaction of impurities in ferromagnets.

G. Isotope Effect on the Electronic Structure of Hydrogenlike Impurities

Since the proton is about 2000 times heavier than an electron and its jump rate ($\sim 10^9/s$) and vibrational frequency ($\sim 10^{13}/s$) are considerably shorter than the plasma frequency of electrons ($\sim 10^{17}/s$), it is generally assumed that the electrons respond to a proton as if it were static. Within this adiabatic approximation, the effect of isotopic mass of hydrogenlike impurities (positive muon, μ^+, proton, deuteron, and trition) on the electronic structure should not exist. On the other hand, if this effect is nonzero, hydrogenlike impurities are the best candidate to see such an effect since the mass of the heaviest (trition) and lightest (μ^+) point charge differ by a factor of about 27. Since these effects are expected to be small, though significant, it requires precision measurements as well as sophisticated theories for their study. Recent advances in experimental techniques and theoretical calculations have enabled one to study the isotope effect of hydrogenlike impurities on the host electronic structure. In the following section we discuss the isotope effect on the charge distribution and scattering properties of screening electrons.

1. Nuclear Spin–Lattice Relaxation Rate $1/T_1$

The measured values (Jena et al., 1978d; Wiley and Fradin, 1978) of nuclear spin-lattice relaxation rates of ^{1}H and ^{2}D in stoichiometric PdH and PdD are

$$\left(T_1 T \gamma^2/4\pi^2\right)_{\mathrm{H}}^{-1} = 0.115(1) \times 10^{-8}\ \mathrm{Oe^2 Hz^{-2}}\ \mathrm{s^{-1}},$$

$$\left(T_1 T \gamma^2/4\pi^2\right)_{\mathrm{D}}^{-1} = 0.121(1) \times 10^{-8}\ \mathrm{Oe^2 Hz^{-2}}\ \mathrm{s^{-1}}. \tag{56}$$

Using Eq. (50) this corresponds to

$$[N_s(E_F)B_{hf}]_H = 4.00(2) \times 10^{15} \quad G/erg,$$
$$[N_s(E_F)B_{hf}]_D = 4.09(2) \times 10^{15} \quad G/erg. \tag{57}$$

Since the hyperfine field is proportional to the electron spin density at the nuclear site, the results in Eq. (57) clearly indicate a significant increase in the electron spin density at the deuteron site over that at the proton site. The lattice constants of PdH and PdD differ by 0.006 Å and thus if ^{1}H and ^{2}D were static, the above results cannot be explained by band structure effects on the s density of states or electron spin density. Jena *et al.* (1978d) have also ruled out the quadrupolar interaction of the deuteron with the electric field gradient (that may be caused by any defects) to be the likely cause for the above result.

It is tempting to assign the results in Eq. (57) to be due to the mass difference of ^{1}H and ^{2}D. This becomes more transparent from studies of neutron inelastic scattering (Yamada *et al.*, 1976; de Graaf *et al.*, 1972; Gissler *et al.*, 1970; Bergsma and Goldkoop, 1960) on metals containing hydrogen. A localized mode for hydrogen in Pd occurs at 56 meV. Assuming that a proton moves in a harmonic potential well, this localized mode corresponds to a root-mean-square vibrational amplitude of $0.45a_0$.

Jena *et al.* (1978d) have studied the effect of this zero-point vibration of hydrogen on the electronic structure. They have shown, using a local density approximation, that the response of the electrons to the vibrating proton is time dependent since the proton vibrates in an ambient inhomogeneous electron cloud (see Fig. 14, for example). Using both real-space and momentum-space (Debye–Waller model) arguments, they successfully account for the results in Eq. 57 to be due to a more efficient screening of

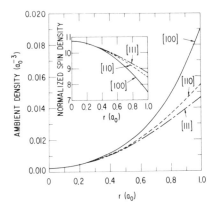

Fig. 14. Ambient electron charge density $n_0(\mathbf{r})$ around octahedral site along (100), (110), and (111) directions obtained by a linear combination of atomic charge densities centered at Pd atom sites. The inset shows the normalized spin density at the nuclear site of hydrogen as a function of displacement. (From Jena *et al.*, 1978d.)

the proton over that of the deuteron. [The readers are referred to Jena *et al.* (1978d, 1979) for a more comprehensive discussion.]

2. MUON HYPERFINE FIELDS IN FERROMAGNETIC METALS

Since the isotope effect of ^1H and ^2D on the electronic structure is significant, it is natural to assume that the effect would be much larger for an interstitial muon. All previous calculations (Petzinger and Munjal, 1977; Jena *et al.*, 1978c; Keller, 1979; Katayama *et al.*, 1979a; Jepsen *et al.*, 1980) of muon hyperfine fields have neglected this effect. Rath *et al.* (1979) have studied this effect in Ni by following the real-space analysis (Jena *et al.*, 1978d). They have found that the muon zero-point vibration reduces the static hyperfine field by as much as 30%. This is too large a correction to be neglected in any quantitative analysis of muon magnetic interactions.

3. TEMPERATURE DEPENDENCE OF IMPURITY RESISTIVITY DUE TO HYDROGEN

For small impurity concentrations, the impurity resistivity $\Delta\rho$ in Eq. (43) is generally assumed to be independent of temperature. Manninen and Jena (1980b) have analyzed theoretically the deviation from this general rule of $\Delta\rho$ due to hydrogenlike interstitials in terms of their temperature-dependent vibrational amplitude.

Following the "real-space analysis," Manninen and Jena (1980b) have computed the ambient electron density around interstitial sites in two test metals—palladium and tantalum. The impurity resistivity $\Delta\rho$ in Eq. (44) was then calculated for several displaced configuration of hydrogen by using phase shifts $\delta_l(E_F)$ from self-consistent density-functional approach. The temperature dependence of the proton distribution was calculated by assuming that it moves in a spherically symmetric harmonic potential well, the strength of which can be estimated from the experimental Einstein temperature θ_E.

The proton/deuteron distributions at several temperatures were convoluted with the calculated $\Delta\rho[n_0(\mathbf{r})]$ according to the prescription by Jena *et al.* (1978d). The temperature dependence of the proton/deuteron resistivity (normalized to the proton resistivity at $T = 0$ K) in Pd and Ta is shown in Fig. 15. The decrease in the normalized impurity resistivity as temperature increases, as well as the isotope effect, are interpreted as a direct consequence of the different screening of proton and deuteron.

There have been many experimental investigations (Peterson and Jensen, 1978; MacLachlan *et al.*, 1975; Verdini, 1976; Westlake and Ockers, 1975; Pryde and Tsong, 1971; Haywood and Verdini, 1968; Westlake, 1967) of the

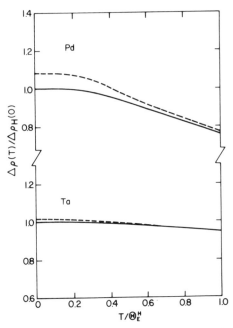

Fig. 15. Temperature dependence of the resistivity of hydrogen (solid lines) and deuterium (dashed lines) in Pd and Ta. The temperature is scaled to Einstein temperature of hydrogen (θ_E^H) in both metals. The result of the hydrogen resistivity at $T = 0$ K is 2.8 $\mu\Omega$ cm/at.% for Pd and 0.58 $\mu\Omega$ cm/at.% for Ta. (From Manninen and Jena, 1980b.)

impurity resistivity due to hydrogen and deuterium in transition-metal systems as a function of temperature. These experiments have mainly concentrated on the study of phase relations at low temperatures where an ordered metal hydride precipitates. Unfortunately several experimental problems, such as poor hydrogen–solubility and precipitation of ordered hydride phases at low temperatures, desorption of hydrogen at elevated temperatures, and imprecise determination of hydrogen concentration and geometry of the specimen, prohibit a precision measurement of the temperature dependence and isotope effect of $\Delta\rho$. Verdini (1976) has observed, in agreement with our results, a decrease in $\Delta\rho$ in Ta as temperature increases. However, Verdini's (1976) results should be analyzed with caution since the slope of $\Delta\rho$ in his measurements depends strongly on hydrogen content even at low concentrations indicating that the resistivity is not proportional to concentration at all temperatures. This is in sharp contradiction with other measurements (Westlake and Ockers, 1975; Haywood and Verdini, 1968; Westlake, 1967). Recently Wenzl (unpublished) and Westlake and Miller (1980) have observed the effect in Fig. 15 in FeTi and NbMo alloys containing hydrogen, respectively.

The results on the temperature dependence of $\Delta\rho$ is not expected to be model dependent since the ambient density would increase as one goes away from the interstitial site irrespective of what model one uses. It is this feature of the calculation that predicts a decrease of $\Delta\rho$ with increasing temperature. It is worth mentioning that the temperature dependence of $\Delta\rho$ for metal–hydrogen systems containing additional impurities may be somewhat more complicated than that discussed above. In certain systems hydrogen may be trapped by impurities in a metal. If this binding energy is small, hydrogen may be detrapped at higher temperatures and thus move to a region of different electronic environment, thus altering the temperature dependence of $\Delta\rho$.

4. Reverse Isotope Effect on Superconducting Transition Temperature T_c

The observed change in the superconducting transition temperature (Skoskiewicz, 1972; Stritzker and Buckel, 1972; Miller and Satterthwaite, 1975; Nakajima, 1976) of PdH and PdD is another evidence of the effect of isotope on the electronic properties of metals. To analyze this, it is convenient to start with the McMillan (1968) formula in the strong coupling limit, namely,

$$T_c = \frac{\theta_D}{1.45} \exp\left[-\frac{1.04(1+\lambda)}{\lambda - \mu^*(1+0.62\lambda)}\right], \qquad (58)$$

where θ_D is the Debye temperature and μ^* is the conventional Coulomb pseudopotential. The electron–phonon coupling parameter λ is given by

$$\lambda = (\eta/M)\langle\omega^2\rangle, \qquad (59)$$

with

$$\eta = N(E_F) \sum_{\substack{\mathbf{k},\mathbf{k}' \\ =k_F}} \langle\mathbf{k}|\nabla V(\mathbf{r})|\mathbf{k}'\rangle \cdot \langle\mathbf{k}'|\nabla V(\mathbf{r}')|\mathbf{k}\rangle. \qquad (60)$$

Here M is the atomic mass and $\langle\omega^2\rangle$ the corresponding second moment of the phonon density of states. $N(E_F)$ is the total density of states of the electrons at the Fermi surface and $V(\mathbf{r})$ the effective potential seen by the electrons. The parameter λ in Eq. (59) has two components—an acoustical contribution primarily due to the heavier Pd atoms and an optical contribution primarily due to the lighter hydrogen atoms. Since we are only interested in the change in T_c between PdH and PdD, it is convenient to deal with the differential form of Eq. (58) in which the contribution of the electron–ion interaction at the Pd atoms would cancel out. Thus,

$$\frac{\Delta T_c}{T_c} = \frac{\Delta \theta_D}{\theta_D} + \frac{1.04(1 + 0.38\mu^*)\lambda}{[\lambda - \mu^*(1 + 0.62\lambda)]^2} \frac{\Delta \lambda}{\lambda}. \tag{61}$$

Defining $\Delta T_c / T_c = [T_c(\text{PdH}) - T_c(\text{PdD})]/T_c(\text{PdH})$, we see from experiment (Wiley and Fradin, 1978) that $\Delta T_c / T_c = -0.22$. Since the isotope effect on the Debye temperature yields $\Delta \theta_D / \theta_D$ to be positive, the observed effect in T_c has been termed the reverse isotope effect. It is obvious that this effect has to originate from the isotope effect on $\Delta \lambda / \lambda$. Using Eq. (59) we can write

$$\frac{\Delta \lambda}{\lambda} = -\frac{\Delta(M\langle \omega^2 \rangle)}{M\langle \omega^2 \rangle} + \frac{\Delta \eta}{\eta}. \tag{62}$$

Considerable theoretical work (Papaconstantopoulos and Klein, 1975; Klein et al., 1977; Papaconstantopoulos et al., 1978; Ganguly, 1973; Hertel, 1974; Rietschel, 1975; Ganguly, 1976; Burger and McLachlan, 1976) has been done to understand the origin of this reverse isotope effect. These calculations have attributed the reverse isotope effect to the effective increase (Rahman et al., 1976) of the PdH force constant over that of PdD due to the larger anharmonicity of the hydrogen motion. Explicit band structure calculations (Papaconstantopoulos and Klein, 1975; Klein et al., 1977; Papaconstantopoulos et al., 1978) using APW scheme and phonon spectra (Rahman et al., 1976) extrapolated from substoichiometric to stoichiometric PdH have been successful in explaining the observed $\Delta T_c / T_c$. The effect due to zero-point vibration on T_c via a redistribution of electrons around ^{1}H and ^{2}D as pointed out by Miller and Satterthwaite (1975) has not yet been properly treated.

IV. Monovacancies and Microvoids in Metals

A monovacancy that corresponds to the absence of one atom from the bulk of a solid is probably the simplest form of an intrinsic point defect. The vacancies may migrate and form vacancy clusters or voids inside the sample. Although microvoids cannot be classified as point defects, we shall discuss their effects as a logical extension of our understanding of vacancies. The vacancies, like impurities, act as external perturbation on the conduction electrons of the host and affect a large class of solid-state properties, e.g., electron redistribution, vacancy formation energy, trapping of point charges, change in the electrical resistivity of the material, quadrupolar interaction, etc. Theoretically, two distinctly different models have been used to treat the vacancies in a metal. The canonical jellium model treats the metal by a uniform positive background together with an interacting electron gas, and the vacancy by a spherical hole in the positive background. The point ion model, on the other hand, models the vacancy by the negative of the ion field of the missing atom. In the next section we discuss the results of various model calculations and compare them with experimental data where available.

A. Electron Distribution around Vacancies

In Fig. 16 we compare the normalized electron-charge density distribution, $n(r)/n_0$ around a monovacancy in Al ($r_s = 2.07$) obtained from various theories based on the jellium model. The inset shows Friedel oscillations on a magnified scale. There are several points to note:

(i)   The linear response theory (dash-dot curve) overestimates the repulsion of electrons from within the vacancy, whereas the nonlinear response theory of Sjölander and Stott (1972) (dashed curve) underestimates the electron repulsion.

(ii)   The results of generalized nonlinear response theory (Gupta et al., 1978), on the other hand, agrees very well with that obtained from self-consistent density-functional approach.

(iii)   All these models exhibit Friedel oscillations that differ from each other in magnitude and phase.

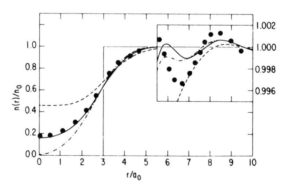

Fig. 16.   The normalized electron density $n(r)/n_0$ around an Al vacancy in the spherical hole model as a function of r/a_0. The radius of the vacancy is $3a_0$. The dash-dot curve is due to the linear theory, while dashed curve corresponds to Sjölander–Stott theory ($a_{12} = 0$). The result of a fully self-consistent nonlinear theory with $a_{12} = 0.88$ is shown by the solid curve. Values denoted by solid circles are based on a fully self-consistent Kohn–Sham calculation. The inset shows the Friedel oscillations on a magnified scale in each case. (From Gupta et al., 1978.)

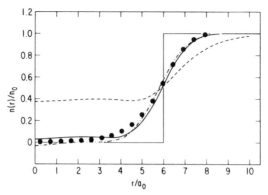

Fig. 17. Normalized electron density $n(r)/n_0$ around a spherical hole model of "8-atom void" in Al (radius $= 6a_0$). Different curves are labeled in the same way as in Fig. 16. (From Gupta *et al.*, 1978.)

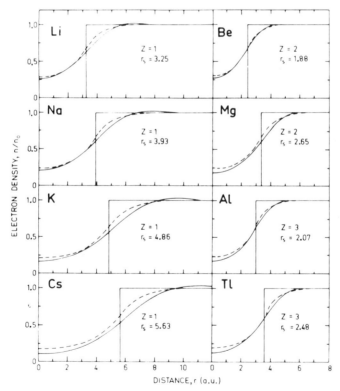

Fig. 18. Electron density profiles at vacancies in eight "simple" metals. The solid lines are the self-consistent Kohn–Sham electron densities, and the dashed ones are the Thomas–Fermi results. (From Manninen *et al.*, 1975.)

The small difference in $n(r)/n_0$ between the generalized nonlinear response theory and Kohn–Sham theory can be attributed to the neglect of the gradient corrections in the exchange-correlation energy functional in the Kohn–Sham method and to an approximate treatment of the kinetic energy in the generalized nonlinear response theory for a two-component plasma. The deviation of the linear response theory result from the Kohn–Sham formalism indicates that the vacancy-electron potential is not very weak. The failure of the linear response theory becomes even more apparent when one computes the electron distribution around an 8-atom spherical void in Al (see Fig. 17). The linear theory gives an unphysical result [$n(r)/n_0$ is negative] near the center of the void. In Fig. 18 we compare the normalized electron density distribution around monovacancies in several metals obtained from Kohn–Sham theory (solid curve) and Thomas–Fermi approach (dashed curve). While both results are in reasonable agreement, it should be emphasized that the Thomas–Fermi model does not give rise to Friedel oscillations. The discrepancy between the results in Fig. 18 is a reflection of the differences between the self-consistent effective potential V_{eff} and Thomas–Fermi potential V_{TF}. The electron density around a vacancy in Al has also been calculated (Gupta and Siegel, 1977) using the supercell APW method. The density at the center of the vacancy in the supercell model is roughly a factor of 3–4 smaller than that obtained from Kohn–Sham theory based on the jellium model. While the band structure results of Gupta and Siegel are not self-consistent, it is probably reasonable to expect that the effect of including the ionic lattice is to decrease the electron density in the vacancy. This assessment is consistent with the spherical solid model calculation (Perrot, 1977a).

In Fig. 19 we have compared the total effective potential and normalized electron density for various defect sizes (monovacancy, 8- and 27-atom voids) in Al. The range as well as the magnitude of the effective potential at the origin grow with the increasing defect size. At this point, it is interesting to compare the quantity

$$W = V_{\mathrm{eff}}(r = 0) - E_{\mathrm{F}}, \qquad (63)$$

which is -1.85, 2.20, and 3.98 eV, respectively, for a monovacancy, 8-atom, and 27-atom void. W in the surface problem has the significance of work function. Lang and Kohn's (1971) value of W for $r_s = 2$ is 3.89 eV. It is also interesting to note that $n(r)/n_0$ is about 0.2, 0.1, and 0.0 for a monovacancy, 8- and 27-atom voids, respectively, indicating that the strength of the potential increases with the size of the defect. Furthermore, the electron density profile near the surface of the defect gets increasingly sharper as the defect gets larger. Comparison with the calculations of Lang and Kohn (1970) for a jellium of $r_s = 2$ indicates that for a void of radius 15 Å or

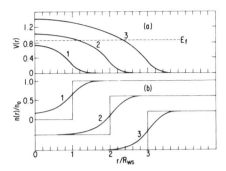

Fig. 19. (a) The electron-defect potential for a monovacancy (curve 1), and 8-atom void (curve 2), and a 27-atom void (curve 3) in Al versus r/R_{WS}. $R_{WS} = 3a_0$ is the radius of the monovacancy. E_f is the Fermi energy. (b) The normalized electron charge density $n(r)/n_0$ around a monovacancy (curve 1), an 8-atom void (curve 2), and a 27-atom void (curve 3) in Al versus r/R_{WS}. The curves have been displaced with respect to each other vertically for clarity. (From Jena *et al.*, 1978b.)

more, the electron profile for all practical purposes would resemble that for a surface. In the following section we make use of these charge densities to study various electronic properties.

B. *Vacancy Formation Energy*

The vacancy formation energy E_v^f at constant volume is defined as the work done when one atom is removed from the metal interior and replaced on the surface. It is written as a sum of two contributions,

$$E_v = E_{\Delta SCF} + E_b, \qquad (64)$$

where E_b is the binding energy of an atom in the perfect solid. $E_{\Delta SCF}$ is the difference between the energy of the defect lattice E_d [containing $(N - 1)$ atoms and a vacancy at the origin] and the perfect lattice E_p (containing N atoms) in the volume Ω:

$$E_{\Delta SCF} = E_d(N - 1, \Omega) - E_p(N, \Omega). \qquad (65)$$

There are several approaches for studying the energetics of vacancy formation. The jellium model where the electron gas is allowed to relax in the presence of a hole in the positive background has been used by many authors (Fumi, 1955; Stott *et al.*, 1970; Robinson and de Chatel, 1975; Evans and Finnis, 1976; Gyemant and Solt, 1977; Perrot, 1977a,b, 1978; Manninen and Nieminen, 1978; Manninen *et al.*, 1977; Nieminen, 1978; Finnis, 1978; Kahn and Rasolt, 1978).

The second scheme, suggested by Harrison (1966), is based on the pseudopotential treatment of the electron–ion interaction, and many calculations have been performed using this approach (Ho, 1971; Chang and Falicov, 1971; Benedek and Baratoff, 1971; Ducharme and Weaver, 1972; Popović et al., 1974; Minchin et al., 1974; Yoshioki and Mori, 1976). While this theory explicitly takes into account discrete lattice effects, it assumes that the perturbation in the conduction-electron distribution due to the vacancy is sufficiently weak to be treated in a linear response formalism. The jellium-based models, on the other hand, treat this response using more sophisticated nonlinear theories and account for the discrete lattice using perturbation theory (Manninen and Nieminen, 1978; Perrot, 1977b). The use of semiempirical interatomic potentials for calculating formation energies of vacancies has been discussed by Johnson (1973). Recently Singhal and Callaway (1979) have attempted to calculate the energetics of the vacancy formation from first principles using band structure methods. The reader is referred to the review article of Evans (1977) for a critical comparison of these various approaches and of Siegel (1978) for experimental studies.

In many models, the pressure p of the system due to the creation of a vacancy is nonzero and must be taken into account while calculating vacancy formation energy E^f,

$$E^f = E_v^f - p \, \Delta V, \tag{66}$$

where ΔV is the volume increase in the metal due to vacancy formation. If the system is in equilibrium, $p = 0$ and E^f can be calculated at constant volume.

Using the jellium model in the self-consistent solution of the Kohn–Sham equations, the vacancy formation energy is given by (Manninen and Nieminen, 1978)

$$E_0^f = \sum \varepsilon_i - \sum \varepsilon_i^0 - \int d\mathbf{r} \, \phi(\mathbf{r}) n(\mathbf{r})$$

$$+ \int d\mathbf{r} \{ [\varepsilon_{xc}(n) - \mu_{xc}(n)] n(\mathbf{r}) - [\varepsilon_{xc}(n_0) - \mu_{xc}(n_0)] n_0 \}$$

$$+ \frac{1}{2} \int d\mathbf{r} \, \phi(\mathbf{r}) [n(\mathbf{r}) - n_+(\mathbf{r})] - p_0 \Omega_0, \tag{67}$$

where ε_i and ε_i^0 are the single-particle energy eigenvalues of Eq. (13) with and without the defect, $\varepsilon_{xc}(n)$ the exchange-correlation energy per particle in a homogeneous electron gas, and p_0 the pressure in the jellium metal.

In Fig. 20 we compare the constant-volume vacancy-formation energies calculated in the jellium model using linear ($\times$) and self-consistent nonlin-

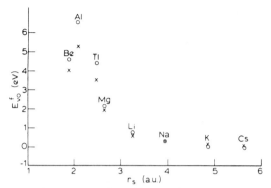

Fig. 20. Constant volume vacancy formation energies calculated for the jellium model; ○ labels the self-consistent, nonlinear results and × labels the linear screening result. (From Evans, 1977.)

ear response (○) results for the screening charge density $n(\mathbf{r})$. The linear screening theory gives a good qualitative account of the variation of E_0^f between different metals. There are, of course, important quantitative discrepancies especially for the polyvalent metals.

In Table III we compare the vacancy formation energies calculated in the jellium model, modified jellium model (including effect of discrete lattice through a variational procedure), and the pseudopotential model (Popović *et al.*, 1974) for several mono-, di-, and trivalent metals. In the jellium models, the effect of relaxation around the vacancies is neglected. Such

TABLE III

VACANCY FORMATION ENERGIES CALCULATED FROM VARIOUS THEORETICAL MODELS

Metal	Jellium	Modified jellium (variational) (Manninen and Nieminen, 1978)	Experimental	Pseudopotential discrete lattice
Li	0.19	0.51	0.34^a–0.40^b	0.29
Na	0.32	0.54	0.39^b–0.42^c	0.41
K	0.38	0.54	0.39^b	0.35
Rb	0.38	0.44	0.27^d	0.31
Cs	0.36	0.38	0.28^e	0.29
Mg	−0.06	1.00	0.79^f–0.89^g	
Ca	0.47	1.13		
Zn	−0.64	0.23	0.54^h	
Cd	−0.12	0.97	0.39^h–0.52^h	
Al	−1.74	0.78	0.62^i–0.75^h	0.86
Pb	−0.95	0.25	0.49^j–0.54^k	

[a] Feder (1970); [b] McDonald (1953); [c] Feder and Charbnau (1966); [d] Górecki (1974); [e] Kraftmacher and Strelkov (1970); [f] Tzanetakis *et al.* (1976); [g] Beevers (1963); [h] March (1973); [i] Fluss *et al.* (1978); [j] McKee (unpublished); [k] Trifthäuser (1975).

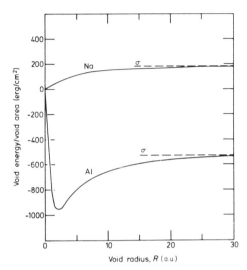

Fig. 21. Formation energies of voids divided by the void area in jellium metals corresponding to the densities of Na and Al as a function of the void radius. The asymptotic limit in each case is denoted by the surface energy value σ. (From Manninen and Nieminen, 1978.)

effects are expected to lower the formation energy. Calculations (Ho, 1971; Finnis and Sachdev, 1976) based on the local pseudopotential theory indicate that relaxation energies are less than about 0.1 eV (Feder, 1970).

In Fig. 21 we plot the formation energies of voids divided by the void area in jellium model for Na and Al as a function of the void radius. The energy reaches its asymptotic surface value only when the void radius is about 15 Å. This should be expected since the electron density profiles for voids with a radius of 15 Å (or larger) approach the surface density profile. The behavior of void formation energy is monotonic in low-density metals, but in high-density metals like Al the (negative) energies in the jellium model have a minimum around monovacancies.

While the theories provide an adequate description of vacancy formation energy in Table III, the discrepancies grow for polyvalent metals. The results are sensitive to the nature of exchange-correlation potential, to the form of pseudopotentials, and to the inclusion of discrete lattice. Since none of these quantities are unique, the quantitative understanding of vacancy formation energies must await sophisticated first-principles calculations based upon the response of interacting Bloch electrons to the defect.

C. *Positron Annihilation Characteristics in Metal Defects*

Since the observation by MacKenzie (1969) that at elevated temperatures the lifetimes of positrons increased in metals and the interpretation of this result by Bergersen and Stott (1969) in terms of the trapping model, considerable amount of work has been done on positron annihilation in metal defects. The positrons are expected to be attracted and trapped by metal vacancies, since the latter correspond to a localized deficiency of positive charge. Thus a localized positron samples the electron density variation around the defect, and its annihilation with these electrons provides information on the electronic structure of point defects. The usefulness of this probe in studying point defects is well established (West, 1973; West, 1979; Seeger, 1973; Doyama and Hasiguti, 1973; Ya, 1974). We next discuss the positron annihilation lifetimes and angular correlation between the photons produced as the positron and electrons annihilate inside the defect.

The positron annihilation rate λ is given by (West, 1973)

$$\lambda = \pi r_0^2 c \int n(\mathbf{r}) n_p(\mathbf{r}) \, dr^3, \tag{68}$$

where r_0 is the classical electron radius, c the velocity of light, $n(\mathbf{r})$ and $n_p(\mathbf{r})$ are the electron and positron densities in the crystal, respectively. The integral in Eq. (68) is evaluated over the unit cell. Gupta and Siegel (1977, 1980) have calculated the spatial distribution of the positron density in the Bloch state in both pure and defective (containing monovacancy) Al using non–self-consistent APW band structure scheme. Figure 22 shows the positron density distribution in perfect Al for the (001) plane. The repulsion of the positron by the ion cores and its accumulation in the interstitial

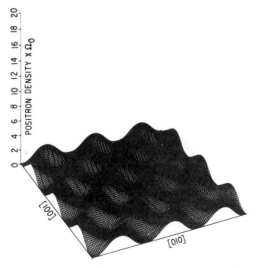

Fig. 22. The spatial distribution of the positron density in the Bloch state of the positron in Al for the (001) plane. (From Gupta and Siegel, 1980.)

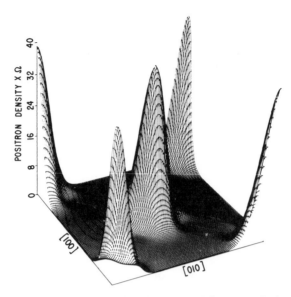

Fig. 23. Positron density $\times \Omega$ (where Ω is the volume of the super cell) shown on a network of 25 atomic sites in the (001) plane. The vacancies are situated at the center and the four corners of this network. (From Gupta and Siegel, 1980.)

regions is clearly seen. Figure 23 shows the spatial distribution of the positron inside an Al vacancy in the (001) plane. The positron was found to be bound to the vacancy with a binding energy of 3.31 eV. The calculated lifetime τ_v for the positron inside an Al vacancy (Gupta and Siegel, 1977) is 231 ps at 250°C, in good agreement with the experimental value of 244 ps (Fluss *et al.*, 1978). This has to be compared with the lifetime of 172 ps (Fluss *et al.*, 1978) for positrons annihilating in pure Al at 250°C. The increased positron lifetimes in vacancies compared to that in perfect metal is understandable since the vacancy contains fewer electrons than the interstitial sites.

While ab initio self-consistent supercell band structure calculations of positron lifetimes are desirable, the enormous computational effort involved in such calculations prohibit a systematic study of positron annihilation characteristics in a large number of metals. We now describe a simple scheme based on the jellium model that has been quite successful in explaining the experimental data.

In the jellium model, the annihilation rate λ of positrons in an inhomogeneous electron gas can be written (Brandt and Reinheimer, 1971) as

$$\lambda = \int dr^3 \, n_p(\mathbf{r}) \Gamma[n(\mathbf{r})], \qquad (69)$$

where

$$\Gamma(n) = (2 + 134n) \times 10^9 \, s^{-1} \tag{70}$$

is the annihilation rate for a homogeneous electron gas of density n. In the jellium model, the positron distribution $n_p(\mathbf{r})$ in the perfect metal is assumed to be uniform. This is, of course, an approximation to that shown in Fig. 22. For a positron in a defect lattice, $n_p(\mathbf{r})$ is computed from

$$n_p(\mathbf{r}) = |\Psi^+(\mathbf{r})|^2, \tag{71}$$

where $\Psi^+(\mathbf{r})$ is the wave function of the positron,

$$\Psi^+(\mathbf{r}) = [u_{nl}^+(r)/r]Y_{lm}(\mathbf{r}). \tag{72}$$

$u_{nl}^+(r)$ is obtained by solving the following radial Schrödinger equation for positron bound states,

$$\left[-\frac{d^2}{dr^2} + V_+(\mathbf{r}) - \frac{l(l+1)}{r^2}\right]u_{nl}^+(r) = -\varepsilon_{nl}^+ u_{nl}(r). \tag{73}$$

$V_+(\mathbf{r})$ and ε_{nl}^+ are, respectively, the effective positron defect potential and the binding energy of the positron in the nl quantum state. The effective positron potential $V_+(\mathbf{r})$ is given by

$$V_+(\mathbf{r}) = -V_0\theta(\mathbf{r} - \mathbf{a}) + \left[V_{corr}^+(n(\mathbf{r})) - V_{corr}^+(n_0)\right] + \phi[n(\mathbf{r})], \tag{74}$$

where V_0 is the zero-point energy of the positron in a perfect lattice and V_{corr}^+ is the positron–electron correlation energy in the local density approximation. The positron-defect electrostatic potential ϕ is simply the negative of the electron-defect potential in Eq. (16). The positron zero-point energies have been tabulated by Hodges and Stott (1973a) for simple metals and by Nieminen and Hodges (1976) for transition metals. The positron–electron correlation energy has been discussed by Bhattacharya and Singwi (1972b). Thus with all the terms in Eq. (74) specified, the positron binding energy and distribution can be easily calculated from Eq. (73).

In Fig. 24, the effective positron-defect potential and the corresponding wave function of the positron in the lowest bound state in Al are shown as a function of defect size. The positron-defect potential is attractive and grows in strength with increasing defect radius. Integrating the positron density inside the defect, we find that 35% of the positrons lie outside the vacancy radius, whereas only 5 and 1% of the positrons lie outside the radius of 8- and 27-atom voids, respectively. Thus in situations where positrons are trapped in shallow bound states, a significant portion of the positrons may annihilate with core electrons of the nearest-neighbor atoms.

The annihilation of positrons with electrons outside the defect radius can be taken into account by scaling the electron density outside the defect in a

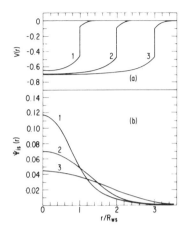

Fig. 24. The positron-defect potential for a monovacancy (curve 1), an 8-atom void (curve 2), and a 27-atom void (curve 3) in Al versus r/R_{WS}. (b) The 1s bound-state positron wave function $\Psi_{1s}(r)$ in a monovacancy (curve 1), an 8-atom void (curve 2), and a 27-atom void (curve 3) in Al versus r/R_{WS}. (From Jena *et al.*, 1978b.)

manner suggested by West (1973), namely,

$$n'(r) = n(r)[1 + \Gamma_c/\Gamma_v], \tag{75}$$

where Γ_c and Γ_v are annihilation rates with core and valence electrons, respectively, and are estimated from angular correlation measurements,

$$\Gamma_c/[\Gamma_c + \Gamma_v] = 0.8A_c/(A_c + A_v), \tag{76}$$

where A_c and A_v are experimental areas under core and valence electron parts of the angular correlation curve, respectively.

In Table IV we compare the positron binding energies and lifetimes obtained from several model calculations with the supercell band structure calculation of Gupta and Siegel (1977). Hodges (1970) used the pseudoatom picture in estimating the electrostatic part of the positron potential. His two values for the binding energy in Table IV are calculated with and without including electron–electron correlations into the screening dielectric function. Arponen *et al.* (1973) used the jellium model and calculated electron density in the nonlinear Thomas–Fermi approximation. Mori (1977a, b)

TABLE IV

CALCULATED POSITRON BINDING ENERGIES E_b AND LIFETIMES τ_t IN AN AL VACANCY

Authors	τ_t (ps)	E_b (eV)
Hodges (1970)		3.81, 2.04
Arponen *et al.* (1973)	243	2.62
Manninen *et al.* (1975)	237	1.75
Mori (1977a)	273	0.8
Jena *et al.* (1978)	243	1.75
Gupta and Siegel (1977)	231	3.31

used the Thomas–Fermi model and took into account the discrete lattice structure around the vacancy. Manninen *et al.* (1975) used the jellium model and electron density obtained from self-consistent density functional formalism. Jena *et al.* (1978b) used the point-ion model for the vacancy and density functional method for their calculation of the electronic structure. As mentioned earlier, the electron density distribution calculated using these models are quite different from each other. The positron distributions, as can be seen from very different binding energies in Table IV, are also different in these models. However, the positron lifetime in Eq. (68) or Eq. (69) is a result of overlap between electron charge and positron distribution and is fairly insensitive to the choice of the model (see Table V).

In Table V we compare the positron lifetimes obtained in the density functional formalism (based on the jellium model) with experiment for several metals. The overall agreement with experiment is good, although the theoretical lifetimes tend to be larger than the experimental in general. So far, the effect of lattice relaxation around a vacancy on the positron lifetime has been neglected. Although this is difficult to treat from first principles, Jena *et al.* (1978b) have estimated its effect by following the prescription of Blatt (1957). According to this procedure, the effective charge of an ion due to lattice relaxation is reduced by the same amount the volume contracts. In

TABLE V

POSITRON LIFETIMES (PS) IN METAL VACANCIES[a]

	Theoretical	Experimental
Li	325	
Na	391	
K	436	
Rb	458	
Cs	452	
Be	177	
Mg	307	255
Zn	233	240
Cd	276	250
Hg	266	165
Al	237	244
Ga	248	260
In	289	240
Tl	302	230
Sn	282	
Pb	291	280
Cu	205	
Ag	227	
Au	212	210

[a]From Manninen *et al.* (1975).

other words,

$$Z^*/Z_v = \Omega/\Omega_0 \tag{77}$$

where Z_v and Ω_0 are the valence and atomic volume in the unperturbed state, Ω the volume occupied by the vacancy after lattice relaxation, and Z^* the charge contained in the relaxed vacancy. This can be taken into account by simply reducing the size of the hole or by reducing the strength of the bare-ion pseudopotential by Ω/Ω_0 and repeating the self-consistent calculation for both electron and positron distribution. Jena et al. (1978b) have found that the lattice relaxation effects could reduce the positron lifetimes in vacancies by about 15% — toward the right direction with experiment in most systems.

The results in alkali metals deserve some special consideration, since the measured lifetimes of positrons in vacancies are found to be almost the same as those in perfect metals, while the theoretical values, even including lattice relaxation effects, indicate that the lifetimes in vacancies should be higher than those in the bulk. To understand this anomalous behavior, we note that for the lifetime of positrons in vacancies and bulk to be different, the following three conditions have to be satisfied: (i) the positron should be bound to the vacancy; (ii) the trapping rate of positrons should exceed the annihilation rate; and (iii) the rate of escape of positrons from the bound state should be less than the annihilation rate.

The first two conditions are satisfied for all metals we have studied. In alkali metals, the positrons lie in a very shallow bound state with binding energy of the order of 0.01 eV. Using the uncertainty principle and the Boltzmann distribution, it can be shown that the escape rate from this potential well is 6×10^{12} s^{-1}. This is much larger than the annihilation rate, which is typically 3×10^9 s^{-1}. Thus in alkali metals the positrons escape the vacancy before they have an opportunity to annihilate. An alternate explanation for positrons in alkali metals based upon the positron detrapping due to vacancy migration has been proposed by Tam and Siegel (1977).

The dependence of positron lifetimes on the defect size in aluminum and molybdenum is plotted in Fig. 25. It is interesting to note that the lifetime of positrons increases markedly with the increase in the size of a void. This is an important observation since it links the size of small voids to the positron life. Positron-annihilation characteristics can therefore be used to monitor the nucleation and growth of small voids where conventional electron microscopy is found to be inadequate. For large voids, the positron lifetime as computed from Eq. (69) would tend to saturate to a value of 500 ps (spin-averaged positronium lifetime). The use of the Brandt–Reinheimer formula (1971) to compute the positron lifetime in the inhomogeneous electron gas implies that the positronium would be formed in large metal

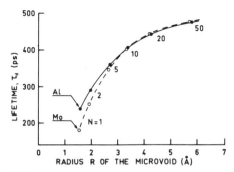

Fig. 25. Positron lifetimes in microvoids of aluminum and molybdenum as a function of void radius *R*. *N* is the number of vacancies in the void. (From Hautojärvi *et al.*, 1977.)

voids. Although there is fragmentary experimental evidence (Trifthäuser *et al.*, 1974; Sen *et al.*, 1975; Gustafson and Barnes, 1973) that the trapped state in large voids has some positroniumlike properties, and recent calculations indicate (Nieminen and Hodges, 1976) that such a state is energetically favorable, experiments done under strong magnetic fields (Cotterill *et al.*, 1972) fail to show any sign of positronium formation. MacKenzie and Sen (1976) have recently cautioned their readers against using the later experimental result as conclusive evidence that positroniumlike states are not possible in large metal voids.

For very large voids (mean diameter 40 Å) lifetimes up to 595 ps have been measured by Cotterill *et al.* (1972) in irradiated molybdenum. Recently, Cheng *et al.* (1976) have measured positron lifetimes in voids of molybdenum by varying the temperature and dosage of irradiation, and they found that the lifetimes saturated to a value of about 453 ps. The discrepancy between these two sets of data in molybdenum is now believed to be due to the presence of impurities around which vacancies tend to cluster to form voids.

As shown earlier, the electron density profile at the internal surface of voids (whose diameter $\gtrsim 20$ Å) resembles closely that of a metal vacuum surface. Thus for voids as large as 20 Å in diameter, the image potential would form a major fraction of the positron trapping potential in Eq. (74). Calculations (Hodges and Stott, 1973b) show that positrons are trapped at internal surfaces of large metal voids. Thus for voids whose internal surfaces resemble metallic surfaces, the positron lifetime would be smaller than 500 ps (spin-averaged positronium lifetime) not only because the electron densities in the region of void surface are larger than those in the center, but the core electrons of the host ions would participate in the annihilation process. The relative importance of image potential as a function of void size is an interesting problem for a future study since it would provide an understanding of how positron lifetime behaves as we approach the critical size at which the void resembles a surface.

Interesting experimental information on the electronic structure of point defects can also be obtained from the study of angular correlation between two annihilation γ-ray quanta. Since the linear momentum of the thermalized positrons is small compared with the Fermi momentum of electrons, this measurement furnishes information on the distribution of electron momentum. In the independent particle model, the momentum density of the electron–positron pair is

$$\rho(\mathbf{p}) \propto \sum_i \left| \int d\mathbf{r} \exp(-i\mathbf{p}\cdot\mathbf{r})\Psi_+ (\mathbf{r})\Psi_i(\mathbf{r}) \right|^2, \tag{78}$$

where $\Psi_i(\mathbf{r})$ is the electron wave function in the occupied state i. The angular correlation curve can then be computed from the expression

$$I(p_z) = \int_{p_z}^{\infty} dp\,\rho(p)p, \tag{79}$$

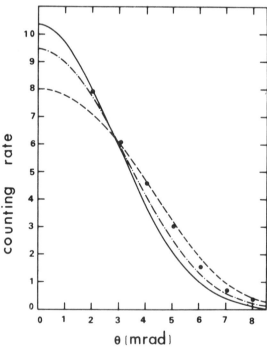

Fig. 26. The angular correlation curve for a monovacancy (dashed curve), an 8-atom void (dash-dot-dash curve), and a 27-atom void (solid curve) in Al versus θ (mrad). All of the curves have been normalized to the same area. The vertical scale is in arbitrary units. Experimental points from Sen *et al.* (1975) are shown as solid circles. (From Jena *et al.*, 1978b.)

where p_z is given by

$$\theta = p_z / m_0 c, \qquad (80)$$

θ being the angle between the two γ-ray quanta.

In Fig. 26 we present the angular correlation curves normalized to equal areas for a monovacancy, an 8-atom void, and a 27-atom void in arbitrary units. We find that the full widths at half maximum (FWHM) of these curves are 8.8, 7.3, and 6.5 mrad, respectively. Trifthäuser *et al.* (1974) have measured the angular correlation curves in a monovacancy and in a large void in Al. Experimental values of FWHM for a monovacancy and a large void are, respectively, 8.8 and 6.5 mrad, which are in good agreement with the calculated values. Not much significance can be attached to this agreement for the large void, since we have not considered surface trapping. However, the narrowing of the peak of the angular correlation curves with the size of the void is consistent with the experimental data in Al (Trifthäuser *et al.*, 1974). The decrease in FWHM with increasing size of the void in molybdenum (Mackenzie, unpublished) is also consistent with our results in Al, although the broadening of the angular correlation peak (Cotterill *et al.*, 1972; Mogenson *et al.*, 1972) with increasing void diameter still remains a puzzling result.

D. Residual Resistivity due to Vacancies and Vacancy Clusters

The scattering of conduction electrons from vacancies and vacancy clusters causes an increase in the electrical resistivity of the metal just as described for hydrogen in the preceding section. There are commonly two methods employed in the calculation of residual resistivity $\Delta\rho$. One of them, described earlier, involves the computation of scattering phase shifts off a vacancy in the self-consistent density functional formalism based on the jellium model. Manninen *et al.* (1977) have followed this procedure and calculated $\Delta\rho$ as a function of valence and electron density of the metal. The second method based upon the pseudopotential approach treats the scattering from a vacancy as the negative of the linearly screened pseudopotential of the missing atom. While the effect of lattice relaxation and applicability to more complex systems can be easily incorporated into the pseudopotential approach, the uncertainty in the choice of pseudopotential and inadequacy of the linear response theory remain as shortcomings of this later model. Popović *et al.* (1973) have used this approach to compute $\Delta\rho$ in the alkali metals.

In Fig. 27 the residual resistivities of vacancies are plotted as a function of the square of the electron density parameter for elements with different valence Z_v. Note that $\Delta\rho$ is proportional to r_s^2, whereas it depends weakly

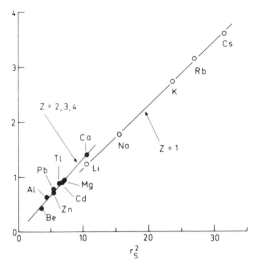

Fig. 27. The calculated residual resistivity of monovacancies as a function of the square of the electron density parameter r_s. Z is the valency of the metal. (From Manninen *et al.*, 1977.)

on the valence of the metal. Since the Wigner–Seitz radius, $R_{WS} = Z_v^{1/3} r_s$, the above result implies that for a fixed Z_v, the density dependence of $\Delta\rho$ is simply related to the geometrical scattering cross section of a vacancy (αR_{WS}^2).

In Table VI we compare the theoretical predictions of $\Delta\rho$ obtained by various authors with the available experimental data. The second column is the result due to Manninen *et al.* (1977) in the jellium model. Manninen and Nieminen (1978) have improved upon this result by incorporating the effect of ion pseudopotential in the density functional formalism through a variational scheme proposed by Perdew and Monnier (1976). The significant difference between the results of columns 2 and 3 indicates the effect of the discrete lattice. For comparison we have also given results of Popović *et al.* (1973) and Stott *et al.* (1970) using pseudopotential approach. The effect of lattice relaxation has been found (Popović *et al.*, 1973; Benedek and Baratoff, 1971) to reduce the magnitude of $\Delta\rho$ by about 30%.

The available experimental results are given in the last column of Table VI. The experimental techniques suffer from difficulties associated with the determination of vacancy concentration, the effect of divacancies and deviations from Matthiessen's rule. A new method based on the current-noise measurement have been presented by Celasco *et al.* (1976). New experimental results of $\Delta\rho$ due to metal vacancies are eagerly awaited.

TABLE VI

RESIDUAL RESISTIVITIES $\Delta\rho$ ($\mu\Omega$ CM/AT.%)

Metal	Theoretical				Experimental
	Manninen et al. (1977)	Manninen and Nieminen (1978)	Popović et al. (1973)	Stott et al. (1970)	
Li	1.22	1.84	1.01		1.9–2.1
Na	1.78	1.67	1.21		
K	2.73	2.25	1.35		
Rb	3.23	1.46	1.43		
Cs	3.57	1.07	1.56		
Be	0.45				
Mg	0.94	0.87		1.21	
Zn	0.71	0.59			
Cd	0.89	1.06			
Ca	1.39	0.84			
Al	0.62	0.86		0.88	
Tl	0.87				1.1–3.0
Pb	0.77	1.35		1.01	2.5

A somewhat related problem is the scattering of electrons from N independent vacancies versus N vacancies forming a single cluster. Hautojärvi *et al.* (1977) have calculated the residual resistivity for this system in Al in the jellium model assuming that even small vacancy clusters (such as di- and trivacancies) are spherical. Their result is shown in Fig. 28. Note that the resistivity rises by about 10% in clustering of two or three vacancies and then gradually starts to fall off. Experimental confirmation of this effect, although difficult, would be worthwhile.

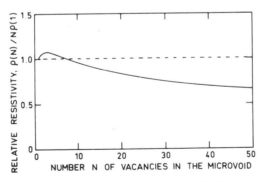

Fig. 28. The change of the total resistivity in Al owing to the clustering of vacancies. $\rho(N)$ is the residual resistivity of the microvoid containing N vacancies and $N\rho(1)$ is the resistivity of N single vacancies. (From Hautojärvi *et al.*, 1977.)

E. Nuclear Quadrupole Interaction due to Vacancies in Cubic Metals

The vacancies in a cubic metal can give rise to electric field gradients at host nuclear sites similar to that discussed in the previous section for impurities. Experimentally the nuclear quadrupole couplings around vacancies are difficult to measure because of low vacancy concentration. Recently Minier et al. (1978) have succeeded in measuring the electric field gradients at four nearest-neighbor sites in Al due to vacancies created by electron irradiation. The efg at first, third, and fourth neighbor sites are 2.8×10^{23} cm^{-3}, 9.3×10^{22} cm^{-3}, and 6.9×10^{22} cm^{-3}, respectively. Unfortunately, the sign of these efg's cannot be determined in conventional nuclear magnetic resonance techniques. These field gradients can be related to the charge density perturbation around the vacancy through the use of Eq. (40). Jena et al. (1978b), using the Bloch enhancement factor of Holtham and Jena (1975) in Al, have computed the first near-neighbor electric field gradient to be -1.5×10^{23} cm^{-3} in semiquantitative agreement with experiment.

This calculation does not take into account the contribution to efg due to the lattice strain caused by relaxation around the vacancy. Sagalyn and Alexander (1977) have shown, through use of approximate charge densities, that this effect can be large and is needed to explain the large asymmetry parameter η:

$$\eta = (V_{yy} - V_{xx})/V_{zz}. \tag{81}$$

V_{xx}, V_{yy}, and V_{zz} are the principal values of the efg tensor. Minier et al. (1978) conclude that $\eta = 0.65$ for first near-neighbor in Al is due to inward relaxation of the nearest-neighbor ions around the vacancy (Singhal, 1973). More experimental data and calculations including lattice relaxation and response of Bloch electrons to vacancies are needed to make useful analysis of systematic trends in nuclear quadrupole interaction.

F. Vacancy Trapping of Hydrogen

This is a rather virgin field and has the potential for being an important field for investigation in the future. Since positrons are known to be trapped by vacancies, it is natural to assume that hydrogen and its various isotopes should be attracted by the vacancies. Several recent experiments tend to support this assumption. Lengeler et al. (1978) have studied this problem in their Doppler broadening measurements on Cu samples containing vacancies and/or hydrogen. They find trapping of hydrogen by vacancies to occur above $\sim$ 150 K, and detrapping of hydrogen does not take place below 450 K. As can be expected, vacancies occupied by hydrogen show a strongly reduced positron-trapping probability as compared to hydrogen-free vacancies. This effect can influence positron lifetimes as well as angular correlation between the annihilation quanta.

Trapping of positive muons (a light isotope of hydrogen) by metal defects has also been observed (Seeger, 1979). In the muon spin rotation (μSR) experiments, the depolarization of the muon spins due to dipolar interactions with the lattice is studied. The experimental data are analyzed in terms of the nuclear and electronic magnetic moments of the neighboring atoms. Thus information on the symmetry and other physical characteristics of the defects acting as traps can be deduced.

The experiment on neutron-irradiated single crystals of Al shows muons trapped by monovacancies (Herlach *et al.*, 1979). Popović *et al.* (1976) had predicted (on the basis of their jellium calculation) the trapping of hydrogen by Al vacancies. However, these authors did not take into account the redistribution of electrons around a vacancy. Manninen and Nieminen (1979a) have shown that when the latter is taken into account, it is not energetically favorable for hydrogen to be in a trapped state with the vacancy in Al. They also reached similar conclusions when the discrete lattice was replaced by a spherical solid instead of a jellium model.

Norskov (1977) and Jena (1978) have studied the electron distribution around a vacancy-trapped hydrogen. They find substantial difference between the electronic structure of interstitial and vacancy-trapped hydrogen. In Fig. 29 we plot the electron charge and spin density distribution around an interstitial muon, a vacancy-trapped muon, and a monovacancy in Al.

The electron charge pileup at an interstitial μ^+ site is 50% higher than that at a muon situated at the center of a monovacancy. This reduction in the electron charge density at the vacancy-trapped muon is understandable since the vacancy provides an additional repulsive potential for the electrons. It is for this reason that the electron density at the center of a monovacancy is only 20% of its average bulk value. From the inserts in Fig. 29, also note that there is a significant spatial shift in the first node of the Friedel charge oscillations. The nature of the normalized electron spin response to the interstitial or vacancy-trapped hydrogen is different from that discussed above. For example, the electron spin density at the vacancy-trapped muon site is about a factor of 3 larger than the corresponding result due to an interstitial hydrogen. This striking result is due to the disappearance of bound states for a vacancy-trapped hydrogen (Jena, 1978).

The electron charge and spin distributions around a monovacancy in Al are also shown in Fig. 29 for reference. The above result implies that the

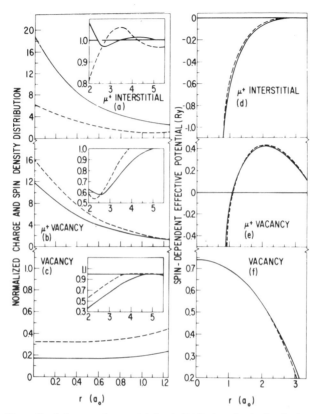

Fig. 29. Normalized electron charge $n(r)/n_0$ (solid line) and spin density $[n(r)\zeta(r)/n_0\zeta_0]$ distribution (dashed line) around (a) an interstitial muon, (b) vacancy-trapped muon, and (c) a monovacancy in Al. The corresponding effective potentials for both spin electrons (solid line for spin ↑ and dashed line for spin ↓) are shown in (d), (e), and (f). The Friedel oscillations of electron charge and spin densities are plotted in the inserts of (a), (b), and (c) on a different scale. (From Jena, 1978.)

Larmor frequency of precession of the positive muons stopped at different locations in a lattice are different and the Knight shift at a vacancy-trapped muon site in Al would be three times larger than that at an interstitial muon. This may be a difficult experiment to perform since the diffusion of μ^+, vacancy concentration, and muon-vacancy binding would be important factors in the success of the experiment. However, with present-day experimental sophistication the prospect of performing the above experiment looks good. Such an investigation would be worthwhile since the positive muons are capable of providing local information regarding the magnetic structure of the electrons around point defects that cannot be obtained by any other known experimental technique.

V. Summary and Conclusions

In this chapter we have tried to review some of the recent developments in our understanding of the electronic structure of point defects in metals. To illustrate the various experimental and theoretical methods, we have chosen two of the simplest form of point defects—an extrinsic defect, such as hydrogen and its isotopes, and an intrinsic defect, such as vacancy and vacancy clusters. From the experimental side, different techniques were shown to provide information on different aspects of the electronic structure. For example, Mössbauer spectroscopy, Compton scattering, and photoelectron spectroscopy shed light on the general nature of the electronic environment around hydrogen; electrical conductivity experiments measure the scattering of electrons from the impurity; nuclear magnetic resonance and muon spin-rotation experiments provide information on the charge and spin density distribution around impurities and vacancies; and positron-annihilation experiments educate us regarding the charge and momentum distribution of electrons around defects in metals. From the theoretical side, several models with varying approximations are introduced to study the electronic properties of point defects. The simplest model based on the jellium approach provides qualitative and, in some cases, even semiquantitative description of the electronic properties. While, in principle, the supercell band structure model is capable of providing quantitative understanding of the electron-defect interaction, it poses tremendous numerical complications. Therefore, it may not be very suitable for systematic analysis. The molecular cluster model is somewhat intermediate between the above two and can be applied to any complicated matrix. The comparison of results between molecular cluster and band models show promise for the former to be a useful tool in providing quantitative description of the electronic structure of point defects.

The efficient screening of point defects by conduction electrons in metals implies that the electronic properties of the defects are governed by local environment. In the case of hydrogen in metals calculations show that there is a net charge transfer from the vicinity of metal ions to hydrogen. This result is consistent with experiments involving Mössbauer and x-ray photoelectron spectroscopy. Theory also accounts for the spin and charge density perturbations around a proton in metallic environment, and the results are consistent with nuclear magnetic resonance and muon spin-rotation experiments. The effect of isotope on the electronic structure of hydrogen can be understood in terms of the zero-point vibration of the point charge. This contributes to our understanding of proton and deuteron spin–lattice relaxation rate, temperature dependence of impurity resistivity due to hydrogen, and reverse isotope effect on superconducting transition in metal hydrides.

The electron charge and momentum distribution around vacancies and vacancy clusters can be understood by studying the annihilation characteristics of positrons in metal defects. Positron lifetimes τ_v are insensitive to the nature of the electron density inside the vacancy and any simple model provides adequate description of the systematics of τ_v in metals. However, quantities such as vacancy formation energy are far from being understood and require a detailed knowledge of the response of Bloch electrons to the defect in a real discrete lattice in the presence of relaxation.

Recent advances in our understanding of interacting electron system in an inhomogeneous environment and availability of high-speed computers make it feasible to venture realistic calculations of the electronic structure of point defects. Problems such as vacancy and solute trapping of hydrogen, study of small nonspherical vacancy clusters, microscopic understanding of hydrogen-induced embrittlement, and kinetics of diffusion are expected to be important problems in the next decade. It is hoped that well-coordinated experimental and theoretical efforts would provide us with a much deeper understanding of the electron-defect interactions and their consequences.

Acknowledgments

It is a great pleasure for me to thank Professor K. S. Singwi for not only introducing me to the field of point defects but also for his collaboration, advice and encouragement. I am grateful to my colleagues with whom I collaborated on various topics discussed in this review. They are Dr. S. G. Das, Dr. F. Y. Fradin, Dr. A. K. Gupta, Dr. M. Manninen, Dr. R. Nieminen, Dr. J. Rath, and Dr. C. S. Wang. I am thankful to Prof. T. Courtney for a careful reading of the manuscript. I would also like to thank Mrs. B. Huff, Mrs. D. Reich, and Miss C. Heimke for their patience in typing this manuscript. The work was supported in part by the National Science Foundation.

References

Abragam, A. (1961). "The Principles of Nuclear Magnetism." Oxford Univ. Press, London and New York.

Adachi, H., Imoto, S., Tanabe, T., and Tsukada, M. (1978). *J. Phys. Soc. Japan* **44**, 1039.

Alefeld, G., and Völkl, J. (eds.) (1978). *In* "Topics in Applied Physics," Vol. 28. Springer-Verlag, Berlin–Heidelberg–New York.

Allen, P. B., and Heine, V. (1976). *J. Phys. C* **9**, 2305.

Almbladh, C. O., and von Barth, U. (1976). *Phys. Rev. B* **13**, 3307.

Almbladh, C. O., von Barth, U., Popović, Z. D., and Stott, M. J. (1976). *Phys. Rev. B* **14**, 2250.

Antonageli, F., Balzarotti, A., Bianconi, A., Burattini, E., and Perfetti, P. (1975). *Phys. Lett. A* **55**, 309.

Arponen, J., Hautojärvi, P., Nieminen, R., and Pajanne, E. (1973). *J. Phys. F* **3**, 2092.

Averill, F. W., and Ellis, D. E. (1973). *J. Chem. Phys.* **59**, 6412.

Baraff, G. A., and Schlüter, M. (1979). *Phys. Rev. B* **19**, 4965.

Beeby, J. L. (1967). *Proc. R. Soc. London, Ser. A* **302**, 113.

Beevers, C. J. (1963). *Acta Metall.* **11**, 1029.

Benedek, R., and Baratoff, A. (1971). *J. Phys. Chem. Solids* **32**, 1015.

Bergersen, B., and Stott, M. J. (1969). *Solid State Commun.* **7**, 1203.

Bergsma, J., and Goldkoop, J. A. (1960). *Physica* **26**, 744.

Bhattacharya, P., and Singwi, K. S. (1972a). *Phys. Rev. Lett.* **29**, 22.

Bhattacharya, P., and Singwi, K. S. (1972b). *Phys. Lett.* **41A**, 457.

Birnbaum, H. K. (1978). Environmental Effects on Fracture, AIMME (A. A. Foroulis, ed.).

Blandin, A., and Friedel, J. (1960). *J. Phys. Radium* **21**, 689.

Blatt, F. J. (1957). *Phys. Rev.* **108**, 285.

Brandt, W., and Reinheimer, J. (1971). *Phys. Lett.* **35A**, 109.

Brewer, J. H., Crowe, K. M., Gygax, F. N., and Schenck, A. (1975). *In* "Muon Physics" (V. W. Hughes and C. S. Wu, eds.). Academic Press, New York.

Burger, J. P., and McLachlan, D. (1976). *J. Phys. (Paris)* **37**, 81.

Burger, J. P., MacLachlan, D. S., Mailfert, R., and Souffaché, B. (1975). *Solid State Commun.* **17**, 277.

Caglioti, G. (ed.) (1976). "Atomic Structure and Mechanical Properties of Metals." North Holland, Amsterdam.

Camani, M., Gygax, F. N., Rüegg, W., Schenck, A., and Schilling, H. (1977). *Phys. Rev. Lett.* **39**, 836.

Camani, M. Gygax, F. N., Rüegg, W., Schenck, A., Schilling, H., Klempt, E., Schulze, R., and Wolf, H. (1979). *Phys. Rev. Lett.* **42**, 679.

Celasco, M., Fiorillo, F., and Mazzetti, P. (1976). *Phys. Rev. Lett.* **36**, 38.

Chang, R., and Falicov, L. M. (1971). *J. Phys. Chem. Solids* **32**, 465.

Cheng, L. J., Sen, P., Mackenzie, I. K., and Kissinger, H. E. (1976). *Solid State Commun.* **20**, 953.

Clementi, E., and Roetti, C. (1974). *At. Data Nucl. Data Tables* **14**, 177.

Cohen, M. L. (1979). *Phys. Today* **32**, 40.

Cohen, M. L., and Heine, V. (1970). *In* "Solid State Physics" (H. Ehrenreich, F. Seitz, and D. Turnbull, eds.), Vol. 24. Academic Press, New York.

Coleridge, P. T., Holzwarth, N. A. W., and Lee, M. J. G. (1974). *Phys. Rev. B* **10**, 1213.

Cotterill, R. M. J., Mackenzie, I. K., Smedskjaer, L. G., Trumpy, G., and Traft, J. (1972). *Nature* **239**, 99.

Crawford, J. H., and Slifkin, L. M. (eds.) (1972–1977). "Point Defects in Solids," Vols. 1–3. Plenum, New York.

Daniel, E. (1967). *In* "Hyperfine Interactions" (A. J. Freeman and R. B. Frankel, eds.), p. 712. Academic Press, New York.

Daniel, E., and Friedel, J. (1963). *J. Phys. Chem. Solids* **24**, 1601.

de Graaf, L. A., Rush, J. J., Flotow, H. E., and Rowe, J. M. (1972). *J. Chem. Phys.* **56**, 4574.

Dekhtyar, I. Ya. (1974). *Phys. Rep. C* **9**, 243.

Doyama, M., and Hasiguti, R. (1973). *Crystal Lattice Defects* **4**, 139.

Ducharme, A. R., and Weaver, H. T. (1972). *Phys. Rev. B* **5**, 330.

Dupree, T. H. (1961). *Ann. Phys.* **15**, 63.

Eastman, D. E., Cashion, J. K., and Switendick, A. C. (1971). *Phys. Rev. Lett.* **27**, 35.

Eichenauer, W. (1968). *Z. Metallkd.* **59**, 613.

Ehrenreich, H., and Schwartz, L. M. (1976). *In* "Solid State Physics" (H. Ehrenreich, F. Seitz, and D. Turnbull, eds.), Vol. 131, p. 149. Academic Press, New York.

Ellis, D. E., and Painter, G. S. (1970). *Phys. Rev. B* **2**, 2887.

Ellis, D. E., Benesh, G. A., and Byron, E. (1977). *Phys. Rev. B* **16**, 3308.
Ellis, D. E., Benesh, G. A., and Byron, E. (1978). *J. Appl. Phys.* **49**, 1543.
Evans, R. (1977). *In* "Vacancies '76," p. 30. Metal Society, London.
Evans, R., and Finnis, M. W. (1976). *J. Phys. F* **6**, 483.
Faulkner, J. S. (1976). *Phys. Rev. B* **13**, 2391.
Feder, R. (1970). *Phys. Rev. B* **2**, 828.
Feder, R., and Charnau, H. P. (1966). *Phys. Rev.* **149**, 464.
Fermi, E. (1928). *Z. Physik* **48**, 73.
Finnis, M. W. (1978). *J. Nucl. Mater.* **69**, 638.
Finnis, M. W., and Sachdev, M. (1976). *J. Phys. F* **6**, 965.
Fluss, M. J., Smedskjaer, L. C., Chason, M. K., Legnini, D. G., and Siegel, R. W. (1978). *Phys. Rev. B* **17**, 3444.
Fluss, M. J., Gupta, R. P., Smedskjaer, L. C., and Siegel, R. W. (1979). *Adv. Chem. Series,* **175**, 243.
Flynn, C. P. (1972). "Point Defects and Diffusion." Clarendon, Oxford.
Flynn, C. P., and Stoneham, A. M. (1970). *Phys. Rev. B* **1**, 3966.
Foy, M. L. G., Heiman, N., and Kossler, W. J. (1973). *Phys. Rev. Lett.* **30**, 1064.
Freeman, A. J. (1981). This volume.
Friedel, J. (1952). *Philos. Mag.* **43**, 153.
Fukai, Y., and Watanabe, K. (1970). *Phys. Rev. B* **2**, 2353.
Fukai, Y., and Watanabe, K. (1974). *Phys. Rev. B* **10**, 3015.
Fukai, Y., Kazama, S., Tanaka, K., and Matsumoto, M. (1976). *Solid State Commun.* **19**, 507.
Fumi, F. G. (1955). *Philos. Mag.* **46**, 1007.
Ganguly, B. N. (1973). *Z. Phys.* **265**, 433.
Ganguly, B. N. (1976). *Phys. Rev. B* **14**, 3848.
Gaspari, G. D., Shyu, W. M., and Das, T. P. (1964). *Phys. Rev.* **134**, A852.
Gauster, W. B., Heffner, R. H., Huang, C. Y., Huston, R. L., Leon, M., Parkin, D. M., Schillaci, M. E., Trifthäuser, W., and Wampler, W. R. (1977). *Solid State Commun.* **24**, 619.
Gehlen, P. C., Beeler, Jr., J. R., and Jaffe, R. I. (eds.) (1972). "Interionic Potentials and Simulation of Lattice Defects." Plenum, New York.
Gelatt, Jr., C. D., Ehrenreich, H., and Weiss, J. A. (1978). *Phys. Rev. B* **17**, 1940.
Gissler, W., Alefeld, G., and Springer, T. (1970). *J. Phys. Chem. Solids* **31**, 2361.
Górecki, T. (1974). *Z. Metallkd.* **65**, 426.
Graf, H., Kundig, W., Patterson, B. D., Reichart, W., Roggwiller, P., Camani, M., Gygax, F. N., Rüegg, W., Schenck, A., Schilling, H., and Meier, P. F. (1976). *Phys. Rev. Lett.* **37**, 1644.
Graf, H., Hofman, W., Kündig, W., Meier, P. F., Patterson, B. D., and Reichart, W. (1977). *Solid State Commun.* **23**, 653.
Guenzer, C. S., and Binenstock, A. (1973). *Phys. Rev. B* **8**, 4655.
Gunnarson, O. (1978). *In* "NATO Advanced Study Institutes Series (B-Physics)" (P. Phariseau and B. L. Gyorffy, eds.), Vol. 42. Plenum, New York.
Gunnarson, O., Lundquist, B. I., and Wilkins, J. W. (1974). *Phys. Rev. B* **10**, 1319.
Gupta, A. K., and Singwi, K. S. (1977). *Phys. Rev. B* **15**, 1801.
Gupta, A. K., Jena, P., and Singwi, A. K. (1978). *Phys. Rev. B* **18**, 2712.
Gupta, M., and Freeman, A. J. (1978). *Phys. Rev. B* **17**, 3029.
Gupta, R. P., and Siegel, R. W. (1977). *Phys. Rev. Lett.* **39**, 1212.
Gupta, R. P., and Benedek, R. (1979). *Phys. Rev. B* **19**, 583.
Gupta, R. P., and Siegel, R. W. (1980). *Phys. Rev. B* **22**, 4572.
Gurevich, I. I., Klimov, A. I., Maiorov, V. N., Meleshko, E. A., Nikolskii, B. A., Selivanov, V. I., and Suetin, V. A. (1976a). JETP **42**, 222.

Gurevich, I. I., Klimov, A. I., Maiorov, V. N., Meloshko, E. A., Nikolskii, B. A., Pirogov, A. V., Roganov, V. S., Selivanov, V. I., and Suetin, V. A. (1976b). JETP **42**, 741.

Gustafson, D. R., and Barnes, T. G. (1973). *J. Nucl Mater.* **48**, 79.

Gyemant, I., and Solt, G. (1977). *Phys. Status Solidi.* **B82**, 651.

Haas, M., King, H. T., Robbins, A. B., Ventura, E., and Kalish, R. (1976). *Bull. Am. Phys. Soc.* **21**, 51.

Hamagaki, H., Nakai, K., Nojiri, Y., Tanihata, I., and Sugimoto, K. (1976). *Hyperfine Int.* **2**, 187.

Hannay, N. B. (ed.) (1975). "Treaties in Solid State Chemistry," Vol. 2. Plenum, New York.

Harmon, B. N., and Freeman, A. J. (1972). *Phys. Rev. B* **10**, 1979.

Harris, R. P. (1970). *J. Phys. C* **3**, 172.

Harrison, W. A. (1966). "Pseudopotentials in the Theory of Metals." W. A. Benjamin, New York.

Harrison, W. A. (1970). "Solid State Theory." McGraw Hill, New York.

Hartmann, O. (1977). *Phys. Rev. Lett.* **39**, 832.

Hartmann, O., Karlsson, E., Norlin, L. O., Richter, D., and Niinikoski, T. O. (1978). *Phys. Rev. Lett.* **41**, 1055.

Hautojärvi, P., Heiniö, J., Manninen, M., and Nieminen, R. (1977). *Philos. Mag.* **35**, 973.

Haywood, C. T., and Verdini, L. (1968). *Can. J. Phys.* **46**, 2065.

Hedin, L. (1965). *Phys. Rev.* **139**, A796.

Hedin, L., and Lundquist, S. (1969). *Solid State Phys.* **23**, 1.

Herlach, D., Decker, W., Gladisch, M., Mansel, W., Metz, H., Orth, H., Putlitz, C. Zu, Seeger, A., Wahl, W., and Wignad, M. (1979). *Hyperfine Int.* **6**, 323.

Hertel, P. (1974). *Z. Phys.* **268**, 111.

Ho, P. S. (1971). *Phys. Rev. B* **3**, 4035.

Hodges, C. H. (1970). *Phys. Rev. Lett.* **25**, 284.

Hodges, C. H., and Stott, M. J. (1973a). *Phys. Rev. B* **7**, 73.

Hodges, C. H., and Stott, M. J. (1973b). *Solid State Commun.* **12**, 1153.

Hohenberg, P., and Kohn, W. (1964). *Phys. Rev.* **136**, B864.

Holtham, P. M., and Jena, P. (1975). *J. Phys. F* **5**, 1649.

Jena, P. (1976a). *Solid State Commun.* **19**, 45.

Jena, P. (1976b). *Phys. Rev. Lett.* **36**, 418.

Jena, P. (1978). *Solid State Commun.* **27**, 1249.

Jena, P. (1979). *Hyperfine Int.* **6**, 5.

Jena, P., and Singwi, K. S. (1978a). *Phys. Rev. B* **17**, 3518.

Jena, P., and Singwi, K. S. (1978b). *Phys. Rev. B* **17**, 1592.

Jena, P., and Geldart, D. J. W. (1978). *J. Mag. Mag. Materials* **8**, 99.

Jena, P., Das, S. G., and Singwi, K. S. (1978a). *Phys. Rev. Lett.* **40**, 264.

Jena, P., Gupta, A. K., and Singwi, K. S. (1978b). *Phys. Rev. B* **18**, 2723.

Jena, P., Singwi, K. S., and Nieminen, R. M. (1978c). *Phys. Rev. B* **17**, 301.

Jena, P., Wiley, C. L., and Fradin, F. Y. (1978d). *Phys. Rev. Lett.* **40**, 578.

Jena, P., Fradin, F. Y., and Ellis, D. E. (1979). *Phys. Rev. B* **20**, 3543.

Jepsen, O., Nieminen, R. M., and Madsen, J. (1980). *Solid State Commun.* **34**, 575.

Johnson, K. H. (1968). *Phys. Lett.* **27A**, 138.

Johnson, K. H., Vvedensky, D. D., and Messner, R. P. (1979). *Phys. Rev. B* **19**, 1519.

Johnson, R. A. (1973). *J. Phys. F* **3**, 295.

Jones, W., and Young, W. H. (1971). *J. Phys. C* **4**, 1322.

Kagan, Y., and Klinger, M. I. (1974). *J. Phys. C* **7**, 4965.

Kane, E. O., and Schlüter, M. (1979). *Phys. Rev. B* **19**, 5232.

Kasawoski, R. V. (1969). *Phys. Rev.* **187**, 891.

Katayama, H., Terakura, K., and Kanamori, J. (1979a). *Solid State Commun.* **29**, 431.

Katayama, H., Terakura, K., and Kanamori, J. (1979b). *J. Phys. Soc. Jpn.* **46**, 822.

Kazama, S., and Fukai, Y. (1976). *J. Less Common Met.* **53**, 25.

Kazama, S., and Fukai, Y. (1977). *J. Phys. Soc. Jpn.* **42**, 119.

Keller, J. (1973). *In* "Computational Methods for Large Molecules and Localized States in Solids" (F. Herman, A. D. McLean, and R. K. Nesbet, eds.), p. 341. Plenum, New York.

Keller, J. (1979). *Hyperfine Int.* **6**, 15.

Keller, J., and Schenck, A. (1979). *Hyperfine Int.* **6**, 39.

Kahn, L. M., and Rasolt, M. (1978). *J. Phys. F* **7**, 1993.

Kirzhnits, D. A. (1957). *Sov. Phys.—JETP* **5**, 64.

Kittel, C. (1963). "Quantum Theory of Solids." Wiley, New York.

Klein, B. M., Economou, E. N., and Papaconstantopoulos, D. A. (1977). *Phys. Rev. Lett.* **39**, 574.

Kohn, W., and Sham, L. J. (1965). *Phys. Rev.* **140**, A1133.

Kohn, W., and Vosko, S. H. (1960). *Phys. Rev.* **118**, 912.

Kolachev, B. A. (1968). "Hydrogen Embrittlement of Nonferrous Metals." Israel Program for Scientific Translation.

Kowalczyk, S. P., Ley, L., McFeely, F. R., Pollak, R. A., and Shirley, D. A. (1974). *Phys. Rev. B* **9**, 381.

Kraftmacher, Ya. A., and Strelkov, P. G. (1970). *In* "Vacancies and Interstitials in Metals" (A. Seeger, D. Schumacher, W. Schilling, and J. Diehl, eds.). North-Holland, Amsterdam.

Lang, N. D., and Kohn, W. (1970). *Phys. Rev. B* **1**, 4555.

Lang, N. D., and Kohn, W. (1971). *Phys. Rev. B* **3**, 1215.

Lang, N. D., and Williams, A. R. (1977). *Phys. Rev. B* **16**, 2408.

Langreth, D. C., and Perdew, J. P. (1979). *Solid State Commun.* **31**, 567.

Lässer, R., and Lengeler, B. (1978). *Phys. Rev. B* **18**, 637.

Lasseter, R. P., and Soven, P. (1973). *Phys. Rev. B* **8**, 2476.

Lea, K. R., Leask, M. J. M., and Wolf, W. P. (1962). *J. Phys. Chem. Solids* **23**, 1381.

Lengeler, B. (1977). *Phys. Rev. B* **15**, 5504.

Lengeler, B., Mantl, S., and Trifthäuser, W. (1978). *J. Phys. F* **8**, 1691.

Lewis, F. A. (1967). "The Palladium-Hydrogen Systems." Academic Press, New York.

Louie, S. G., Schlüter, M., Chelikowsky, J. R., and Cohen, M. L. (1976). *Phys. Rev. B* **13**, 1654.

McDonald, D. K. (1953). *J. Chem. Phys.* **21**, 177.

McKee, B. T. A. (1978). Unpublished.

MacKenzie, I. K. (1969). *Phys. Lett. A* **30**, 115.

MacKenzie, I. K., and Sen, P. (1976). *Phys. Rev. Lett.* **37**, 1296.

MacLachlan, D. S., Mailfert, R., Burger, J. P., and Souffaché, B. (1975). *Solid State Commun.* **17**, 281.

McLellan, R. B., and Oates, W. B. (1973). *Acta. Metall.* **21**, 181.

McMillan, W. L. (1968). *Phys. Rev.* **167**, 331.

Manninen, M., and Nieminen, R. M. (1978). *J. Phys. F* **8**, 2243.

Manninen, M., and Nieminen, R. M. (1979a). *J. Phys. F* **9**, 1333.

Manninen, M., and Jena, P. (1979b). Unpublished.

Manninen, M., and Jena, P. (1980a). *Phys. Rev. B* **22**, 2411.

Manninen, M., and Jena, P. (1980b). *Solid State Commun.* **34**, 179.

Manninen, M., Nieminen, R., Hautojärvi, P., and Arponen, J. (1975). *Phys. Rev. B* **12**, 4012.

Manninen, M., Hautojärvi, P., and Nieminen, R. M. (1977). *Phys. Lett. A* **63**, 60.

March, N. H. (1957). *Adv. Phys.* **6**, 1.

March, N. H. (1973). *J. Phys. F* **3**, 233.

Meier, P. F. (1975). *Helv. Phys. Acta.* **48**, 227.

Miller, R. J., and Satterthwaite, C. B. (1975). *Phys. Rev. Lett.* **34**, 144.

Minchin, P., Meyer, A., and Young, W. H. (1974). *J. Phys. F* **4**, 2117.

Minier, M., Andreani, R., and Minier, C. (1978). *Phys. Rev. B* **18**, 102.

Mogenson, O., Petersen, K., Cotterill, R. M. J., and Hudson, B. (1972). *Nature* **239**, 97.

Mook, H. A. (1966). *Phys. Rev.* **148**, 495.

Moon, R. M. (1964). *Phys. Rev.* **136**, A195.

Moon, R. M., Koehler, W. C., Cable, J. W., and Child, H. R. (1972). *Phys. Rev. B* **5**, 997.

Mori, G. (1977a). *J. Phys. F* **7**, L7.

Mori, G. (1977b). *J. Phys. F* **7**, L89.

Mueller, W. M., Blackledge, J. P., and Libowitz, G. G. (1968). "Metal Hydrides." Academic Press, New York.

Mueller, F. M., Freeman, A. J., Dimmock, J. O., and Furdyna, A. M. (1970). *Phys. Rev. B* **1**, 4617.

Mulliken, R. S. (1955a). *J. Chem. Phys.* **23**, 1833.

Mulliken, R. S. (1955b). *J. Chem. Phys.* **23**, 1841.

Munjal, R., and Petzinger, K. (1978). *Hyperfine Int.* **4**, 301.

Murnick, D. E., Hass, M., King, H. T., and Ventura, E. (1976). *Hyperfine Int.* **1**, 367.

Myles, C. W., and Dow, J. D. (1979). *Phys. Rev. B* **19**, 4939.

Nakajima, T. (1976). *Phys. Status Solidi* **B77**, K147.

Nieminen, R. M. (1978). *J. Nucl. Mater.* **69**, 633.

Nieminen, R. M., and Hodges, C. H. (1976). *Solid State Commun.* **18**, 1115.

Nieminen, R. M., Manninen, M., Hautojärvi, P., and Arponen, J. (1975). *Solid State Commun.* **16**, 831.

Nieminen, R. M., and Manninen, M. (1979). *In* "Topics in Current Physics, Vol. 12: Positrons in Solids" (P. Hautojärvi, ed.), p. 145. Springer-Verlag, Berlin–Heidelberg–New York.

Nishida, N. *et al.* (1977). *Solid State Commun.* **22**, 235.

Norskov, J. K. (1977). *Solid State Commun.* **24**, 691.

Papaconstantopoulos, D. A., and Klein, B. M. (1975). *Phys. Rev. Lett.* **35**, 110.

Papaconstantopoulos, D. A., Klein, B. M., Economou, E. N., and Boyer, L. L. (1978). *Phys. Rev. B* **17**, 141.

Patterson, B. D., and Falicov, L. M. (1974). *Solid State Commun.* **15**, 1509.

Perdew, J. P., and Monnier, R. (1976). *Phys. Rev. Lett.* **37**, 1286.

Perdew, J. P., Langreth, D. C., and Sahni, V. (1977). *Phys. Rev. Lett.* **38**, 1030.

Perrot, F. (1977a). *Phys. Status Solidi* **B81**, 205.

Perrot, F. (1977b). *Phys. Rev. B* **16**, 4335.

Perrot, F. (1978). *Phys. Status Solidi* **B87**, 707.

Peterman, D. J., Harmon, B. N., Marchiando, J., and Weaver, J. H. (1979). *Phys. Rev. B* **19**, 4867.

Peterson, D. T., and Jensen, C. L. (1978). *Metall. Trans.* **9A**, 1673.

Petzinger, K. G., and Munjal, R. (1977). *Phys. Rev. B* **15**, 1560.

Pryde, J. A., and Tsong, I. S. T. (1971). *Acta Metall.* **19**, 1333.

Popović, Z. D., Carbotte, J. P., and Piercy, G. R. (1973). *J. Phys. F* **3**, 1008.

Popović, Z. D., Carbotte, J. P., and Piercy, G. R. (1974). *J. Phys. F* **4**, 351.

Popović, Z. D., Stott, M. J., Carbotte, J. P., and Piercy, G. R. (1976). *Phys. Rev. B* **13**, 590.

Rahman, A., Sköld, K., Pelizzari, C., and Sinha, S. K. (1976). *Phys. Rev. B* **14**, 3630.

Rajagopal, A. K., and Callaway, J. (1973). *Phys. Rev.* **137**, 1912.

Rasolt, M., and Geldart, D. J. W. (1975). *Phys. Rev. Lett.* **35**, 1234.

Rath, J., Manninen, M., Jena, P., and Wang, C. S. (1979). *Solid State Commun.* **31**, 1003.

Reilly, Jr., J. J. (1978). *In* "Synthesis and Properties of Useful Metal Hydrides" Proceedings of International Symposium on Hydrides for Energy Storage. Geilo, Norway, August 14–19. Pergamon, Oxford.

Rietschel, H. (1975). *Z. Phys. B* **22**, 133.

Robinson, G. G., and de Chatel, P. F. (1975). *J. Phys. F* **5**, 1502.

Rosén, A., and Ellis, D. E. (1976). *J. Chem. Phys.* **65**, 3629.

Rowland, T. J. (1960). *Phys. Rev.* **119**, 900.

Sagalyn, P. L., and Alexander, M. N. (1977). *Phys. Rev. B* **15**, 5581.

Satterthwaite, C. B., and Toepke, I. L. (1970). *Phys. Rev. Lett.* **25**, 741.

Schenck, A. (1976). *In* "Nuclear and Particle Physics at Intermediate Energies" (J. B. Warren, ed.). Plenum, New York.

Seeger, A. (1973). *J. Phys. F* **3**, 248.

Seeger, A. (1978). *In* "Hydrogen in Metals" (G. Alefeld and J. Völkl, eds.), Topics in Applied Physics, Vol. 2. Springer, Berlin–Heidelberg–New York.

Seeger, A. (1979). *Hyperfine Int.* **6**, 313.

Sen, P., Cheng, L. J., and Kissinger, H. E. (1975). *Phys. Lett. A* **53**, 299.

Shenoy, G. K., Dunlap, B. D., Westlake, D. G., and Dwight, A. E. (1976). *Phys. Rev. B* **14**, 41.

Shull, C. G. (1967). *In* "Magnetic and Inelastic Scattering of Neutrons by Metals" (T. J. Rowland and P. A. Beck, eds.). Gordon and Breach, New York.

Shull, C. G., and Mook, H. A. (1966). *Phys. Rev. Lett.* **16**, 184.

Siegbahn, K. *et al.* (1967). "ESCA: Atomic, Molecular, and Solid State Structure Studied by Means of Electron Spectroscopy," Nova Acta. Regial. Soc. Sci. Upsaliensis, Series VI, Vol. 20.

Siegel, R. W. (1978). *J. Nucl. Mat.* **69**, **70**, 117 (Proc. Int. Conf. on the Properties of Atomic Defects in Metals, Argonne, 1976).

Simpson, R. W., Lane, N. F., and Cloney, R. C. (1978). *Nucl. Mater.* **69**, **70**, 582.

Singhal, S. P. (1973). *Phys. Rev. B* **8**, 3641.

Singhal, S. P., and Callaway, J. (1979). *Phys. Rev. B* **19**, 5049.

Singwi, K. S., and Tosi, M. P. (1981). *In* "Solid State Physics" (H. Ehrenreich, F. Seitz, and D. Turnbull, eds.), Academic Press, New York, to be published.

Singwi, K. S., Tosi, M. P., Land, R. H., and Sjölander, A. (1968). *Phys. Rev.* **176**, 589.

Sjölander, A., and Stott, M. J. (1972). *Phys. Rev. B* **5**, 2109.

Skoskiewicz, T. (1972). *Phys. Status Solidi* **A11**, K123.

Slater, J. C. (1974). "The Selfconsistent Field for Molecules and Solids," Vol. 4. McGraw Hill, New York.

Slichter, C. P. (1963). "Principles of Magnetic Resonance." Harper and Row, New York.

Soven, P. (1970). *Phys. Rev. B* **2**, 4715.

Stoneham, A. M. (1972). *Ber. Bunsenges. Phys. Chem.* **76**, 817.

Stoneham, M. (1975). "Theory of Defects in Solids." Clarendon, Oxford.

Stott, M. J., Baranovsky, S., and March, N. H. (1970). *Proc. R. Soc. London* **A316**, 201.

Stritzker, B., and Buckel, W. (1972). *Z. Physik* **257**, 1.

Suits, B., Shenoy, G. K., Dunlap, B. D., and Westlake, D. G. (1977). *J. Mag. Mag. Mat.* **5**, 344.

Sussmann, J. A. (1971). *Ann. Phys.* **6**, 135.

Switendick, A. C. (1972). *Ber. Bunsenges. Phys. Chem.* **76**, 535.

Takahashi, J., Ozawa, K., Yamaguchi, S., Fujino, Y., Yoshinari, O., and Hirabayashi, M. (1978). *Phys. Status Solidi* **A46**, 217.

Tam, S. W., and Siegel, R. W. (1977). *J. Phys. F* **7**, 877.

Terakura, K. (1976a). *J. Phys. Soc. Jpn.* **40**, 456.

Terakura, K. (1976b). *J. Phys. F* **6**, 1385.

Thomas, L. H. (1927). *Proc. Cambridge Phil. Soc.* **23**, 542.

Trifthäuser, W. (1975). *Phys. Rev. B* **12**, 4634.

Trifthäuser, W., McGervey, J. D., and Hendricks, R. W. (1974). *Phys. Rev. B* **9**, 3321.

Tzanetakis, P., Hillairet, J., and Revel, G. (1976). *Phys. Status Solidi* **B75**, 433.

Van Vleck, J. H. (1948). *Phys. Rev.* **74**, 1168.

Vashishta, P., and Singwi, K. S. (1972). *Phys. Rev. B* **6**, 875.

Veal, B. W., Lam, D. J., and Westlake, D. G. (1979). *Phys. Rev. B* **19**, 2856.

Veal, B. W. (1981). This volume.

Verani-Borgucci, M., and Verdini, L. (1965). *Phys. Status Solidi* **9**, 243.

Verdini, L. (1976). *J. Less Common Met.* **49**, 329.

Viccaro, P. J., Shenoy, G. K., Dunlap, B. D., Westlake, D. G., Malik, S. K., and Wallace, W. E. (1979). *J. Phys. (Paris)* **40**, C2-157.

von Barth, U., and Hedin, L. (1972). *J. Phys. C* **5**, 1629.

Wampler, W. R., and Lengeler, B. (1977). *Phys. Rev. B* **15**, 4614.

Wang, C. S., and Callaway, J. (1977). *Phys. Rev. B* **15**, 298.

Wang, C. S., and Freeman, A. J. (1979). *J. Appl. Phys.* **50**, 1940.

Watson, R. E., Perlman, M. L., and Herbst, J. (1976). *Phys. Rev. B* **13**, 2358.

Weizsäcker, C. F. V. (1935). *Z. Phys.* **96**, 431.

Wenzl, H. (1978). Private communication.

West, R. N. (1973). *Adv. Phys.* **22**, 264.

West, R. N. (1979). *In* "Positrons in Solids" (P. Hautojärvi, ed.), p. 89. Springer-Verlag, Berlin–Heidelberg–New York.

Westlake, D. G. (1967). *Trans. Metall. Soc. AIME* **239**, 1341.

Westlake, D. G., and Miller, J. F. (1980). *J. Phys. F* **10**, 859.

Westlake, D. G., and Ockers, S. T. (1975). *Metall. Trans.* **6A**, 399.

Westlake, D. G., Satterthwaite, C. B., and Weaver, J. H. (1978). *Phys. Today, November*, p. 32.

Wiley, C. L., and Fradin, F. Y. (1978). *Phys. Rev. B* **17**, 3462.

Williams, A. R., and Lang, N. D. (1978). *Phys. Rev. Lett.* **40**, 954.

Williams, B. (1977). "Compton Scattering." McGraw Hill, New York.

Ya, I. Dekhtyar (1974). *Phys. Rep. C* **9**, 243.

Yamada, R., Watanabe, N., Sato, K., Asano, H., and Hirabayashi, M. (1976). *J. Phys. Soc. Jpn.* **41**, 85.

Yoshioki, S., and Mori, G. (1976). *J. Phys. F* **6**, 1743.

Zaremba, E., Sander, L. M., Shore, H. B., and Rose, J. H. (1977). *J. Phys. F* **7**, 1763.

Zeller, R., and Dederichs, P. H. (1979). *Phys. Rev. Lett.* **42**, 1713.

Zunger, A., and Freeman, A. J. (1977). *Phys. Rev. B* **16**, 2901.

Index

Contents of Previous Volumes